GENES, ENVIRONMENT, AND BEHAVIOR

Genes, Environment, and Behavior

An Interactionist Approach

JACK R. VALE

University of California, Berkeley

HARPER & ROW, PUBLISHERS, New York
Cambridge, Hagerstown, Philadelphia, San Francisco,
London, Mexico City, São Paulo, Sydney

1817

To Patti
Chris
Jamie
and Michelle

The bright stars in my firmament

Sponsoring Editor: George A. Middendorf
Project Editor: Pamela Landau
Senior Production Manager: Kewal K. Sharma
Compositor: Maryland Linotype Composition Co., Inc.
Printer and Binder: The Maple Press Company
Art Studio: Vantage Art, Inc.

GENES, ENVIRONMENT, AND BEHAVIOR: An Interactionist Approach

Library of Congress Cataloging in Publication Data

Vale, Jack R
    Genes, environment, and behavior.
    Includes bibliographical references and index.
    1. Behavior genetics.   2.   Behavior evolution.
3.   Nature and nurture.   I.   Title.
QH457.V34   591.51   80-13097
ISBN 0-06-046758-4

Contents

Preface

What place have the biological and behavioral sciences in the construction of the human self-image? Through the popular media all of us are familiar with the modern scientist's role as consultant to government and hence are aware that there is presumably at least a pro forma relationship between the enactment of social legislation and some scientific predication of success. Thus, we are all aware of the notion that at one level what scientists think about human behavior affects sociopolitical attempts to influence that behavior.

However, a less familiar but much deeper and more pervasive relationship between science and society derives from the scientist's role as producer of knowledge and purveyor of thought. In this the scientist joins philosophers, clerics, and other sages who would tell us what we are. What we think we are—our image of ourselves—substantially contributes to what we do. Thus the scientist, like his or her predecessors, far from simply reporting upon behavior, actually determines it.

All human social action presupposes a theory of humanity. As cognizing, speaking, remembering, anticipating animals we can deal with one another only when we can predict, and we can predict only within some reasonably comprehensive behavioral framework. More and more, as society's vanguard of wisdom, scientists are the architects of that framework.

Is behavior learned? Is it inborn? Are these even proper questions? What of aggression, intelligence, madness? Whence have we evolved and into what? What of our diversity? These and other issues, in plain and subtle guises, burden our conduct beyond the mere application of intellect. Their occasional resolutions have provided all the emotional ingredients attendant upon an outlook upon the world.

There is thus a crucial relationship between the behavior of humans toward their own kind and the view of life they hold. It is a relationship too little pondered. Yet the place we reserve for certain of us in the scheme of things is directly and dramatically affected by what passes for truth at the time. It will be our purpose first to examine specific aspects of recent scientific contributions to such truth, particularly as they have emanated from our investigations of the provenance of behavior.

It is, then, one thesis of this book that there is much in science that is essential to any appraisal of the human condition, that this loses little in nontechnical transcription, and that hence there are insights to be gained, and perhaps biases to be surrendered, by a perusal of past and present scientific thought. Thus, in early chapters we shall emphasize the profound connections between conventional wisdom and human relations using some examples from the history of biology and experimental psychology. At the same time we shall briefly examine the development of two historically disparate attempts to find the origins of behavior.

It is a second thesis here that an organism-environment interaction approach to the study of behavior is possible both conceptually and in practice, and that it is within this context that we see ourselves in neither the most comely nor the most forbidding, but in the most revealing and therefore in the most sympathetic, light. In later chapters, then, we shall examine the theory and evidence supporting the interaction approach.

Writing is an exhilarating, often lonely, sometimes melancholy chore. And, particularly with a book of this sort, only does the reader's nod of recognition satisfactorily punctuate the sentences so arduously wrought. It is with the hope of earning that recognition that this work is offered.

To a degree we write for ourselves. Probably it is true in this case more than others. I wrote the book I would liked to have read in the formative stages of my own intellectual development. This may seem dangerous, but I think not. The issues treated are broad and run deeply through the human condition, framing its boundaries and events like a membrane on the nerve of history. This book therefore is for all who with rigor and sympathy would begin to face the querulous complexities of the provenance of behavior. To them goes my respect and, I trust, some small assistance.

Jack R. Vale

Chapter 1
Organism and Behavior:
Retrospect

THE PRE-DARWINIAN VIEW

Few ideas have, at a stroke, so altered the thoughts of man about himself as those expressed by Charles Darwin in *The Origin of Species*. There had been other revolutions: Copernicus changed our place in the astronomical universe, and Newton described physical laws for the objects that surround us. But it was Darwin who, finally and irrevocably, connected humans to nature by showing them to be a part of all life, past and present. Consequently, humanity's existence took on a new dimension—indeed, a new reality. Darwin gave us perpetual change. This accomplishment may seem less now because, as noted by Dobzhansky (1962), the concept of evolution has long since become an integral part of our inward vision. It is thus something of a shock to look back to the times before 1859 and see what state the intellectual world was in.

The Grand Ascending Chain of Perfection

Darwin did not invent the concept of evolution. He discovered the principle by which it works, namely, natural selection. It was for a principle such as this, resting upon demonstrable, natural causes, that the scientific community had to wait before it could accept the theory of evolution itself. When it was finally presented, the farsighted biologist Thomas Huxley (Kogan, 1960) would remark: "How extremely stupid not to have thought of that!" And 20 years later Huxley would write with obvious satisfaction that he doubted whether there was a single responsible representative of the old view remaining, so rapid and sweeping had the transition been.

The reason for the swift change is that the old view had reached the limits of credulity. It was not nearly so homogeneous as presented here, but did generally include these features (Eiseley, 1956, 1959; Krutch, 1957; Laughlin & Osborne, 1968; Lovejoy, 1936): (1) The world was as full of diverse forms as it could be; no species were being made anew, and none being destroyed. (2) Species existed because of acts of special creation by God. They were therefore not "transmutable"; that is, one species could not over time have derived from another. (3) Each species was a preordained type; variation among individuals within a species was thus considered to be due to error, and in some instances pathology. (4) Extinct species were recognized, but they had been destroyed in one of numerous world cataclysms thought to have occurred in the short history of the earth. New species were then created and lasted until the next cataclysm. (5) There was a continuity of existing forms, reflected in a gradual transition of complexity from the inanimate to the animate, from plants to animals, and among animals. Continuity was maintained between land, sea, and air animals by intermediate groups, such as seals and bats. (6) Not only was the world as full as it could be of immutable species, and not only did these species show continuity, but they could be arranged *hierarchically* to form a *Grand Ascending Chain of Perfection.* In the Grand Chain each link had its place, and it was separated from the next lower or higher link by the least possible difference in complexity. Thus the hierarchical differentiation of superior and inferior could be, and was, applied not only to vastly different forms but also to extremely similar ones. (7) At the highest portion of the Chain was man, the Lord of Creation. But the Chain concept applied equally to humans, so that some forms were slightly but unchangeably more perfect. (The seeds of rational racism thus were ably sown.) (8) Atop the Grand Chain, in the opinion of its describers (the Western Europeans) were, of course, the Western Europeans, and their descendants. The evidence was clear and everywhere, in terms of who had progressed farthest and contributed most to civilization.

Given the above assumptions, a major task of the eighteenth- and nineteenth-century biologist was to inventory nature—to locate and describe all the links, from highest to lowest. Where continuity did not appear to exist, where one link was not separated from the next by the least possible difference, a link was missing. Much of the naturalists' time was thus taken up in the search for missing links in order to better substantiate the Grand Chain. Such indeed was the task intended by Robert Fitzroy, captain of *H.M.S. Beagle*, for young Charles Darwin.

Naturalist supporters of the Chain had an obvious ally in Christian theology. Belief in one clearly reinforced belief in the other. In fact, the earthly Chain may be considered to have been modeled upon a vision of the heavenly one, with humans at the top of the former and God at the top of the latter, and all the lesser creatures in each—apes and angels, snakes and spirits—below. Doubt about the earthly version could then be taken to imply doubt about the other, a case of no mean consequence at the time and a partial explanation for the equivocation of a number of otherwise committed naturalists. Altogether then the Grand Ascending Chain was a blend of Western religious orthodoxy, science, and, clearly, the self-serving beliefs of the well-to-do. For them at least, it seemed to present a cosmology that was orderly on all fronts.

Some fronts were little more than facades, however, and none more than in the discipline of geology. Theological authority had been strong here at the beginning of the nineteenth century, and there was pressure upon geologists to avoid direct conflict with the church. Indeed, some geologists were themselves primarily theologians. But genuine discovery continued, a situation that led to the existence of a body of fact and interpretation that could be maintained consonant with scripture only by the exercise of great ingenuity.

Then, in 1830, Charles Lyell adduced evidence against the theory of successive catastrophes and the short age of the earth envisaged by theology. The Lyell school was not evolutionary in the biological sense, but nevertheless made a strong imprint in favor of long-term, natural causes in geological history, preparing the way for the evolutionary view. More and more data contradicting the theological version subsequently became available, but there was no interpretatively comfortable place to put them, so they remained, contradictory, on the mind. People were forced into differential public and private positions on matters of science. Clearly, the strain had begun to show.

Into this charged atmosphere strode Darwin (1859) with a lighted match. Aware of the potential impact of his theory, he envisaged a poor reception from established biologists:

> Although I am fully convinced of the truth of the views given in this volume. . . , I by no means expect to convince experienced naturalists whose minds are stocked with a multitude of facts all viewed, during a long course of years, from a point of view directly opposite to mine. It is easy to hide our ignorance under such expressions as the "plan of

creation," "unity of design," etc. and to think that we give an explanation when we only restate a fact. A few naturalists, endowed with much flexibility of mind, and who have already begun to doubt the immutability of species, may be influenced by this volume; but I look with confidence to the future—to young and rising naturalists, who will be able to view both sides of the question with impartiality. Whoever is led to believe that species are mutable will do good service by conscientiously expressing his convictions; for thus only can the load of prejudice by which this subject is overwhelmed be removed. (p. 444)

It was only later that he speculated upon the fertile, if fitfull, ground upon which his ideas had fallen, remarking at last: "innumerable well-observed facts were stored in the minds of naturalists ready to take their proper places as soon as any theory which would explain them was sufficiently explained" (Eiseley, 1956).

The reason Thomas Huxley could make the boast noted earlier now seems clear. Seldom had a change of such magnitude occurred in so short a time. It is instructive to study briefly the conditions and implications of that change.

Lyell and the New Geology

"I feel as if my books came half out of Sir Charles Lyell's brain," Darwin once confessed. What had given the great biologist cause to speak so of a geologist? It was Lyell who began the revolution brought to a thunderous and irrevocable conclusion by Darwin; it was Lyell who took the first step toward making evolution scientifically feasible by giving it time.

The earlier, theologically dominated, theory of the earth's history had little regard for time or natural cause. First, the earth was held to be, in keeping with the biblical description, relatively young. Second, the fossil record, particularly of the land vertebrates, was quite incomplete, and the idea arose that the apparent breaks in the record were real, that there truly had been an interruption between the life of one age and that of another. Each geological period was thus seen as having its own flora and fauna, distinct from those that preceded and followed it. What had intervened to cause the breaks? Catastrophes, presumably of divine inspiration, which had successively destroyed all life. This was replaced by new life, as living proof of creative interference. The religious appeal of such a proposition is obvious. It was, in Eiseley's (1956) words, "a geotheological drama [which] retained both the creative excess and fury of an Old Testament Jehovah."

A third feature of this view was the doctrine of Progression. Paleontologists had discovered the increased complexity among life forms of successive geological ages, which they then explained by the theory of a *necessary* advance in the organization of life as it was created and recreated from the beginning to the present. All life forms moved toward perfection, namely humanity, after which there were to be no new developments. Progressionism was thus evolution by miracle, which was to cease only when the human level had been attained.

Lyell, in his *Principles of Geology* in 1830, proposed instead a uniformitarian geology, which emphasized that the same natural forces seen currently at work on the earth were also those that had shaped it in the past. Thus volcanic activity, earthquakes, wind, frost, and rain were in the "uniform" sources of geological history. Most important, these sources were available for study. The castastrophic view presented a scientific cul-de-sac, for who could know what was on the Creator's mind, or when He would strike again? A theory of natural causes, on the other hand, offered one the opportunity of discovering something.

If the uniformitarian hypothesis removed one partner from the geological scene, however, it added another, perhaps even more terrifying, one: time, endless time. The comfort of circumscribed and near events was to be lost in the abyss of an earthly infinity in which there was, in the words of Lyell's intellectual predecessor, James Hutton, "no vestige of a beginning, no prospect of an end." Small wonder that people turned their eyes from Lyell's apparition.

Lyell, nevertheless, persisted. He was, wrote one of his contemporaries, "deficient of power in oral discourse, and was opposed by men who were his equals in knowledge, his superiors in the free delivery of their opinions. But in resolute combats, yielding not an inch to his adversaries, he slowly advanced upon the ground they abandoned, and became a conqueror without ever being acknowledged as a leader" (Eiseley, 1959). He documented the case for geological change due to natural causes over long periods so effectively that its validity could not be ignored. By degrees the uniformitarian hypothesis prevailed, and the interrupted periods of the catastrophists were seen as continuous, extending into a still awe-inspiring past.

Naturally the doctrine of progressionism, in which life forms were seen as progressing in perfection through the successive cataclysms, was rejected by Lyell. But he found no set of natural causes, such as those at work in geology, to explain organic change. He therefore remained in something of a quandary over the question of the transmutation of species. What was not natural was miraculous, was tainted with progressionism, and was unacceptable. On the other hand, the fossils did show an increase in organization over time. The question of organic evolution therefore remained unresolved.

The answer was to come later. But it was Lyell who both provided the foundation for that answer and set the stage for its ultimate triumph. The intellectual tension necessary for a revolution in thought would never have surrounded the question of the origin of species had not Lyell generated it with his new geology. If the earth were old, and its features due to natural forces, what of its inhabitants? Here lay Darwin's contribution. Lyell's was no less, but men above all value that which relates directly to their own condition, and when the wave of organic evolution broke over the world from England, it raised Darwin above all others. Lyell remained his steadfast friend and counselor. Yet still, a few years before his death, the onetime idol of British natural science wrote somewhat sadly to biologist Ernst Haeckel: "Most of the zoologists forget that anything was written between the time of Lamarck and the publication of our friend's *Origin of Species*" (Eiseley, 1959). History shows it was not so.

The Theory of Natural Selection

Charles Darwin, as noted, was not responsible for the concept of evolution. Ideas involving the transmutation of species, the "derivative" hypothesis, had long opposed those of species fixity and special creation. Darwin's grandfather, Erasmus, had himself been a transmutationist. But the best known evolutionary system prior to Darwin was that of Jean Baptiste de Monet, Chevalier de Lamarck.

Lamarck's system represented something halfway between the scientific and the spiritual—in other words, that which was explicable on the basis of natural cause and that which was not. There were four features in particular that excited his critics: (1) the inheritance of acquired characters; that is, the passing to one generation, through the germ plasm, of structures or habits acquired during the lifetime of the previous generation. This implied the modifiability of the germ plasm through use or disuse of the somatic structures or of habits, an issue that went unresolved for many years. (2) Changes in physical form through conscious will, with the changes then being inherited. Cats, for instance, may have gradually obtained claws by desire, in order better to catch mice.

The above two points are clearly intended to explain evolution by means of adaptation. Traits acquired through use or will were thought to be passed on to the progeny, leaving them better suited to their environments. The idea of adaptation to the environment of course remains central to evolutionary theory today. Finally, Lamarck believed in (3) progressionism, and because all life forms were thought to progress to the more complex, in order to explain the continued existence of simple forms, he was forced to postulate (4) spontaneous generation—that is, the production of simple forms from inorganic materials.

Such proposals won Lamarck few converts among natural scientists, and, in fact, left them ill disposed toward the concept of evolution in general. He became something of a whipping boy, as evidenced by Thomas Huxley's remarks (Kogan, 1960): "Since Lamarck's time, almost all competent naturalists have left speculation on the origins of species to . . . dreamers. . . . The Lamarckian hypothesis has long since been justly condemned, and it is the established practice for every tyro to raise his heel against the carcase of the dead lion" (p. 45). (He continues, however: "But it is rarely either wise or instructive to treat even the errors of a really great man with mere ridicule . . .")

Lamarck's treatment of the problem was followed by that of Robert Chambers, a bookseller and journalist, who in 1844 published *The Vestiges of the Natural History of Creation* in an attempt to support Lamarck. However, *Vestiges*, in the opinion of Huxley (Kogan, 1960), was an effort by which "the Lamarckian hypothesis received its final condemnation in the minds of sound thinkers" (p. 45). Nevertheless, it went through several editions, and was widely read. It is acknowledged by Darwin (1859) in later editions of *Origin*, not because of theoretical agreement but because "it has done excellent service in this country in calling attention to the subject, in removing prejudice, and in thus preparing the ground for the reception of analogous views" (p. 4).

Darwin was once again not only generous, but insightful. The utility of the exposition of a concept, even incorrectly, before its time, cannot be overvalued. The theories of the early evolutionists hung nagging in the air, and the just but hard reviews they received from scientific critics armed with the knowledge of the day served not to eliminate them, but to send them underground, whence they could emerge most acceptably in a new and proper atmosphere. An important proposition seems all the more convincing, once its workings are understood, if one has labored alongside it, than if it is presented fresh. It is almost revelational, perhaps because one has had to think about it more.

Besides religious objections, the chief difficulty with the evolutionary view was the lack of a feasible scientific explanation for its mechanism. Most relevant was that no one had ever *seen* evolution; the phenomenon had a name, but it had never been observed. This, coupled with the ease of thinking by types, which was both reflected in and resulted from the classificatory work of Linnaeus, led to a superficially comfortable understanding of the provenance and destiny of life.

Darwin changed all that. He demonstrated the *fact* of evolution by showing its *effect*, and he *accounted* for evolution with his principle of natural selection. The old reasons for rejecting the evolutionary view were therefore no longer valid. Suddenly it was possible. The key of course was the new principle—so obvious with hindsight, but so elusive at the time.

Darwin was himself unsure just how or when the idea of natural selection occurred to him. At the age of 22 he had been, according to Eiseley (1956), "a gentleman idler after hounds who had failed at medicine and whose family, in desperation, hoped he might still succeed as a country parson." In the autumn of 1831 he met Captain Robert Fitzroy, a religious man who intended to take *H.M.S. Beagle* on a long voyage of exploration. He was seeking a naturalist (whose report, he hoped, would refute Lyell's heretical geology). Darwin, recommended by one of his former professors, accepted, with little thought of scientific upheaval. "If it is desirable to see the world," he wrote, "what a rare and excellent opportunity this is. Perhaps I may have the same opportunity of drilling my mind that I threw away at Cambridge" (Eiseley, 1956).

The fact that Darwin had indeed been a poor student of the classical education may have served him well later, for he departed the shores of England with a mind healthily open and uncluttered. He also departed into a ferocious storm, which nearly capsized the ship and gave the young naturalist a solid hint that he was vulnerable to seasickness. But with the *Beagle* driven back into Plymouth Harbor and given an opportunity to quit the voyage at hand, Darwin found in himself a stubborn determination to see it through. It was a trait that later would also serve him well.

The journey lasted five years, and included a trip down the coastline of South America, with numerous stops at islands. Darwin observed, noted, recorded, thought. Then, as he reported in the introduction to the *Origin:*

> I was much struck with certain facts in the distribution of the organic beings inhabiting South America, and in the geological relations of the present to the past inhabitants of that continent. These facts . . . seemed to throw some light on the origin of species. On my return home, it occurred to me, in 1837, that something might perhaps be made out on this question by patiently accumulating and reflecting on all sorts of facts which could possibly have any bearing on it. (p. 27)

The observation most striking to Darwin was the extraordinary amount of variation, even in "important organs," among individuals of the same species. Individuals differed from one another, and many of the differences were inherited. This could be demonstrated in domestic animals by the success of breeders, who were able to change both anatomical and behavioral traits by selecting for parents of the next generation the animals having the traits desired. Small differences among individuals could be used to alter the characteristics of a species by humanly imposed selection over relatively short periods of time. Could the same thing not have occurred in nature over much longer periods of time? If Lyell's ideas were correct, the earth was very old, and one could think in terms not of tens or hundreds of generations, but of thousands, perhaps tens of thousands. Under such conditions, with some sort of selective principle in operation, it was possible that the graded differences that meant separate species to the catastrophists actually indicated relationship, that species were derived one from the other, and that similar species were derived from a common ancestor. But what sort of selective principle could have operated in wild nature from the dawn of living things?

In 1838 Darwin chanced to read an essay by Thomas Malthus, and found the essence of his principle. Malthus had written on humanity, glumly noting that our population tended to increase geometrically, whereas the food supply did not; moreover, were it not for war, famine, and disease, we would soon increase to such a point that none could live. These checks on the population imposed a continuous struggle for existence among individuals for a place in the world. Darwin saw immediately that the same condition obtained throughout the whole of nature. The natural checks on population growth acted analogously to the selection techniques used by humans on domestic species, in that some parents were favored to leave more progeny than others. Any wild species has the capacity to produce far more offspring than it numbers, and in fact it does, in the normal course of events, produce as many as it can. Relatively few survive. Those that do, however, tend to come from parents slightly better adapted to the environment. These slight genetic advantages are passed on in turn to future generations of the favored lines, and over a long period the species becomes modified to the extent that it may appear to be a form different from its ancestor. The struggle for existence created by an excess population tends to favor those best adapted and eliminates those less well adapted. The variety among individuals within a species, so annoying to the Linnaean typologists, was the raw material of evolution.

Darwin called his principle natural selection, and he hoarded it for 20 years. This was not capriciousness on his part; he well knew the fate of the earlier evolutionists. Perhaps, like Roentgen when he first saw through his own flesh to his living skeleton in the experiment in which X rays were discovered, he was unnerved by the power of what he had found. In addition, he correctly predicted that his materialistic ideas would create havoc in some quarters. He therefore found it prudent to retire and gather enough facts to make his theory irrefutable. There is also little question that such a course suited his personality. He was not contentious, not given to public debate; indeed, he was not up to it. He had come off the *Beagle* ill, perhaps with a tropical fever, never to recover fully.

He therefore isolated himself and his family on an estate in a small village in Kent, where he worked on his experiments and put together the cohesive argument that would eventually culminate in *The Origin of Species*. It was not until 1842 that he briefly outlined his theory. In 1844 he wrote a larger version, which he showed to Charles Lyell and to a famous botanist, Joseph Hooker. Thereafter he labored in relative silence, thinking, building his case. He was apparently anxiety ridden and suffered from insomnia; some of his illness may in fact have been psychosomatic. But he persevered, no doubt perplexing a good many, who were unable to comprehend the intangible character of his work. His gardener is supposed to have told a visitor once who inquired about his master's health: "Poor man, he just stands and stares at a yellow flower for minutes at a time. He would be better off with something to do" (Eiseley, 1956).

Darwin continued at much the same pace, in contact with Lyell and Hooker but few others about his project, for 14 years. In 1856 Lyell warned him that he should publish or that he would be anticipated, but even these words, which would numb the heart of any modern scholar, failed to move him. Two years later another man independently discovered the principle of natural selection.

Alfred Wallace was a young naturalist on a collecting trip in Indonesia:

> The question of *how* changes of species could have been brought about was rarely out of my mind, but no satisfactory conclusion was reached till February 1858. At that time I was suffering from a rather severe attack of intermittent fever . . . and one day while lying on my bed . . . the problem presented itself to me, and something led me to think of the "positive checks" described by Malthus in his *Essay on Population,* a work I had read several years before, and which made a deep and permanent impression on my mind. These checks—war, disease, famine and the like—must, it occurred to me, act on animals as well as on man. Then I thought of the enormously rapid multiplication of animals, causing these checks to be much more effective in them than in the case of man; and while pondering vaguely on this fact there suddenly flashed upon me the *idea* of the survival of the fittest—that the individuals removed by these checks must be on the whole inferior to those that survived. In the two hours that elapsed before my ague fit was over I had thought out almost the whole of my theory, and that same evening I sketched the draft of my paper, and in the two succeeding evenings wrote it out in full, and sent it in the next post to Mr. Darwin. (Kogan, 1960, pp. 65–66)

To Mr. Darwin! It was one of the splendid ironies in the history of science, and one of the splendid transactions. Darwin was not unknown as a naturalist, having published some of his observations made on the *Beagle*, and he had previously corresponded with Wallace on the topic of the origin of species—without, however, mentioning his own views. It was not unusual that Wallace should have sent him his paper for criticism.

Darwin was, of course, quite shaken. The great secret of his work of 20 years lay in his hands, over another man's signature. The question was, what should he do, and the answer came quickly: the honorable thing. He would withdraw completely in favor of Wallace. "I would far rather burn my whole book," he insisted, "than that he or any other man should think that I behaved in a paltry spirit" (Eiseley, 1956). Fortunately, Lyell and Hooker, who knew of Darwin's priority, found a compromise by arranging to have Wallace's paper and one hastily written by Darwin presented jointly at the meetings of the Linnaean Society in 1858. Thereafter Darwin prepared what he considered to be an abstract of his final work and published it in 1859 as *The Origin of Species*. Even then, with his usual unpretentiousness, he wrote to the publisher of the book (which Ashley Montagu decribed as the most influential in human thought outside of the Bible): "Accept your offer. But I feel bound for your sake and my own to say in clearest terms that if after looking over part of the MS you do not think it likely to have a remunerative sale I completely and explicitly free you from your offer" (Huxley, 1958). The first edition sold out in a day.

Wallace, upon learning the actual state of affairs, minimized his own contribution as but one week's work to that of 20 years and became a champion of the theory, which he advanced as "Darwinism." The two men remained cordial friends, with Wallace carrying a part of the public burden that Darwin could not.

Thus the theory of evolution was well launched at last. Hooker, Wallace, and particularly Thomas Huxley, entered into a long series of public debates with those who held opposing views, and eventually they carried the day. It was not easy, for the complexities and subtleties of the theory could be simplistically misrepresented. *Origin*, for instance, had very little to say about human evolution, but it was upon this topic that much public interest centered: "Descended from the apes! My dear, we hope it is not true. But if it is, let us pray that it may not become generally known" (Dobzhansky, 1962, p. 5). So spoke a Victorian lady (who no doubt had witnessed the low and vulgar behavior of monkeys in the London zoo), epitomizing a common misinterpretation and reaction to Darwin's theory. That humans and apes might have a common ancestor became garbled into: People descended from the apes. It was with this as much as anything else that the new evolutionists were forced to contend.

That they did well enough is a matter of record. Huxley, the keenest scientific debater of them all, probably delivered the most telling riposte in a confrontation with Samuel Wilberforce, bishop of Oxford, himself a formidable if unctuous orator (known to his detractors as Soapy Sam). Wilberforce had summed his arguments; he then turned and sarcastically asked whether it was from his grandmother's or grandfather's side that Huxley claimed descent from an ape. Huxley answered the bishop's other objections; he then remarked that he would be less ashamed to have descended from an ape than from a man who, unacquainted with the genuine issues, introduced aimless rhetoric and mockery into a serious scientific discussion. The immediate audience was won, and so, in time, was the world.

THE POST-DARWINIAN VIEW

Darwin's was a true conceptual revolution, in Kuhn's (1962) sense: it so restructured the mind that nothing of the old position remained. Previous observations were of course not discarded, but were interpreted from a perspective so different that they actually appeared not the same. How often this happens in a mature science may be open to question (Feuer, 1974), as for instance when Einstein's theory made Newton's laws a limiting case for everyday objects at low velocities. Newton's laws still remained valid for that segment of experience in which they were tested. No such circumstance obtained in the immature natural science of 1859. There were no limited cases of immutability or special creation. The theory of evolution through natural selection simply replaced all ideas before it. Within five years the honorable and reclusive Darwin had been given England's highest scientific award, the Copely Medal of the Royal Society, and all of biological science turned to nature with his eyes.

The importance of a change of this magnitude cannot be overstated, and was well understood at the time, as the following remark by Wallace indicates:

> Let us consider for a moment the state of mind induced by the new theory and that which preceded it. So long as men believed that every species was the immediate handiwork of the Creator, and was therefore absolutely perfect, they remained altogether blind to the meaning of the countless variations and adaptations of the parts and organs of plants and animals. They who were always repeating, parrotlike, that every organism was exactly adapted to its conditions and surroundings by an all-wise being, were apparently dulled or incapacitated by this belief from any inquiry into the inner meaning of what they saw around them, and were content to pass over whole classes of facts as inexplicable, and to ignore countless details of structure under vague notions of a "general plan," or of variety and beauty being "ends in themselves"; while he whose teachings were first stigmatised as degrading . . . was enabled to bring to light innumerable hidden adaptations, and to prove that the most insignificant parts of the meanest living things had a use

and a purpose, were worthy of our earnest study, and fitted to excite our highest and most intelligent admiration. (Kogan, 1960, p. 68)

One sees what one wants to see, and ordinary scientists are no better about this (perhaps they are worse) than laypersons. It behooves us, therefore, if we cannot be open-minded, to be prepared. Such is the value of a great theory. Biologists spent their days differently after *Origin*. The task now was to find evidence of evolution—not, as before, meticulously to catalog the innumerable links in the Grand Ascending Chain of Perfection.

The Disintegration of the Chain

What of the Grand Chain? It was clearly incompatible with the new view, and nothing basic to it was saved. The chief assumption, that of the immutability of species, was demolished by the weight of evidence and reason in *Origin*. Progressionism had been confronted by Lyell, but Darwin disposed of it neatly by showing that life forms *could* evolve into more highly organized states, but that it was not *necessary:*

> . . . it may be objected that if all organic beings thus tend to rise in the scale, how is it that throughout the world a multitude of the lowest forms still exist; and how is it that in each great class some forms are more highly developed than others? Why have not the more highly developed forms everywhere supplanted and exterminated the lower? On our theory, the continued existence of lowly forms offers no diffi-culty; for natural selection . . . does not necessarily include progressive development—it only takes advantage of such variations as arise and are beneficial to each creature under its complex relations to life. And it may be asked what advantage . . . to an earthworm to be highly orga-nized? (1859, pp. 123–124)

The same reasoning applies down to the level of viruses. Small size and fast reproduction may be more important in certain environ-ments than a high level of complexity.

The related doctrine of ascending perfection among presently existing forms fell to a similar argument. Although it was obvious that species do differ in complexity, Darwin showed that their differences were multidimensional rather than unidimensional, and that there was no series whereby each could be considered just a little more perfect than the next. He saw populations of individuals instead of immutable types, with natural selection working to produce diversity rather than uniformity.

> As each species tends by its geometrical rate of reproduction to increase inordinately in number; and as the modified descendants of each species will be enabled to increase by as much as they become more diversified in habits and structure, so as to be able to seize on many and widely

different places in the economy of nature, there will be a constant tendency in natural selection to preserve the most divergent offspring of any one species. (1859, pp. 434–435)

Differences among individuals were therefore not due to error deviations from the species type, but were part of both the product and the process of evolution.

But there was still the problem of the links. Links were missing in the Ascending Chain, true; however, Darwin's theory also required links. Where were the intermediates between species? If species were related, where was the proof? "The links," the critics would rail. "Show us the links!" Darwin could only answer that this was the point at which his theory was most vulnerable, and that proof would have to wait upon a more extensive paleontology. "With respect to existing forms, we should remember that we have no right to expect (excepting in rare cases) to discover directly connecting links between them, but only between each and some extinct and supplanted form" (1859, p. 429). His prediction was once again correct.

With the disintegration of the Grand Ascending Chain of Perfection, the most prominent Western view of nature at the time had suddenly vanished. The enormity of Darwin's scientific achievement, if appreciated then, and it was (he was openly referred to as the Newton of biology), is perhaps even more appreciable today. It must be recalled that in 1859 biological ignorance was large in all areas. We knew nothing of the mechanisms of fertilization, heredity and variation, or of embryonic differentiation. The study of animal behavior, biogeography, and ecology had hardly begun. No good paleontological evidence of evolution, such as was later discovered for the horse and elephant, was available, and there was absolutely nothing bearing upon the ancestry of humans. Yet in *Origin* Darwin gave an excellent general account of the evolutionary process as we know it now, and further, accurately deduced the implications of his theory. *Origin* is a book, notes Julian Huxley (1958), "which after a century of scientific progress, can still be read with profit by professional biologists."

Darwin was aware of the changes he had wrought, and of their necessary conditions:

> Why, it may be asked, until recently did nearly all the most eminent living naturalists and geologists disbelieve in the mutability of species? The belief that species were immutable productions was almost unavoidable as long as the history of the world was thought to be of short duration. But the chief cause of our unwillingness to admit that one species has given birth to clear and distinct species, is that we are always slow in admitting great changes of which we do not see the steps. (1859, p. 443)

It was Darwin's genius that he could *imagine* the steps.

The theory of evolution through natural selection is not a philosophy, but, like all scientific positions, lends itself to philosophic interpretation. In that role it left nothing untouched. The world as it is must be known before we can philosophically connect it with the world as we want it to be. Darwin's theory thoroughly changed our vision of the world as it is. First, humans lost their special place; a smug, anthropocentric system, with all lower forms serving perfected humanity, was no longer defensible. The human state was not an ordained finality, but only a stage in a natural process. Second, perhaps even more frightening, was the manner of our derivation. Up from the slime, through the struggle for existence? It was not the portrait one would prefer. Human origins were different from what we had desired them to be. No longer a mystical child of the universe, the human being was a thinking animal, suspended in an organic web binding past and future, beyond sight of a benevolent creator, alone with his personality.

The New Natural Science

THE INHERITANCE OF ACQUIRED CHARACTERS

A great theory is predictive as well as conclusive; it offers new questions as well as settling old ones. Darwin's exceeded in this as in all other features. If the world were seen differently, there would be many areas, previously overlooked, to explore:

> When we regard every production of nature as one which has had a long history; when we contemplate every complex structure and instinct as the summing up of many contrivances, each useful to the possessor, in the same way as any great mechanical invention is the summing up of the labour, the experience, the reason, and even the blunders of numerous workmen; when we thus view each organic being, how far more interesting—I speak from experience—does the study of natural history become! (1859, pp. 448-449)

Among the most important issues foreseen by Darwin were the causes of heritable variation among individuals, the effects of use and disuse of anatomical structures and habits upon heritable variation, the psychology of organisms, and the origin of humans. The first two were clearly of importance to the theory of evolution itself. In a reductionistic science every mechanism must in turn have a mechanism; as, for instance, the molecules of chemistry are composed of the atoms of physics. Once evolution was admitted, the burning question centered upon the mechanisms by which the individual differences upon which natural selection depends are generated. These concerns were to occupy biologists for several decades.

The effect of use and disuse upon heritable traits, which we consider first, was not a new issue. As noted earlier, it was included in the evolutionary system of Lamarck, and, in fact, it is this limited aspect that is historically most associated with his name. Earlier opponents of Lamarck, repelled by its ethereal qualities, had rejected evolution altogether, thereby throwing out the baby with the bath. When evolution was accepted after publication of *The Origin of Species*, the baby was welcomed back, but it was not entirely dry. A part of Lamarck's system came with it. Thomas Huxley, it seems, was premature in pronouncing the Lamarckian hypothesis unfit for sound thinkers. The proposition that acquired characters may be inherited apparently exerts a certain sorcery on the mind or fulfills a profound psychological need, for it was defended vehemently by some post-Darwinian evolutionists at the time and has persisted in one form or another in the biobehavioral sciences to this day. (For an example involving Lysenko in the Soviet Union see Dobzhansky (1962), p. 26.)

Lamarckism remained an issue in immediate post-Darwinian biology chiefly because biologists were ignorant of the mechanisms of inheritance. They were even unsure how the germ plasm—the sperm and the egg in animals—is constituted. It was, in fact, around this question that the debate centered. If the germ plasm were modifiable in a predictable way over the lifetime of a parent, then that parent could pass the modification on to its offspring. If stretching the neck to reach higher and higher up a tree could somehow alter germ plasm, for instance, the potential for a longer neck would be inherited, and one had the beginning of a giraffe. If, on the other hand, the constituents of a sperm or egg could not be changed by the actions of its possessor, no such inheritance was possible. Because not enough was known physiologically to settle the affair, there ensued a long season of argument by negative evidence.

Charles Darwin was a Lamarckian in the later, limited sense because he could not convince others, and was himself not convinced, that natural selection was the sole agent of evolution. If this seems somewhat improbable and contradictory given his commitment to natural selection, one must recall the then tenuous condition of the pre-science of what would be called genetics. Darwin supposed natural selection to use small individual differences, but he was unsure as to how those differences originally arose. He saw two sources: direct action by the environment, and intrinsic variation, that is, variation among organisms due to causes originating in the organisms themselves. He was much more impressed with intrinsic variability, for he found individual differences in domestic breeds even when the animals were maintained under the same conditions. But the basic causes of such variability remained a mystery, acknowledged so by him. Under such circumstances, it was impossible to dismiss any hypothesis for which a reasonable argument could be made, and he saved a small place for Lamarckism in evolutionary theory. It was different in effect from that of natural selection, although the two might work in conjunction.

Birds that cannot fly, for example, seemed to Darwin an illuminating anomaly. He could think of no selective advantage in *not* flying, and so concluded that flightless birds had found themselves under circumstances where their wings were rarely needed, and through many generations of disuse had lost the power of flight. Or consider certain eyeless crabs living deep in caves: "In some of the crabs the footstalk for the eye remains, though the eye is gone. . . . As it is difficult to imagine that eyes, though useless, could be in any way injurious to animals living in darkness, their loss may be attributed to disuse" (1859, p. 135).

Perhaps more often, however, the effects of use or disuse were combined with natural selection. In such instances the inheritance of acquired characters provided the individual differences upon which natural selection would operate. For example, there was the case of the wingless beetles on the windswept plain off Spain:

> . . . certain large groups of beetles, elsewhere excessively numerous, which absolutely require the use of their wings, are here almost entirely absent; these several conditions make me believe that the wingless condition of so many Madiera beetles is mainly due to the action of natural selection, combined probably with disuse. For during many successive generations each individual beetle which flew least, either from its wings having been ever so little less perfectly developed or from indolent habit, will have had the best chance of surviving from not being blown out to sea; and, on the other hand, those beetles which most readily took to flight would oftenest have been blown to sea, and thus destroyed. (1859, p. 134)

Inferences such as these, coupled with expositions of genetical ignorance, and, later, with appeals back to the authority of *Origin* itself, were used by the neo-Lamarckians to further their cause. Opposition was provided by the ultra-Darwinians, so called because of their fervid devotion to the principle of natural selection. There followed then behavior normal in a victorious postrevolutionary camp; various factions fell to squabbling, each righteously waving the same book at the other.

Among the neo-Lamarckians were the philosopher Herbert Spencer and the biologist and early defender of Darwin on the continent, Ernst Haeckel. Each had his own reason for espousing the cause, but neither seemed able to present strong evidence on its behalf. On the ultra-Darwinian side were August Weismann and his supporters. Weismann had developed and promulgated the idea that the germ plasm was *not* modifiable by normal environment agencies, and therefore inheritance of acquired characters was quite impossible. It followed that natural selection alone was responsible for evolution. Weismann, however, was unable to obtain convincing evidence for his position either. It developed that he was right, but it took the discovery of Mendel's hereditary units to prove it. Meanwhile, there was a lively disagreement between two other groups, the mutationists and the biometricians.

CONTINUOUS VERSUS DISCONTINUOUS VARIATION

In the first 25 years after the publication of *The Origin of Species* there had been little open opposition to Darwin's assumption that natural selection worked upon the very small differences in a trait exhibited among individuals. The gradual accumulation of differences over many generations was considered to be the chief (if not the only, à la Lamarck) method of evolution. To the classifiers, the biological systematists, this was entirely satisfactory, for their role now was to collect the best evidence of evolution, which would necessarily come from the past. In fact, Darwin's careful habits of inquiry, reflected in his zeal for collection, had by then permeated the universities. But to the experimentalist, using materials of the present, the weight of time was crushing. How does one do work on evolution when it may require thousands of generations to demonstrate the expected change? Many experimentalists had turned to embryology, as suggested by Darwin, and it was from this field that some of the major mutationists came.

William Bateson was a young British embryologist who had come to the United States to work with a newly discovered marine invertebrate. Here he began an association with an American, W. K. Brooks, which would change his life.

> For myself I know that it was through Brooks that I first came to realize the problem which for years has been my chief concern. At Cambridge in the eighties morphology held us like a spell. It scarcely occurred to us that the supply for that particular class of facts was exhaustible, still less that facts of other classes might have a wider significance. In 1883 Brooks was just finishing his book *Heredity* and naturally his talk used to turn largely on this subject. He used especially to recur to his ideas on the nature and causes of variation. . . . The leading thought was that . . . "the obscurity and complexity of the phenomena of heredity afford no ground for the belief that the subject is outside the legitimate province of scientific inquiry." To me, the whole province was new. Variation and Heredity with us had stood as axioms. For Brooks they were problems. (Carlson, 1966, pp. 2–3)

Brooks also believed that the small variations of Darwin might not be the only sources upon which natural selection operated. He felt that much larger hereditary differences were possible and would be found if properly sought. Bateson spent the next ten years searching for and amassing numerous cases of marked differences in varieties and individuals. These differences he called "discontinuous variations." Meanwhile, he angrily noted that the collectors were adding nothing to the study of evolution. They were in the habit of saving only their best specimens and destroying those that were "imperfect," thereby destroying evidence of variation at the same time.

Discontinuous variations offered experimental evolutionists hope for something of which they had little before: a phenomenon with which to work in their own lifetime. If variation occurred in large steps, it could be possible over a manageably short period of time to manipulate the forces of evolution in order to learn about them. Perhaps even more exciting at the moment, however, the concept of discontinuous variation could lead to a new view of evolution. Such indeed was the position of Hugo de Vries, Bateson's contemporary in Holland.

Since de Vries was much impressed with discontinuous variation, he sought an organism with which it could be well demonstrated. After trying more than 100 different plants, he found the one he wanted in the evening primrose, *Oenothera lamarckiana*. It was a lovely irony, for the mechanism of evolution that the study of the primrose supported was quite the opposite from that advocated by the individual after whom it was named.

> In these cultures the species is seen to be very pure and uniform in the large majority of its offspring, but to produce an average of one or two aberrant forms in every hundred of its seedlings. It is obviously a constant and inheritable condition which is the cause of these numerous and repeated jumps. By means of isolation and artificial fecundation, these races are easily kept during their succeeding generations. (Carlson, 1966, p. 40)

The jumps were called by de Vries "mutations." A mutation occurred when in an apparently true-breeding line there was a sudden (i.e., in one generation) and very marked difference in some of the offspring. Further, by breeding the mutated individuals to one another, the mutated state could be maintained and the new line would breed true for the trait.

To de Vries, the existence of the mutated condition seemed to be evidence for a different mechanism of evolution from that envisaged by either Lamarck or Darwin. He developed a *Mutationstheorie*, in which he postulated that "new species and varieties are produced from existing forms by sudden leaps. The parent-type itself remains unchanged throughout this process, and may repeatedly give birth to new forms" (Kogan, 1960, p. 120). The role of natural selection in developing new species was thus greatly reduced, putting a gentler frame around the evolutionary process in the early opinion of T. H. Morgan, who would go on to be a leading proponent of the chromosomal theory of inheritance:

> Nature makes new species outright. From this point of view the process of evolution appears in a more kindly light than when we imagine that success is only attained through the destruction of all rivals. Evolution is not a war of all against all, but is largely a creation of new types for the unoccupied, or poorly occupied, places in nature. (Carlson, 1966, p. 41)

Meanwhile, in England, Bateson was still evangelically pursuing the sources of discontinuous variation in his own way. He strongly believed that the next great advance would not occur until they were known. Further he believed that experimentation with the crossing of related varieties and a careful accounting of the traits as they appeared in the offspring presented a unique opportunity to study heredity.

> It is perhaps simplest to follow the beaten track of classification or of comparative anatomy, or to make for the hundredth time collections of the plants and animals belonging to certain orders . . . but . . . any one of them may well take a man's life. If all the work which is now being put into these occupations were devoted to the careful carrying out and recording of experiments of the kind we are contemplating, the results . . . would in some five-and-twenty years make a revolution in our ideas of species, inheritance, variation, and the phenomena which go to make up Natural History. (Carlson, 1966, p. 5)

Bateson sought the laws of transmission of the hereditary material from parent to progeny. He had followed his own advice and was engaged in experimentation that probably quite soon would have led him to their discovery. He could not know that he had been preempted 35 years earlier by Gregor Mendel.

Ranged against the mutationists, the disciples of discontinuous variation, were the champions of continuous variation, the biometricians. The biometrical approach was begun by another of Erasmus Darwin's grandsons, Francis Galton. Galton was a spirited, intellectually restless man who already had a small reputation as an explorer, inventor, and geographer. Upon reading *The Origin of Species,* however, he felt he had found his true calling. He wrote later in a letter to his half-cousin:

> I always think of you in the same way as converts from barbarism think of the teacher who first relieved them from the intolerable burden of their superstition. I used to be wretched under the weight of the old-fashioned arguments from design; of which I felt though I was unable

to prove myself, the worthlessness. Consequently the appearance of your *Origin of Species* formed a real crisis in my life; your book drove away the constraint of my old superstition as if it had been a nightmare and was the first to give me freedom of thought. (McClearn & DeFries, 1973, p. 13)

He plunged into biology with characteristic vigor and soon discovered the topic that was to occupy him for the remainder of his life, the inheritance of mental characters. In so doing, he was instrumental in the development of at least four modern scientific disciplines: statistics, mental testing, the study of individual differences, and behavior genetics.

Galton was convinced that mental abilities were inherited in the same manner as were physical traits, and he set out to demonstrate this in a book on *Heredity Genius* (1869), published ten years after Darwin's *Origin*. His basic argument was simple: among the biological relatives of eminent men one would find more persons who had themselves become eminent than would ordinarily be expected if the potential for eminence were not inherited in some degree. However, there was an obvious problem in defining what would be ordinarily expected. What is the probability of eminence in the first place, and how many relatives of the eminent person have to be eminent to give evidence of inheritance of the character? Here lay Galton's innovation. He was aware that Quetelet, the Royal Astronomer of Belgium, had published tables showing that if very large numbers of observations are made, the values of those observations tended to be distributed in a certain way. The higher or lower the value, the less frequently it occurs in the distribution; the closer the value to the average of all values, the more frequently it occurs. Quetelet's distribution was of course the famous "normal" curve, which was described as the "law of deviation from an average."

Most importantly, the normal curve could be used predictively as well as descriptively. If, for instance, in a distribution of height only one man in a hundred were taller than 78 inches, one could predict that of any new observations only 1/100 would fall into that category. One could thus assign a probability to each division under the curve, no matter how large or small one wanted to make it. One could then detect departures in actual data from what would be expected, what was probable, under the curve. If, referring to the above example, one measured the height of all the men in a village and found that 10 out of 100 were taller than 78 inches, study of the values expected under the curve would suggest that there was an abnormal proportion of tall men. There would in fact be ten times as many as expected.

Further, since Quetelet's distribution had its inception at the gambling table, the factor that determined the value of any observation was said to be chance. Play at dice, for instance, must be according to certain rules that allow chance to operate fully in determining the outcome of any throw. Violations of those rules, such as shaving or loading the dice, leads to a game that is literally not "fair," because the operation of chance has been restricted. Improbable events, such as rolling 20 straight sevens, are, if the game is fair, merely improbable—that is, unlikely to occur under the determination of chance alone, but still possible.

Galton used exactly the same logic to apply to the inheritance of mental characteristics:

> Now, if this be the case with stature then it will be true as regards every other physical feature—circumference of head, size of brain, weight of grey matter, number of fibers, etc.; and thence, by a step on which no physiologist will hesitate, as regards mental capacity. This is what I am driving at—that analogy clearly shows there must be a fairly constant average mental capacity in the inhabitants of the British Isles, and that deviations from that average—upwards towards genius and downwards towards stupidity—must follow the law that governs deviations from all true averages. (Jenkins & Patterson, 1961, p. 13)

It is actually only fortuitous that the normal curve applies to the real world; it is not, as Galton thought, a law of nature, but a theoretical probability function that is only approximated by real events. Nevertheless, the approximation serves well, and Galton could argue on this basis, as he did, that departures from a chance distribution of eminence—that is, the tendency of eminence to be concentrated in the relatives of index cases rather than being distributed randomly throughout the population—was evidence supporting the heredity view. Lacking a means of quantifying natural ability, he relied upon the earned reputation of leaders of opinion and originators.

Using published accounts, biographies, and personal inquiry, Galton evaluated the accomplishments of men in many fields. Altogether nearly 1000 eminent men were identified, associated with 300 families. With the incidence of eminence in the population being only 1 in 4000, the result clearly indicated that eminence tended to be a family trait. Furthermore, the results suggested a decrease in the probability of eminence the farther hereditarily removed was the relative from the index case. The son of an eminent person was much more likely to be eminent, for instance, than was the great-grandson.

Galton was not unaware of the possible vitiating effects of the environment. Relatives of eminent persons might share social, educational, financial, and psychological advantages not common in the remainder of the population, and these might account for the concentration he found. In opposition to this interpretation he made three arguments: (1) Many eminent men had risen from humble backgrounds, so such advantages could not have been a necessary precondition for eminence; (2) the proportion of eminent men in England was not less than in the United States, which had a far more egalitarian educational policy; and (3) the adopted kinsmen of Catholic popes, who were given great social advantages, produced proportionally fewer men of eminence than did the sons of his index cases.

So began the biometrical school of heredity. Galton did much more, but chiefly it was his use of the normal curve in the study of heredity which launched the movement. Soon he attracted the mathematically gifted Karl Pearson, and others, to his cause, and precise methods of measuring the degree of relationship of traits in a population, such as Pearson's product-moment correlation coefficient, were developed. Always, however, the biometricians dealt with continuous variation, with differences that are extremely small. There was no room in biometry for discontinuous variations, the mutations of Bateson and de Vries.

Biometry's concern with continuous variation derived from two sources. First, Darwin had emphasized that natural selection worked upon the minute differences between individuals. Second, the kinds of traits that interested the biometricians were naturally continuous. There were no sudden jumps in intellectual ability any more than there were in height, weight, or chest size. Rather, when large numbers of people were considered, they could be arranged in a series in which the difference between any two individuals was almost infinitesimal; it was all a matter of how fine a measurement one cared to make. The mathematics and the biology of the situation seemed to coalesce with purpose.

One foundation for this belief was the concomitant belief in blending inheritance. It appeared to most naturalists, including Darwin, that what they saw in the progeny of crosses between parents differing on a trait was best explained as if the two parental germ plasms had blended together to form an intermediate. A white-blossomed plant crossed to a red-blossomed one thus would produce one with a pink flower. Moreover, the blend was *irrevocable*. The parental germ plasm, once having come together in the offspring, could not then be separated, but was passed on by that offspring as the blend it received, to be blended again with the germ plasm of its mate in their progeny. The germ plasm of an individual was thus inevitably diluted over generations. (It followed that the only way to maintain "good" germ plasm was to keep passing it around among the right people. Little wonder that the choice of mate was a serious affair.)

The blending hypothesis seemed to explain a number of other facts known to naturalists and breeders of domestic animals and plants. For instance, if one bred the progeny of a red × white cross, a pink, back to the red line, the offspring of that cross were a color between pink and red. Clearly the red × white mating had led to a half-red, half-white blend, a pink, whereas the pink × red mating had led to a three-quarters red, one-quarter white blend. Continuing in this procedure, one eventually arrived at progeny that were all red, as red as the original parent. The white had simply been diluted so many times that it was effectively lost. It could not be recovered. The same seemed to apply to animal traits and human characters as well, although for all the exact mechanism remained unspecified.

To Bateson, the elucidation of the *mechanisms* of inheritance was the central problem of biology, not to be solved by biometrical laws, which did not in fact purport to be causal, but were merely statistical regularities derived from large numbers of observations. Thus, in a study claiming to show that anatomical and mental characters are correlated to the same extent among siblings (and hence equally heritable), Pearson remarked:

> Once fully realise that the psychic is inherited in the same way as the physical, and there is no room left to differentiate one from the other in the evolution of man. Realise all this, and two mysteries have been linked into one mystery, but the total mystery is no less in magnitude and no more explicable than it was before. We know not why either physical or psychical characters are inherited, nor wherefore the existence at all of living forms, and their subjection to the great principle of selective evolution. We have learnt only a law common to the physical and the psychical (although) I would emphasize . . . that the mission of science is not to explain but to bring all things, as far as we are able, under a common law. Science gives no real explanation, but provides comprehensive description. (Anatasi, 1965, p. 117)

For Bateson this was hardly enough: "We want to know the whole truth of the matter; we want to know the physical basis, the inward and essential nature, the causes, as they are sometimes called, of heredity. We also want to know the laws which the outward and visible phenomena obey" (Carlson, 1966, p. 5).

On May 8, 1900, on a train to the meetings of the Horticultural Society, he got what he wanted. Bateson was prepared to address the meetings, and had brought some journals to read on the way. In one was a paper by Hugo de Vries recounting his discovery of results reported by Gregor Mendel in 1865 and of their reconfirmation. De Vries, in search of support for his mutation theory, had turned them up. Bateson was thunderstruck. Here was evidence of discontinous variation indeed, as well as evidence against the blending hypothesis. Here were the very laws of the transmission of characters he had known must surely exist. He altered his manuscript at once, and, upon arriving, boldly announced Mendelism to England.

Such actions were not for the uncourageous. Bateson was by this time 40 years old and had as yet to hold a permanent teaching or research position. He worked at Cambridge on an insecure yearly stipend, which he supplemented with occasional small grants. He had been, in effect, a postgraduate on fellowship for nearly 20 years. Further, his militant attitude against continuous variation, by then a tenet of Darwinian orthodoxy, left him few friends. An answer to his espousal of Mendel's rules as the principles of heredity was therefore not long in coming.

Galton and Pearson, hoping thereby to interest biologists in their biometry, had founded a journal, *Biometrika,* in 1901. In the foreword, asserting the purpose and scope of their intention, the editors reaffirmed their faith in population analysis as the source of hereditary laws. (The journal was dedicated to Darwin, and a marble statue of him appeared in a photograph cut to resemble a gothic church window.) *Biometrika* was the logical tool for a counterattack against Mendelism, and the man who began it was Professor W. F. R. Weldon, an embryologist and convert to Galton's views, and the person who, inevitably in the tide of these affairs, had originally arranged for Bateson to study at Brooks's laboratory in the United States.

MENDELISM

Gregor Mendel was an Austrian monk of the Augustinian order who lived in a monastery near Brünn (now Brno, Czechoslovakia). As such, he may seem a figure even less likely to revolutionize biology than was Charles Darwin. However, he shared Darwin's characteristics of care, patience, and modesty. He was, indeed, left unaware of the magnitude of his own discoveries and did not live to see them recognized. He may even have died thinking them a failure. But happily for him there were other events to occupy his life, and we may leave it to historians to ponder, we hope not mootly, the balance of his personal rewards and the languished acknowledgment of his labors.

It was customary in the religious communities of that day for men to be involved in either scientific or artistic projects along with their religious duties; thus Mendel's concern with basic scientific questions is not as strange as it may initially appear. His home was in a gardening and fruit-growing area. He was raised on a fruit tree farm, was much delighted with living things, both plant and animal, and thus turned naturally to experiments on plant hybridization.

Plant hybridization—the deliberate mating of unlike plants to one another—was in fact a most common and intense field of interest at the time. Almost nothing was known of cell division, reproduction, or heredity, however, and it was therefore extraordinarily difficult to find organisms suitable for analysis. Many hybridization experiments went to the extreme of crossing different species, and the offspring tended to be sterile or of very low fertility. Succeeding generations were difficult or impossible to obtain. Also, the difference in number of characters between such crosses was so great, and some of the features so complex, that a clear conception of heredity simply did not emerge.

Crosses of different varieties within the same species produce viable offspring, however, offering greater analytical opportunity (a fact noted by both Darwin and Bateson), and it was to these that Mendel turned. His major innovations were to examine only traits that differed in ways so large that they could be considered qualitative, to study only one or two of these qualitative traits at a time, and to actually count and categorize the progeny rather than relying upon a verbal summary. He was then able to describe his results in simple mathematical ratios, from which he deduced the rules of inheritance.

Mendel chose the garden pea as his experimental material because it had well-defined qualitative characters, could be crossed easily, and because it is annual, growing in the summer, which left him time for teaching during other parts of the year. Moreover, the flower contains both male and female parts, and usually fertilizes itself. Cross-fertilization can be introduced by humans, but self-fertilization is the rule in nature. Therefore, experimenters have more control over matings than they might otherwise have. In addition, the garden pea had other properties, unknown to Mendel, which made it valuable for his purpose. First, through many generations of natural self-fertilization under domestication, garden peas had developed pure lines. There was no variability, except that caused by environmental agents, within the lines, although there were heritable differences between them. Second, in the traits Mendel chose to study, one character in a line was clearly dominant or recessive to that character in another line. Thus, no intermediates were formed, a fact that was crucial for his interpretation.

One trait studied by Mendel, which is sufficiently illustrative here, is height. One line regularly grew to be 6 or 7 feet tall, whereas another, growing in the same environment, reached 9 to 18 inches. This is a difference that may be considered qualitative. What would happen if the two lines were crossed (i.e., hybridized) instead of being allowed to self-fertilize as they usually would? Under the blending hypothesis the two germ plasms would inextricably mix, half tall and half short, and the progeny would be intermediate in height. But this is not what Mendel found. All of the progeny were tall! The condition "tall" seemed to prevail completely over the condition "short." Rather than a blend, what Mendel saw and so described was a case of one condition being *dominant* over its analogous counterpart, which was called *recessive*.

Continuing with the experiment, what would happen if the tall progeny of the cross "tall" × "short" were allowed to self-fertilize? If, under the blending hypothesis, all of this group were tall (no matter how this occurred), then when they were self-fertilized there was no possibility for anything but tall progeny. The germ plasm of the talls contained nothing but the potential for "tall." In contrast, Mendel found that one-fourth of these progeny were short, just like their short grandparents, whereas the other three-fourths were tall! This was a momentous discovery, for it demonstrated clearly for the first time that the potential for a condition could be carried but not expressed in one generation and then find complete expression in a later generation, with no evidence of having been altered by association. Thus, even though the elements for "tall" and "short" were together in the progeny of the "tall" × "short" cross, when those progeny were bred, "short" appeared again in its pure state. This was the first and most important of Mendel's principles. It asserts the *purity of gametes;* that is, although the contributions from the paternal and maternal sides come together in the progeny, they do not blend, but rather retain their integrity, and may thus be expressed separately and intact in later generations. Mendel called it the disjunction of hybrids. It is known now as the law of segregation.

Such experiments were undertaken using monohybrid crosses, that is, hybrids that differed in the expression of only one condition. What would happen in dihybrid crosses? Suppose, taking two other characters, he crossed plants having round, yellow seeds with those having wrinkled, green seeds? The progeny all had round, yellow seeds, indicating dominance of these two conditions over wrinkled and green. When these progeny were self-fertilized, however, there was the expected segregation of the gametes, and something else as well. Segregation of round versus wrinkled had no connection with that of yellow versus green. It was as if two monohybrid crosses were made, neither one affecting the results of the other. This resulted in Mendel's second principle, now known as the law of independent assortment.

THE MEANING OF MENDELISM

Mendel was aware of the implications of his experiments, recognizing that the germ plasm did not blend, and that the "elements" of heredity were and remained particulate. Each element appeared to transmit one and only one form of the "differentiating characters," and which form was expressed depended upon whether the combination of elements contributed by the male and female happened to contain a dominant or were both recessive. The elements were of course unseen; they were abstractions from the observations made on the transmission of the characters from one generation to the next. From such observations it appeared *as if* there were particulate hereditary units that exhibited segregation and independent assortment. This is the common practice of science: One invents an entity that seems to account for the data at hand and that hopefully predicts new data.

Mendel came to his abstraction through experimental discovery. Before his work and while it lay in obscurity there were several other theories of heredity that had as their major premise particulate units serving as hereditary determiners. None had predictive value for laboratory tests, and none came from experimental findings. They did serve, however, as such premature theories do, to so seed the scientific atmosphere that by the time Mendel was recognized in 1900, his work was not only acceptable but comprehensible (Feuer, 1974).

Darwin himself had espoused a theory of particulate inheritance. He suggested that microscopic or ultramicroscopic particles, which he called gemmules, existed in all cells of the body. The gemmules responded to the physiological conditions of the tissues in which they subsisted and thus could be altered by the external environment. They were released at various times into the circulatory system and found their way to the gonads, where they became the germ plasm. The tissue source, the numbers, the time they were released, and the environmental conditions at play all affected the composition of gemmules that would go to make up a particular gamete, thus conferring upon it its uniqueness. Darwin's ingenious system, which he hesitantly called the provisional theory of pangenesis, thus accommodated two seemingly contradictory features: the blending inheritance thought to be true in his day, and the individual differences he required for natural selection. On the blending theory, variations among organisms would tend to be reduced over generations, but if the gametes carried by each individual were unique owing to their particular assortment of gemmules, variation could be maintained. Pangenesis also served to explain such obvious facts as the failure of the same parents to produce exactly the same offspring each time. If changes occurred inside and outside the parents, and these changes affected the gemmules and hence the germ plasm, room for differences was left. Pangenesis also clearly incorporated Lamarckian inheritance, since use or disuse would affect the organs involved, which in turn would affect the gemmules and the germ plasm. (Galton, incidently, disagreed with Darwin on the subject of Lamarckism, and thus the biometricians were not forced to defend it. To them, the physical laws of heredity remained an acceptable mystery, as evidenced by the remarks of Pearson.)

The appeal of the theory of pangenesis was nowhere felt more strongly than by de Vries. However, he too differed with Darwin on the Lamarckian issue, and so combined Weismann's idea of the separateness and integrity of the germ plasm with Darwin's gemmules in a theory of intracellular pangenesis. De Vries renamed the gemmules "pangens" and specified that, since all somatic pangens remained inside the cells and were not carried to the gonads, it was the gonadal pangens that were the units of heredity. The mutations for which he sought evidence in *Oenothera lamarckiana* were intended to supply variation.

The rediscovery of Mendel's work by de Vries (and also simultaneously by Carl Correns and Erich von Tschermak-Seysenegg, the former of whom had at the time been experimenting with garden peas and had arrived at principles similar to Mendel's) thus brought an end to one phase of biological thought and began another. The mutationists seized upon the "differentiating characters" as evidence of the generality of discontinuous variation, and upon Mendel's principles as the only verifiable laws of heredity. Lamarckism once again fell from favor, there being no place in Mendel's rules for the transmission of acquired characters. And finally, the blending model was once and for all discredited, or would be.

Mendel carried out his experiments in the years between 1856 and 1864. He read his results and conclusions before the Brünn Natural History Society in 1865, and they were published under the title "Experiments in Plant Hybridization" in the proceedings of the society in 1866. For the next 34 years they may as well not have existed. We know them now to be fundamentally correct and know they were so regarded upon their rediscovery, at least by some. Certainly they were at the center of biology's concerns. Why did they sleep so long?

There are undoubtedly many reasons. For one thing, science was not Mendel's only professional interest. He was elected abbot of his monastery in 1868, and for nearly all of the last ten years of his life he struggled against the political anticlerics in Austria who were imposing taxes upon religious organizations. This did not leave him entirely free to publicize his scientific views. For another, Mendel's publication in 1866 came only seven years after *The Origin of Species*, Darwin's enormous influence was at its initial height, and the theory of natural selection emphasized small, not large, differences in the forms of characters. Also, the mathematical treatment used by Mendel may have been too alien for naturalists of the time. Darwin was nonmathematical in his approach, as were most of his contemporaries. But Mendel had taken mathematics at the University of Vienna and taught physics for 14 years at the Brünn modern school, and thus he was familiar with mathematical concepts and their application to real objects.

He was also very well aware of the resultant confusion in the study of heredity when one attempted to examine too many characters at once. The potential combination of forms was just too large to reveal the simple laws which underlay them. It was for this reason that he preferred to examine one character at a time. His reasoning is evident in the following passage:

> . . . the series in each separate experiment must contain very many forms, since the number of terms . . . increases with the number of differentiating characters as the powers of three. With a relatively small number of experimental plants the result therefore could only be approximately right, and in single cases might fluctuate considerably. If, for instance, the two original stocks differ in seven characters, and 100–200 plants were raised from the seeds of their hybrids to determine the grade of relationship of the offspring, we can easily see how uncertain the decision must have become since for seven differentiating characters the combination series contains 16,384 individuals under 2,187 various forms; now one and then another relationship could assert its predominance, just according as chance presented this or that form to the observer in a majority of cases. (Bennett, 1965, p. 44)

Such probabilistic considerations, though basic to transmission genetics today, were not to the taste of leading naturalists then. Mendel's ratios, upon which his laws were built, did not please Karl Nägli, an eminent Swiss botanist, who considered the experiments "empirical rather than rational." Nägli held to a grand, speculative vision of a creative force in nature, a *Vervollkommingskraft*, which propelled the variations of species in a particular direction (Feuer, 1974, p. 282). The passing of Mendel's "elements" from generation to generation in a probabilistically determinate way was antithetical to his thinking. He thus informed Mendel in personal correspondence that his experiments were unfinished and in need of further verification. One other person, W. O. Focke, cited Mendel in this period, but to little avail. The consensus now is that his work simply was not, perhaps could not have been, understood.

Finally, there was the question, to be revisited, whether Mendel's conclusions applied generally. His results seemed uncommonly clear, and quite unlike those found in other organisms. Breeders and naturalists had, after all, been experimenting for centuries and had not seen their equal. In addition, Mendel himself became uncertain when upon the advice of Nägli he attempted to replicate his data in different plants. It must be recalled that domestic garden peas have some traits not shared by all other organisms. Thus, when he crossed varieties of hawkweed, the progeny did not give evidence of a segregation of elements from both parents, but rather looked like the female side. We now understand that hawkweeds are apomictic (capable of reproduction without fertilization). Therefore, the offspring that Mendel considered to be hybrids were not hybrids at all. But since he could not have seen these reproductive differences, he may have considered that he had disconfirmed his own results. Little wonder he failed to publicize them further and was never to know of his legacy. Alas, this sort of episode has occurred in intellectual affairs with an alarming frequency. If there is a devil in science his handiwork is most evident, and he must he happiest, at times like these.

BEANS AND WHEAT: CONTINUOUS VARIATION
AND MULTIFACTORAL INHERITANCE

In 1900, however, Bateson was alive and well and anxious to do battle. Weldon, speaking for the biometricians, had studied the Mendelian evidence and concluded that if indeed it were valid, it was anomalous and irrelevant. He first objected to the fact that in Mendel's system the condition of the progeny depended solely upon that of the parents, whereas:

> . . . it is well known to all breeders, and it is clearly shown in a number of cases by Galton and Pearson, that the condition of an animal does not as a rule depend upon the condition of any one pair of ancestors alone, but in varying degrees upon the condition of all of its ancestors in every past generation, the condition in each of the half dozen nearest generations having a quite sensible effect.

Next, he noted that in most organisms and for most characters the distribution is continuous, nondiscrete:

> . . . if Mendel's statements were universally valid, even among peas, the character of the seeds in the numerous hybrid races now existing should fall into one or another of a few definite categories, which should not be connected by intermediate forms. . . . I have carefully examined the seed characters of some twenty named varieties, and the present condition of many I have studied seems to me quite incompatible with the general validity of Mendel's statements. (Carlson, 1966, p. 10)

Weldon then cited his own and the observations of others showing commercial varieties whose seeds were wrinkled (wrinkling of the seeds was a differentiated, that is, discrete, character in Mendel's experiments) in a continuous curve from barely perceptible dents to sharply corrugated troughs. Similarly, he found yellow strains that showed degrees of color from chlorotic tints to full green, often with piebald patches of green interspersed with the yellow. Where, then, was the complete separability of characters upon which the Mendelian laws were based? It was clearly Darwin's and Galton's view of heredity, not Mendel's, that should prevail:

> . . . the fundamental mistake which vitiates all work based upon Mendel's method is the neglect of ancestry, and the attempt to regard the whole effect upon offspring, produced by a particular parent, as due to the existence in the parent of particular structural characters; while the

contradictory results obtained by those who have observed the offspring of parents apparently identical in certain characters show clearly enough that not only the parents themselves but their race, that is their ancestry, must be taken into account before the results of pairing them can be predicted. (Carlson, 1966, p. 11)

This was critical evidence, and it required a substantial reply. Bateson therefore chose to write a book, *A Defense of Mendel's Principles of Heredity*, in which, far from conceding a weakness in Mendelism, its potential was stressed. Included was a detailed report of his own work as further data supporting discontinuous variation. The crucial issue was the possibility of two kinds of inheritance, and upon this Bateson found no compromise: "[T]he Mendelian principle of heredity asserts a proposition absolutely at variance with all the laws of ancestral heredity, however formulated. In those cases to which it applies strictly, this principle declares that the cross-breeding of parents need not diminish the purity of their germ cells or consequently the purity of their offspring." But if there was to be no compromise on rival views, how could continuous variation be explained?

Bateson did not know. It did appear as though the most important part of Mendel's discovery was not dominance, recessivity, and differentiating characters, but the purity of the gametes. Under this interpretation what seemed to be blended inheritance might very well be Mendelian, without dominance. Mendelism might therefore incorporate Galtonism.

> It is difficult to blame those who on first acquaintance concluded Mendel's principle can have no strict application save to alternative inheritance. Whatever blame there is in this I share with Professor Weldon and those whom he follows. Mendel's own cases were almost all alternatives; also the fact of dominance is very dazzling at first. But that was two years ago, and when one begins to see clearly again, it does not look too certain that the real essence of Mendel's discovery, the purity of the germ cells in respect of certain characters, may not apply also to some phenomena of blended inheritance. (Carlson, 1966, p. 12)

With such commitments on the part of the participants, the edge of scientific nicety began to fray. *Biometrika* rejected one of Bateson's papers, and in 1903 the editor of the prestigious journal *Nature* returned another with a note that the editor was "not prepared to continue discussion on Mendel's principles" (Feuer, 1974, p. 279). The Cambridge University Press (Weldon was at Oxford), in turn, arranged for a private printing of one of Bateson's rejected papers, pointedly using the format of *Biometrika*. Things continued in this vein up through the meetings of the British Association in 1904. Bateson, by now in high umbrage, declared that "The imposing Correlation

Table into which the biometrical Procrustes fits his arrays of unanalysed data is still no substitute for the sieve of a trained judgment" (Feuer, 1974, p. 280). As the session ended, Pearson proposed a three-year truce.

More productive events were, fortunately, under way on the continent. In 1903 Danish botanist Wilhelm Johannsen began publishing an analysis of continuous variation in beans. Using seed weight as his character, he first noted that in a mixture of seeds obtained from a number of different plants, there is a continuous, normal distribution (just as for the characters studied by the biometricians). Applying selection over several generations by mating plants producing heavier seed to one another on the one hand and plants producing lighter seed to one another on the other, he first demonstrated that seed weight is heritable. The progeny of the "heavy" × "heavy" seed matings produced heavier seed, whereas the progeny of the "lighter" × "lighter" seed matings produced lighter seed. The seeds in the original mixture, taken from different plants, thus differed from one another at least in part because they contained different heredity factors.

Johannsen next took advantage of the fact that beans can self-fertilize as well as be cross-fertilized by the actions of humans. He allowed 19 plants to self-fertilize. Keeping the seeds from each plant separate, he noted that within each set there was a continuous distribution of seed weight that tended toward normal, just as there had been with the mixture of seeds from many plants. However, when selection was applied as before by mating "heavy" × "heavy" and "light" × "light," but still maintaining each of the 19 lines separately, there was no response! There continued to be the same distribution within each line as when selection was begun.

Johannsen had in these experiments greatly clarified the nature of continuous variation by demonstrating that it was due to *both* heritable and nonheritable factors. In a mixture from many lines there would be heritable differences. In a pure line, achieved by self-fertilization, the differences were due to the effects of the environment. It followed that continuous variation under usual circumstances was due to a combination of heritable and nonheritable factors, in sharp contrast with the discontinuous variation of Mendel, Bateson, and de Vries, wherein the heritable elements alone played the crucial role. Further, Johannsen showed that although heritable effects could be small in terms of visible differences between his beans, the effects of selection were considerable and rapid.

Johannsen also made important contributions to the concepts and terminology of heredity. First, he gave heritable elements the name we know: *genes.* He wanted to dissociate them from any role played in previous theories involving particulate elements. Second, he wished to dissolve any necessary hypothetical ties of a single particulate unit with a single differentiated character. The unit closest to what we had in mind was the pangen of Darwin and de Vries. To avoid the historical connotations of this term, Johannsen chose to abbreviate it simply to "gene": "The word 'gene' is completely free from any hypotheses; it expresses only the evident fact that, in any case, many characteristics of the organism are specified in the gametes" (Carlson, 1966, p. 20). Finally, he made a fundamental distinction, which followed from the separability of heritable and nonheritable effects. The total genetic complement of an individual he called the *genotype;* the part that is visible, measurable, apparent (such as the actual weight of a seed) he called the *phenotype.* Variations in genotype were all due to differences in genes. In contrast, variation in phenotype could be due to variation in genotype *and* variation in environment. One of the central concerns of genetics since its inception has been to infer events at the genotypic level from observations made at the phenotypic level.

Using Johannsen's analyses, Bateson saw immediately that Galton's cases of blending could be incorporated into the Mendelian framework. If what was important about Mendelian inheritance was not discontinuous variation per se (although this originally provided the key for Mendel's rules) but the purity of gametes, and if heritable and nonheritable factors both affect continuous variation, then the apparent blending in Galton's cases could simply be due to a combination of heritable and nonheritable effects. The effects of the genes were not so obvious because of fluctuations in the environment.

The blending hypothesis continued to lose credibility when it was shown that several genes, each having a small effect but the total effect being cumulative and depending upon the number of such genes present, could account for the heritable portion of continuous variation. H. Nilsson-Ehle of Sweden and E. M. East of the United States were the first to demonstrate what came to be known as multifactoral inheritance. Nilsson-Ehle crossed wheat having red kernels with that having white kernels. The progeny were intermediate in color. When the progeny grew and were bred among themselves, they produced a third generation that had the full range of color between red and white. Upon close classification, however, it was revealed that about $\frac{1}{16}$ of these kernels were as red as the original red parent, about $\frac{4}{16}$ had more red than the intermediates, about $\frac{6}{16}$ were just intermediate, about $\frac{4}{16}$ were lighter than the intermediates but not as white as the original white parent, and about $\frac{1}{16}$ were as white as the original parent. This obviously symmetrical but discrete distribution could be explained if in each parent two pairs of genes were segregating, each affecting the color in one direction or the other an equal amount. The more genes determining red a kernel had, the redder it was; the more genes determining white, the whiter it was.

Multifactoral inheritance, plus the effects of environment as demonstrated by Johannsen, meant that all cases of blending could be accounted for by postulating numerous genes operating on the character, with whatever gaps were left in the distribution being filled by environmental fluctuations. A whole new approach to genetics was thereby begun.

In 1918 Ronald Fisher, who was to contribute greatly to the development of both genetics and statistics, showed that with certain assumptions the correlations obtained by the biometricians followed mathematically from the assortment of Mendelian genes. Thus, the statistical regularities adduced by Galton and Pearson (the .50 correlation in mental ability between parents and offspring and between siblings, for example) maintained empirical validity but were subsumed under Mendelian rules. Finally, in 1930 Fisher published *The Genetical Theory of Natural Selection*, demonstrating that natural selection was the only possible evolutionary mechanism, given Mendelian inheritance. Blending would reduce genetic variance, and high mutation rates were required to sustain it, rates much higher than were in fact found to occur. On the other hand, if the genes did not blend, then genetic variance could be maintained under mutation rates far less, particularly since most mutants are recessive, and thus could be stored, ready to provide new combinations of genetic material when environmental conditions changed. Fisher remarked:

> The tacit assumption of the blending theory of inheritance led Darwin, by a perfectly cogent argument, into a series of speculations, respecting the causes of variations, and possible evolutionary effects of these causes. In particular the blending theory, by the enormous mutation rates which it requires, led Darwin to some extent and others still more to attach evolutionary importance to hypothetical agencies which control the production of mutations. A mechanism (Mendelism) of particulate inheritance has since been discovered, requiring mutations to an extent less by many thousandfold. . . . The whole group of theories . . . must be set aside, once the blending theory of inheritance is abandoned. The sole surviving theory is that of Natural Selection. (Kogan, 1960, p. 133)

Fisher's *Genetical Theory* thus eliminated one of the major difficulties that beset Darwinian evolution while at the same time maintaining its integrity. Darwin, that "working dreamer" with the orderly mind, who was never at ease with the sources of variation he felt forced to accept, could be content at last.

Fisher's work was, for our purposes, the capstone of the Darwinian edifice. Much, much more has since been accomplished by those interested in evolution and genetics, but it has been toward refined understanding, intricate instrumentation, and sophisticated experimental design. Nothing has altered the general proposition that evolution occurs through natural selection operating on genes transmitted according to Mendelian principles. Mutation—that is, the alteration of a gene from one state to another—has been found to be of any extent, large or small, and provides new genetic material, but it does not guide evolution. It is thus something of a historical quirk that the mutationists for a time put Darwinism into disfavor, even leading, in Julian Huxley's terms, to its eclipse. The enthusiastic Bateson, indeed, in his relentless pursuit of the biometricians,

> . . . did not hesitate to draw the most devastating conclusions from his reading of the Mendelian facts . . . assuming that change in the germplasm is always by large mutation and secondly that all mutation is loss, from a dominant something to a recessive nothing, he concluded that the whole of evolution is merely an unpacking. The hypothetical ancestral amoeba contained—actually not just potentially—the entire complex of life's hereditary factors. The jettisoning of different portions of this complex released the potentialities of this, that, and the other gorup and form of life. Selection and adaptation were relegated to an unconsidered background. (Huxley, 1964, p. 24)

But Bateson's errors were soon put right, and since that time data from all branches of the life sciences, from biochemistry to zoology, have converged to confirm the correctness of the selectivist approach.

BIOLOGY AND SOCIETY IN THE POST-DARWINIAN ERA

A great change in scientific thought does not necessarily lead to a change in moral philosophy, although there appears to be a correlation. In the initial stages this is probably due less to the conversion of the unlike-minded than to the liberation of those who, for whatever reasons, find the new thought congenial to their prejudices and are eager to see it extrapolated into other realms of discourse. The first friends of a new scientific movement in areas outside (as well as inside) science, then, are likely to be those whose needs it serves.

So it was that *The Origin of Species* was seized upon, and its principles translated and very often overextended, into areas of human enterprise ranging from poetry to statecraft. A timely example is recorded in the biography of the English mathematician William K. Clifford:

> For two or three years the knot of Cambridge friends of whom Clifford was the leading spirit was carried away by a wave of Darwinian enthusiasm: we seemed to ride triumphant on a new life and boundless possibilities. Natural selection was to be the master-key of the universe. . . .

Among other things it was to give us a new system of ethics, combining the exactness of the utilitarian with the poetical ideas of the transcendentalist. We were not only to believe joyfully in the survival of the fittest, but to take an active and conscious part in making ourselves fitter. At one time Clifford held that it was worth while to practise variation of set purpose; not only to avoid being the slaves of custom, but to eschew fixed habits of every kind, and to try the greatest possible number of experiments in living. (Feuer, 1974, p. 331)

Intellectuals found themselves with a new set of clothes and avidly altered their manner to suit them.

In the United States in 1860, however, there were more urgent problems than speculation upon the deportmental implications of *Origin*. An election was beginning that would be followed by the Civil War. Thus, although Darwin's book received wide review, its impact on all but a few professionals was delayed until the postbellum period. Then it swept the land like a firestorm.

The reasons are not far to find. Disaster opens the mind; suffering is, alas, the forge of receptivity. Of suffering there had been plenty, and monstrous hate, not easily ameliorated by the old pieties. It was a time of turmoil, of restlessness, of uncertainty. Fundamentalist faith was shaken, and biblical criticism and the study of comparative religion were on the rise. To Henry Adams, evolution could replace religion. "Unbroken evolution . . . was the very best substitute for religion: a safe, conservative, practical, thoroughly Common-Law deity . . . the idea was only too seductive in its perfection; it had the charm of art" (Hofstadter, 1955, p. 15). It had even more. In a torn and bitter land, it offered a new cause for optimism. Science could do what God would not; it could guarantee the fruition of human virtue, the perfection of man. Charles Brace, a leading social reformer, having read *Origin* 13 times, decided that "if the Darwinian theory be true, the law of natural selection applies to all of the moral history of mankind, as well as the physical. Evil must die ultimately as the weaker element, in the struggle with good" (Hofstadter, 1955, p. 16). Natural selection was to sit in judgment of beneficence and malevolence, and find the former more fit!

Why not? It could even be a part of God's subtle plan. Asa Gray, an American friend and confidant of Darwin's publicly defended the principles in *Origin* but, anticipating a clerical reaction, was careful to note that they were not incompatible with theism. Evolution through natural selection may have been God's order all along. The proposal was in fact put directly to Darwin:

A celebrated author and divine has written to me that "he has gradually learnt to see that it is just as noble a conception of the Deity to believe that He created a few original forms capable of self-development into other and needful forms, as to believe that He required a fresh act of creation to supply the voids caused by the action of His laws." (1895, p. 443)

Such an interpretive retrenchment actually saved a place for something that is antievolutionary: the argument from design. This proposition was popularized by the English theologian William Paley, who declared that one may infer the existence of a designer from his apparent work. Just as one infers a watchmaker from a watch, so the existence of complex forms and relations in the organic and inorganic world requires the inference of a worldmaker. (The idea is still in use, although now the person who approaches you is more likely to hold up a flower.) But it was Darwin's intent to show that complex forms could *evolve*, given time and natural selection, and therefore Gray's placation may seem questionable. His concern, however, was to make no more enemies for Darwin than he had to, and he evidently succeeded.

Another set of conditions was being altered simultaneously, the outcome of which was to promote the evolutionary view. Classical education began to give way to the scientific, as demand for technically trained people grew. The prestige of the scientists in turn increased, and it was the tension and thrill of their affairs which charged the atmosphere of the universities. Evolution and its potential implications permeated faculties and student bodies alike. Thus, in an address at Dartmouth in 1873, Whitelaw Reid remarked: "Ten or fifteen years ago the staple subject here for reading and talk, outside study hours, was English poetry and fiction. Now it is English science. Herbert Spencer, John Stuart Mill, Huxley, Darwin, Tyndall, have usurped the places of Tennyson and Browning, and Matthew Arnold and Dickens" (Hofstadter, 1955, p. 21). Meanwhile, O. C. Marsh of Yale had collected a set of American fossil horses that Darwin later acclaimed as the best evidence supporting evolution that had appeared in the 20 years following *Origin*. And psychologist William James was writing to his brother Henry of the antievolutionist Louis Agassiz: "The more I think of Darwin's ideas, the more weighty do they appear to me, though of course my opinion is worth very little—still, I believe that that scoundrel Agassiz is unworthy either intellectually or morally for him to wipe his shoes on, and I find a certain pleasure in yielding to the feeling" (Hofstadter, 1955, pp. 17–18).

The conversion of scientists promised early success in the universities. At the same time, popular journals opened their pages to the evolution controversy, thereby ensuring a wide hearing. Huxley, on a speaking tour in the United States, found his lectures reprinted in full in the newspapers and was accorded the ceremonial treatment of royalty. Evolution had become de rigueur, so much so that a reporter sketched the members of the Senate in Darwinian terms as bulls, lions, foxes, and rats, and at the latest Mardi Gras one costume motif was the "missing link." Clearly Darwin's ideas were moving swiftly to all levels.

A final reason for the acceptance of the evolutionary view so radidly here lies in our proud mystique of self-sufficiency and opportunity. We had, after all, begun as a nation by wresting away the land in conflict, and had for 100 years continued to expand by much the same practice. Clearly, the measure of a man was what he could carve out, almost in defiance of high-born privilege. If only he would master his tools and himself, there was a decent and independent place for him in the world. We like to believe this today (which is not to say it is altogether wrong), and such faith, following the trial of the war, was even more of value. A scientific proposal that could be taken to mean that the hardworking and adaptable were the most deserving by force of natural law offered cause for renewed optimism.

Such naive faith must be, and was, short-lived. Whatever evolves, something evolves to feed upon it. In this case the principles of *The Origin of Species* were misapplied by sometimes well-meaning but overzealous moralists who sought and found a too-direct relationship between those principles and a behavioral credo. Some were philosophers, some naturalists, sociologists, or businessmen. The less technical aspects of the intellectual effort have come down to us under the horrible misnomer of "Social Darwinism."

"Social Darwinism": The Individual Phase

The most notable of those attempting seriously and systematically to align the new biology with human ethics was the British philosopher Herbert Spencer. Spencer had been an evolutionist before *Origin*. He had published books on psychology and ethics (and later claimed to have anticipated Darwin on natural selection). He nevertheless integrated Darwin's ideas with his own to produce a massive world system that accounted for the behavior of all life under one general process. Using the recently discovered law of conservation of energy, coupled with the proposition of the embryologist von Baer that each organism proceeds from a state of homogeneity to a state of heterogeneity, he inferred that development is inevitable, in organism or society, through a series of equilibrations and disintegrations, until a last stage is reached. In the single organism, this results in death. However, society continues to evolve to a condition of final equilibration wherein all is harmony and stability. "Evolution," wrote Spencer, "can end only in the establishment of the greatest perfection and the most complete happiness" (Hofstadter, 1955, p. 37).

Such a proposal, however befuddling, would not ordinarily engender hostility among the societally oriented. But Spencer combined it with an intense social laissez-faire in which he advocated a continuous economic struggle for existence among individuals as the price of progress. For several decades, particularly in the United States, his doctrine was accepted, and his fame and influence came to exceed that of Darwin.

It is not difficult to understand why the ideas in *Origin* suited Spencer's philosophical predilections so well. The concept of organic evolution through natural selection need only be extrapolated in order to produce an ethic for the industrial state. Make analogous the organic and societal conditions, and progress in both could be determined by the same principle. Interference in the operation of that principle would be, in turn, certain to deter progress. The ethic of laissez-faire was therefore all-subsuming. Success in competition demonstrated a selective advantage and therefore dictated the rights to wordly goods. Industrialization's human waste, unhappily, must be consigned to the dumping ground of social development during the long process to final adaptation. "Nature," Spencer proclaimed, "is a little cruel that it may be very kind."

It would be a mistake, however, to conclude that Spencer or his philosophical followers felt *themselves* to be cruel. It was simply a matter of logic. Although the existing human moral constitution still contained undesirable elements, we would eventually evolve a new morality in keeping with the needs of civilized life. Human perfection was not only possible but inevitable, so long as the natural processes remained unchecked. This would be maintained through continuous adaptation to the conditions of life. The root of evil was failure to adapt. (Spencer's Lamarckism may be seen to play a central, if under-emphasized, role in all of this: Presumably the genetic residual of proper moral behavior would be passed on in the germ plasm, thus inclining the next generation to behave even more properly. But so would the residual of improper or slothful behavior. Such actions were all the more to be discouraged, the progeny of parents so behaving thus to be all the more unprotected from their natural fate.) Spencer was himself a pacifist and an internationalist, believing war to be but a passing stage that would soon lose its utility and die out. Further, he fought for the rights of women and children in an age when it was not popular. His thrusts, rather, were reserved for individuals who he felt detracted from societal advance because they could not or would not successfully adapt.

It was Spencer, not Darwin, who coined the phrase "survival of the fittest," and although Darwin used it in later years, it was much more in keeping with Spencer's interpretation of the Malthusian thesis. Darwin's original choice had been "struggle for existence," which, although it conveys a competitive tone, does so with less emphasis upon direct conflict. Spencer appeared to glorify the victor. This, along with other aspects of his philosophy, made it particularly applicable to the American scene. It was scientific, comprehensive, and (to many) reassuring. And America was ready for it: "With its rapid expansion, its exploitive methods, its desperate competition, and its peremptory rejection of failure, post-bellum America was like a vast human caricature of the Darwinian struggle for existence and survival of the fittest. Successful business entrepreneurs apparently accepted almost by instinct the Darwinian terminology which seemed to portray the conditions of their existence" (Hofstadter, 1955, p. 44). (One might wish to see Spencer's name substituted for Darwin's.)

Guests at plush dinners and public banquets in New York could, and did, consider themselves the evolutionary elite, deserving of their good fortune because their "superior ability, foresight, and adaptability" had been proved in competitive economic combat. John D. Rockefeller served notice that "The growth of a large business is merely a survival of the fittest. . . . It is merely the working out of a law of nature and a law of God." And Andrew Carnegie, troubled over the collapse of Christian theology, recalled: "I remember that light came as in a flood and all was clear. Not only had I got rid of theology and the supernatural, but I had found the truth of evolution. 'All is well since all grows better' became my motto, my true source of comfort" (Hofstadter, 1955, p. 45). Science and business were one.

It did not last. Laissez-faire ethics puts a premium not only on economic but political noninterference, an intolerable condition to the many who saw the need for political and social reform. Spencer, as had Compte before him, built his sociological system upon the premise that universal physical and biological laws are directly applicable to human societies. One could, however, argue that the evolution of the human intellect, with its awareness of its own consciousness and of purpose, had altered the process of evolution itself. We did not have to wait eons for social change; we could make the changes ourselves. Such an argument was made by Lester Ward.

Ward was the son of an itinerant mechanic, brought up in poverty, severely wounded in the war, and chiefly self-educated. He resented the aristocratic aura of Spencerism (to the end of his life he remembered how, as a child in public school, he had felt keen satisfaction whenever the ragged boys of his own class beat the rich men's sons), and in the United States became its most formidable critic. He attacked the assumption that biological and human social phenomena are equatable in natural law. To him sociology became a special discipline dealing with a novel and unique organization, with which biological and physical laws had little to do. By the time his view succeeded, however, Spencerism had passed into a new and modified stage.

"Social Darwinism": The Collective Phase

For a short period, as the social and political ramifications of the ethics of unrestrained competition at the individual level came to be denounced by those seeking reform, Spencerian precepts were applied at the level of international relations. If progress through competition was natural among individual organisms, why not among nations? Had it not always been so? Was it not prudent to be ready? Was it not obvious that certain nations exhibited the energy and organization to dominate, and would it not be unnatural for them to fail to do so? Was there not, for some, written in the selective history of a thousand years, a manifest destiny?

International Spencerism suited the spirit of the times, particularly in the United States. It was thus used as an intellectual vindication for our flirtation with imperialism, lasting roughly from just prior to the Spanish-American War to just before World War I. During that period we took the Philippines from Spain, annexed Hawaii, partitioned Samoa with Germany, and joined other Western powers in suppressing the Boxer Rebellion in China. All the while, we sought justification for our ideas from a presumed cosmic order. We were not alone. The Europeans were hardly idle. Indeed, following the Franco-Prussian War, both sides had invoked Darwinian principles as a higher level pretext for their behavior.

But this had nothing to do with science. The principles of natural selection had simply been taken into the service of the imperial inclination. War and conquest had notable devotees before 1859. Darwin, or more properly Spencer, only provided an excuse for those whose enthusiasms led them to foresee a quick and easy expansion of the nation's influence and territories. That enthusiasm would end in the life-absorbing trenches of France.

Meanwhile, another interpretive retrenchment, similar to that involving the religious argument from design, was under way. Racism was reintroduced as, if not a part of God's plan, at least a part of Nature's. In the Grand Ascending Chain of Perfection each species had had its place, and within each species, each variety, including those of humans. Suppose the chain were invalid. Looking at the progress of each race, was it still not possible to infer that the difference in attainment were due to differences in rate of evolution? Could it not be that some races were selectively more gifted than others? And if that were true, was it not the right and duty of the evolutionarily superior peoples to expand at the expense of those more primitive?

No group has a monopoly upon the underlying bases of such attitudes (although some appear to have an inordinately freakish capacity to publicize them). However, it fell to the world-dominating Europeans, as colonizers from the seventeenth to the nineteenth centuries, and to Americans, as slave holders and Indian fighters during much of the same period, to develop global descriptions and categorizations of peoples of different origins. These could be put to use under almost any set of circumstances, in good times and in bad. We shall therefore examine very briefly previous European and American thoughts on racial differences.

In so doing, however, some indulgence must be given, particularly to the earlier writers. They did not have our information, and what information they did have was viewed from an entirely different perspective. Valid observations were actually so scarce that the first question had to do not with differences among races of people but with which races in fact *were* people. (It is not as eccentric as it sounds. European explorers of the fifteenth and sixteenth centuries had continually brought back excited stories of jungle creatures intermediate between humans and animals—humanoid, but lacking human attributes, particularly the power of speech. Moreover, tales of bizarre (and barbaric) customs of some admitted humans not living on the continent only tended to confirm the suspicion that God had filled the world with an awesome multitude of forms.)

The chief concern of natural scientists of the time was the systematic classification of specimens. Relying on what data they had, they did so classify. But the human data were so poor that Linnaeus, for instance, felt obliged to put humans and chimpanzees into the same genus. In 1758 he distinguished two species, *Homo sapiens* (humans) and *Homo troglodytes* (anthropoid apes). "I shall confess," he had written earlier (1746) "that as a natural historian I have so far been unable to dig up any character by which to distinguish (them)" (Baker, 1974, p. 31). We, who are accustomed to seeing chimps do clever tricks, wear clothes, and communicate using human sign language, should not be too amazed at such an assignment, particularly in view of the accounts then coming from Africa and Borneo of the wild, hairy people who lived in trees. Lord Monboddo, a Scottish lawyer and philologist of the time, did not hesitate to write of the "Ouran Outangs":

> They are exactly of the human form, walking erect, not on all four . . . they use sticks for weapons; they live in society; they make huts of branches of trees, and they carry off negro girls, whom they make slaves of, and use both for work and pleasure. . . . But though from the par-

ticulars mentioned it appears certain that they are of our species, and though they have made some progress in the arts of life, they have not come the length of language." (Baker, 1974, p. 23)

(Actually, Edward Tyson, a British physician, did in 1699 dissect a chimpanzee and concluded that it was not human, although it resembled human beings more than do monkeys. Evidently his report had slight effect.)

Baker (1974), whom it is probably safe to follow in his historical chapters, notes that at this time political radicalism rather than orthodoxy tended to be associated with belief in the inequality of races. Jean Jacques Rousseau, essayist on the evils of civilization, was not an egalitarian:

> From this lack of knowledge there has arisen that fine dictum of morality much bandied about by the philosophical crowd, that men are everywhere the same, and that having everywhere the same passions and the same vices, it is rather useless to attempt to characterize the different races; which is just about as reasonable as if one were to say that one could not distinguish Peter from James, because each of them has a nose, a mouth, and eyes. (p. 16)

Other philosophical and political thinkers of the period made comments, mostly short ones, upon racial differences. David Hume concluded that people living in the tropics or beyond the polar circles were inferior, although there was no relation between intelligence and latitude within the temperate zone. Immanuel Kant produced a rather elaborate discourse upon the mental characteristics of peoples of various European nations, and also the Arabs, Persians, Indians, Japanese, and Chinese. He was particularly vexed at the fetishism of sub-Sahara Africans, and so concluded that they had "received from nature no intelligence that rises above the foolish." Voltaire was also convinced: "The inclinations, the characters of men differ as much as their climates and their governments" (Baker, 1974, pp. 19–20).

Biological writers considered the question at greater length. Linnaeus in fact had listed the mental qualities of each race as distinguishing characters, comparable with physical ones. But perhaps the most influential of the earlier writers was Johann Blumenbach, professor of medicine at Göttingen in Germany. Writing in 1776, his real desire was to show that all races of men are indeed human and to convince others that the reported differences among races had been exaggerated. In the attempt, however, he inadvertently planted as nefarious a seed as exists in racial ground: the unhappy term *degeneration*.

Blumenbach used *degeneration* innocently. He intended to suggest only that differences among human races could be explained as due to the derivation of various peoples from a single original stock. (Indeed the term is itself derived from the Latin meaning "departed from its race.") The only hint at superiority had to do with aesthetics, in that Blumenbach felt that people of the "primary variety," the original, were "of the most beautiful form" (Baker, 1974, p. 26). (He later specified the Caucasus Mountains as both the place of origin of man and the home of the handsomest people—hence the misapplication of "Caucasian" to all people with white skin.)

Of course Blumenbach knew nothing of evolution in the Darwinian sense, but he did recognize that plants and animals had been modified in form as a result of domestication or change of climate, and, like Lamarck, he supposed that acquired characteristics could be inherited. He believed all humans to belong to one species (new species arose by divine creation), and his task therefore was to account for the differences in races. This he imagined to have occurred through changes caused by the agencies mentioned above. Humans were, after all, so widely distributed as to be subject to climatic extremes and were by far the most domesticated of organisms, sometimes even interfering artificially with the structures of their bodies. It followed that people would be different on different parts of the earth. But if there were only a single human species, and the varieties of people arose through "degeneration," which race was the original? Not surprisingly, the European.

The next large step, following Blumenbach, came in 1853 with the publication of the *Essay on the Inequality of Human Races* by Joseph Arthur, Comte de Gobineau. Gobineau was a career diplomat and a novelist and poet who also wrote on history, philosophy, archaeology, and Oriental studies. His approach was neither biological nor anthropological, but historical. Also, he was less concerned with the reasons for racial differences than with the reasons why great past civilizations had fallen. Examining the causes usually cited—decline of religion, corruption of morals, luxury, poor government, despotism—he rejected all on historical grounds. There were too many contradictions: Morally inferior nations had replaced those that were morally superior; the Greeks, Romans, and Persians declined even though luxury never reached so high a peak among them as in France in his own day; and so forth.

It occurred to him instead that the answer might be found in the loss, through hybridization, of what we would call now (Gobineau of course had no knowledge of genes) a genetically based propensity for social development. The first move toward civilization, he believed, was the union of several tribes, by alliance or conquest, to form a nation. Certain races seemed capable to him of forming units large enough for civilization to result; others seemed incapable. Next, nations that were formed became in their ascendant stage genetically homogeneous through intermarriage. But such civilizations attract strangers, among them individuals belonging to races lacking the genes requisite for statehood. The capable bred with the incapable, and after a sufficient number of generations the nations' descent stage began. The genetic capacity for civilization had been washed away in a sea of lesser peoples.

Gobineau very specifically used a more onerous interpretation of "degeneration" to describe this process:

> I think the word "degenerate," applied to a people, should and does signify that the people has no longer the intrinsic worth that it formerly possessed, that it no longer has in its veins the same blood, the worth of which has been gradually modified by successive mixtures; or to put it in other words, that with the same name, it has not retained the same race as its founders. (Baker, 1974, p. 36)

Plainly, Gobineau's criterion for judging the inherent qualities of a race was its capacity to originate a nation. He noted ten such races in the world's history: the Indian, Egyptian, Assyrian, Greek, Chinese, Roman, north European, "Alleghanien" (North American Indian mound culture), Mexican, and Peruvian. No fewer than six of these he described as descendants of the original "Aryans," an ancient race that had presumably spread throughout the world. It was the Northern Europeans, for instance, who, having intermarried to some extent with Slavs and others without losing their initiative too quickly, were responsible for European culture. Where the Aryans themselves originated is left rather vague.

Gobineau's long book ends with characteristic pessimism. Hybridization, he concluded, was destroying the present civilization as it had those of the past; the period upcoming, indeed, would be one of "faltering procession toward decreptitude." Darwin's work, published a few years later, did not measurably improve his prognosis. Upon (mis)reading it he announced: "We are not descended from the apes, but we are rapidly getting there."

Gobineau had given an entirely different treatment to the racial issue than had Blumenbach, while using the same term. Degeneration for Blumenbach meant the *derivation* from an original stock through adaptation. For Gobineau it meant the *degeneration* of *pure stock* through *hybridization*. The Aryans of Gobineau's imagination time and again had their great works destroyed through mixture with inferior peoples. Now it was happening again. Was there cause for alarm? Of course. Concern with racial purity would haunt and demean Western culture for the next 100 years.

Francis Galton, in *Heredity Genius* (1869), did not eschew racial comparisons. He depended upon historical data also, and he used the normal curve to establish grades of intellect by which different peoples could be compared. The grades ranged from illustrious to complete imbecile. The Attic Greeks, who between 530 and 430 B.C. produced Pericles, Thycydides, Socrates, Xenophon, Plato, Euripides, and Phidas, among others, were on the highest plane, with other achieving civilizations ranked below. Galton was nothing if not exhaustive, for even animals came in for their fair share; an illustrious dog was equivalent to a very dull Englishman.

Heinrich von Treitschke was, unlike previous writers mentioned here, a professional historian (and a political activist). He had great admiration for north European culture, and it was his fear that it would be adulterated by Jews. He was particularly disturbed by the influence of Jews on literary criticism—books were praised or abused, so he claimed, according to whether or not they supported the Jewish cause. He deplored the tendency of Jewish pupils to dominate in numbers the higher classes of Berlin colleges. His prescription for all this was, however, "gentle restraint."

Following Treitschke was Houston Chamberlain, son of a British admiral, who was educated by a German tutor. He subsequently studied under Carl Vogt in Vienna, lived in Dresden, and married the daughter of German composer and nationalist Richard Wagner. (Vogt taught that the various races of man had separate origins.) Chamberlain attempted to demonstrate, in his *Foundations of the Nineteenth Century,* that it was to the north Europeans that Europe owed its culture. In order to do so he was required to place the origin of the Middle Ages centuries earlier than was usual, and to lessen the significance of the Renaissance, which began around the Mediterranean. He designated 1200 A.D. as the year the *Germanen* awakened to "their destiny in world history as the founders of an entirely new civilization and an entirely new culture" (Baker, 1974, p. 49). The Middle Ages were in fact not over; the nineteenth century was only a part of the long period of preparation for the new age which would succeed it. (He used the term *Germanen* for the various populations of Northern Europe.)

Chamberlain also devoted considerable effort to detracting from the accomplishments of Greeks, Romans, and Jews. Like Treitschke, he deplored the influence of Jews, and, like Gobineau, he feared for the civilizing race. He regarded Jews as not only different from *Germanen,* but from Indo-European peoples altogether, writing of them as a "strange people . . . quite strange." European Jews he thought to be hybrids between Semites, Syrians, and Indo-Europeans, therefore violating the "sanctity of unmixed race." For the fall of Rome he did not find the Jews culpable, but it was supposed to have occurred because of "raceless chaos." The worst example of race mixing was, nevertheless, the ancient Middle East.

In the United States over this later period the fascination was with Anglo-Saxons, who were supposed to hold the genetic key to good government and appreciation of the arts. But this, as in the European case, originally had little to do with Darwin's theory: "This Anglo-Saxon dogma became the chief element in American racism in the imperial era; but the mystique of Anglo-Saxonism, which for a time had a particularly powerful grip on historians, did not depend upon Darwinism either for its inception or its development. Like other varieties of racism, Anglo-Saxonism was a product of modern nationalism and the romantic movement rather than an outgrowth of biological science" (Hofstadter, 1955, p. 172).

Thus the idea of survival of the fittest was, in the hands of the Anglo-Saxon imperialists, simply a new instrument. The measure of world power already achieved by Anglo-Saxons seemed to some in the latter nineteenth century to demonstrate their fitness, their particular genius being self-government. This was concluded because it was believed that methods of government could be compared as if they were animal forms, some appearing evolutionarily more favored than others.

The leading proposition of the Anglo-Saxon school was that the democratic institutions of Britain and the United States could be traced back to the political institutions of the early Teutonic tribes. Their derivatives alone maintained the genetic capacity for freedom. Wrote James Hosmer in 1890:

> Though Anglo-Saxon freedom in a more or less partial form has been adopted (it would be better to say imitated) by every nation in Europe but Russia, and in Asia by Japan, the hopes for that freedom, in the future, lie with the English-speaking race. By that race alone it has been preserved amidst a thousand perils; to that race alone is it thoroughly congenial. . . . The inevitable issue is to be that the primacy of the world will lie with us. English institutions, English speech, English thought, are to become the main features of the political, social, and intellectual life of mankind. (Hofstadter, 1955, p. 308)

How was this to be accomplished? John Fiske, an early American supporter of Spencer, thought he knew. There would be an inevitable Anglo-Saxon expansion, followed by a *pax Anglo-Saxonia*. Fiske had long believed in Aryan race superiority and also accepted the Teutonic origin of democracy, which he felt justified the expansion. Anglo-Saxons would number more than 700 million in America alone, whereas at the same time the English would cover Africa with cities, farms, railroads, telegraphs—the devices of civilization. All of the world that was already so developed would become English. Four-fifths of the earth would trace its pedigree to English forefathers and hold such sovereignty over commercial enterprises that the states of Europe would be unable to afford armaments and would finally see the advantages of peace and federation. All this was delivered in a series of lectures, to an English audience, and to a warm reception.

It should be noted, though, that however unsavory the main elements of Fiske's supposition appear today, they accrued from benevolent conviction rather than personal malice. Fiske was in fact intellectually gifted, a devourer of knowledge who had mastered eight languages and begun six others by the time he was 20. He was also a pacific Spencerian who believed the practice of war to indicate a lesser evolutionary status. Expansion of the Anglo-Saxon peoples was in his view a necessary precondition to world harmony. For other peoples it would be as a small penance to the expurgation of sin: Only take Nature's way, and its blessings would follow. (He was not the first gentle person to be run down by a vision.)

The whole Anglo-Saxon fascination, indeed, had about it an air of benign despotism:

> One outgrowth of the Anglo-Saxon legend was a movement toward an Anglo-American alliance which came to rapid fruition in the closing years of the nineteenth century. In spite of its unflagging conviction of racial superiority, this movement was peaceful rather than militaristic in its motivation; for its followers generally believed that an Anglo-American understanding, alliance, or federation would usher in a "golden age" of universal peace and freedom. (Hofstadter, 1955, p. 182)

It would be, according to one senator, an "English-speaking people's league of God for the permanent peace of this war-worn world."

And it passed quickly. The ethnic composition and cultural background of the United States made it in reality an unlikely place to sell Anglophilia. An attempt to revive it during World War I was unsuccessful, and then the powerful isolationist sentiment that settled over Americans following that war swept it away altogether.

During the same period, however, there developed quite a different strain of white racism, one far more virulent and durable, because it came not from hope but from fear. Suppose there were an Anglo-Saxon destiny? It could not be expected that other peoples would concede. Conflict was therefore inevitable. But were the Anglo-Saxons ready? There was genuine concern. Wrote Theodore Roosevelt: "The twentieth century looms before us big with the fate of many nations. If we stand idly by, if we seek merely swollen, slothful ease and ignoble peace, if we shrink from hard contests where men must win at hazard of their lives and the risk of all they hold dear, then the bolder and stronger peoples will pass us by, and win for themselves the domination of the world." And said John Barrett, a former minister to Thailand: "The rule of the survival of the fittest applies to nations as well as to the animal kingdom. It is a cruel, relentless principle being exercised in a cruel, relentless competition of mighty forces; and these will trample over us without sympathy or remorse unless we are trained to endure and strong enough to stand the pace." (Hofstadter, 1955, p. 180). Was it possible that the Anglo-Saxon race, ordained to rule the earth, felt a trifle flabby?

Most of this talk was of course merely expansionist wind, but it helped to propogate the notion that the presumed struggle between human races had a Darwinian basis, and, most frighteningly to some at the time, that whites were losing. The Rev. Josiah Strong, in his popular *Our Country: Its Possible Future and Present Crisis*, published in 1885, had been clear but positive:

> Then the world will enter upon a new stage of its history—*the final competition of races for which the Anglo-Saxon is being schooled.* If I do not read amiss, this powerful race will move down upon Mexico, down upon Central and South America, out upon the islands of the sea, over upon Africa and beyond. And can anyone doubt that the result of this competition of races will be the "survival of the fittest"? (Hofstadter, 1955, p. 179)

Charles Pearson, an English educator wrote more pessimistically in 1893. The white races, he believed, can live only in the temperate zone and would be barred from colonization of the tropics. Overpopulation and economic pressures would give rise to state socialism, eventually sapping initiative and capacity for independent thought. Meanwhile, other races, not so enervated, would continue to progress, until:

> We shall wake, to find ourselves elbowed and hustled, and perhaps even thrust aside by peoples whom we looked down upon as servile, and thought of as bound always to minister to our needs. The solitary consolation will be that the changes will have been inevitable. It has been our work to organize and create, to carry peace and law and order over the world, that others may enter in and enjoy. Yet in some of us the feeling of caste is so strong that we are not sorry to think that we shall have passed away before that day comes. (Hofstadter, 1955, p. 186)

Indeed, as noted in a famous remark by a weary black American soldier sent to suppress a rebellion in the Philippines, "Dis shyar white man's burden ain't all it's cracked up to be."

Meanwhile, scholars and literari took up Pearson's thesis with morbid alacrity. Henry Adams was convinced ("the dark races are gaining on us"), as was his brother Brooks, who in 1896 authored the proposition that highly centralized societies disintegrate under the pressure of economic competition because the energy of the race has been exhausted. And one had only to look around to see yet another foe. The military successes of the Japanese, particularly their defeat of Russia in 1905, brought to our attention the "Yellow Peril" to white hegemony. Could this be only a prelude to the reawakening of the Mongol race?

Hardly scholarly but highly influential were two later books by Americans, Madison Grant's *The Passing of the Great Race or the Racial Basis of European History* (1917) and Lothrop Stoddard's *The Rising Tide of Colour Against White World-Supremacy* (1920). The titles do not disguise the themes. Grant was concerned mainly with the ethnic composition of the United States. The "Great Race" was of course the North European, identifiable as Chamberlain's *Germanen*, and Grant was troubled that the proportion of those of North European extraction was diminishing. He did not happily see his country proclaim itself as an asylum for the world's oppressed. He was unfavorably disposed toward Jews, but particularly toward Asians and those of Mediterranean or East European ancestry. Stoddard's concern was that white civilization might disappear by race mixture with nonwhites. This he thought would be irrevocable calamity, for the greatest creative ability would be destroyed. Grant's and Stoddard's books, along with the policy implications that seemed extensible from the eugenics movement, probably played a role in the passing of the Immigration Restriction Act of 1924, in which strong preference was given to people from North and Western Europe. It has only recently been changed.

Meanwhile, on the Continent, Oswald Spengler was publishing *The Decline of the West*. Trained as a mathematician and philosopher, Spengler, like Gobineau, was interested in the demise of civilizations. He recognized that each had its individual twists, but saw general similarity in the overall process. Decline, he concluded, was inevitable, chiefly because of what we might now call "the urban syndrome," the loss of the "we" identification in the big cities. Cities, according to this thesis, grow so large and polyglot that they no longer reflect the culture of the people who made them. They are inhabited by "factual man," clever but traditionless, appearing in amorphous, fluctuating, unproductive masses. The creative force is lost, and religion replaced by practical substitutes. True abstract thinking disappears and is superseded by "professional scientific lecture-desk philosophy" (Baker, 1974, pp. 53–54).

All of this (some of which has a remarkably modern sociological sound) led in Spengler's view to internationalism, pacifism, and socialism—a loss of whatever originally made the people great. Modern civilization, far from being a *goal*, was a *fate*.

Spengler's work was enormously popular. For one thing, he wrote for the general educated public. For another, the events of 1914–1918 at least put the question to those with a belief in inevitable progress, and the results were disturbing. Along with Grant, Stoddard, and Charles Pearson, Spengler did not offer happy thoughts. Indeed, together they presented to North European and Anglo-Saxon audiences an alarming diagnosis: racial encroachment from all sides, and decay within. There had to be a reaction.

Disgusted (and chastised), America withdrew into itself, making few drastic changes. In Germany, however, forces were being aligned which would generate events unimagined by even the harshest racial polemicist. The Aryan myth, Nordic supremacy, the awakening of *Germanen*, degeneration through racial mixture, the dissolution of Western culture, even the prostitution of Nietzsche's *Ubermensch*— these related themes wound sinuously through those times of intellectual uncertainty and economic havoc, available to those of fertile, if pathological, mind. Such minds are always with us. This one belonged to Adolf Hitler.

Untroubled by the efforts of physical anthropology, Hitler in *Mein Kampf* (1925) nevertheless writes with heavy conviction in the story of his "struggle" that

> . . . the road which the Aryan had to take was clearly marked out. As a conqueror he subjected the lower beings and regulated their practical activity under his command, according to his will and for his aims. But in directing them to a useful, arduous activity, he not only spared the life of those he subjected; perhaps he gave them a fate that was better than their previous so-called "freedom." As long as he ruthlessly upheld the master attitude, not only did he remain master, but also the preserver and increaser of culture. For culture was based exclusively on his abilities and hence upon his actual survival. As soon as the subjected people began to raise themselves up and probably approached the master in language, the sharp dividing wall between master and servant fell. The Aryan gave up the purity of his blood and, therefore, lost his sojourn in paradise which he had made for himself. He became submerged in the racial mixture, and gradually, more and more, lost his cultural capacity, until, at last, not only mentally but physically, he began to resemble the subjected aborigines more than his own ancestors. For a time he could live on the existing cultural benefits, but then petrification set in and he fell a prey to oblivion." ((pp. 295–296)

Another civilization lost. But there was enough Aryan blood remaining in Germany to save the genetic propensity for culture. It wanted only purification and the master attitude. Thereafter, a new order awaited the world, another gift from the Aryans. It would be declined at great cost.

Meanwhile, Nazi theoreticians were very busy keeping peoples straight. Those living on the Mediterranean were inferior until the

Italians became a part of the Axis. The condition of the Slavs was also dependent upon whether there was or was not a pact with Moscow. The Norwegians originally had a large amount of Aryan blood, but they were demoted suddenly when they resisted German invasion. Finally, when Japan was included on the side of the Nazi cause, a high proportion of Aryan genes were found in their ancestry. Politics does indeed make strange bedfellows.

Philosophies that self-consciously exploit the notion of racial destiny were fortunately interred in the rubble of Berlin in 1945. Modern racial concerns have usually concentrated upon political and economic freedom rather than domination. The misuse of Darwin's theory in regard to human racial thought, thus, has long since ceased among reputable students of heredity and evolution, just as it has among sociologists and economists. Appeal to the concept of survival of the fittest for maintenance or extension of the dominion of those already politically powerful has appeared for some time as the thin tactic it is. One could hope then that it has finally been perceived that today's dogma may be tomorrow's joke, and that the rush to turn scientific principles into social policy before their implications are fully understood is not only unjustified but dangerous. However, it is perhaps unhappily true that the only thing we learn from history is that we do not learn from history. Pernicious priggery lives on and one can only suspect from its antiquity and ubiquity that it cannot be killed by an idea but rather must be fought anew each generation.

Finally, lest it be imagined that it was used exclusively in the support of conservative opinion, nationalism, and racism, it should be noted that Darwin's theory was interpreted favorably by the left as well as by the right. The struggle for existence could just as easily be applied to classes as to anything else. Karl Marx recognized this potential, writing to Friedrich Engels: "Darwin's book is very important and serves me as a basis in natural science for the class struggle in history" (Kogan, p. 166). Indeed, Marx wanted to dedicate the second volume of *Das Kapital* to Darwin. (Charles declined. An uncut version of Marx's book still lies in his library at Down.) Meanwhile, in the eventual home of Marx's disciples, the liberals were using evolution to attack orthodox religion, generating Dostoevsky's angry comment that Russians take for unshakable truths what elsewhere are only hypotheses. Everyone, it seems, wants science on his side.

Eugenics

Leonardo da Vinci was the illegitimate child of one Piero, a notary from the village of Vinci, and a peasant girl, Caterina. He was of course also recognized as a giant in his own time. His stepbrother by his father's third wife, Bartolommeo, greatly admired Leonardo; so much so that he determined to make another:

> Bartolommeo examined every detail of his father's association with Caterina and he, a notary in the family tradition, went back to Vinci. He sought out another peasant wench who corresponded to what he knew of Caterina and, in this case, married her. She bore him a son but so great was his veneration for his brother that he regarded it as profanity to use his name. He called the child Piero. Bartolommeo had scarcely known his brother whose spiritual heir he had wanted to produce and, by all accounts, he almost did. The boy looked like Leonardo, and was brought up with all the encouragement to follow his footsteps. Pierino da Vinci, this experiment in heredity, became an artist and, especially, a sculptor of some talent. He died young. (Ritchie-Calder, 1970, pp. 39–40)

This is but one, and, it must be thought, one of the more pleasant, instances in a long history of what might now be described by some stainless steel term such as *genetic engineering*. Actually, it was breeding for a specific purpose, something that in industrialized countries most of us contact very little but in former times immediately affected the lives both of individuals and of empires. The reasons for the change are obvious. First, the proportion of our population that grows plants or animals for sustenance (or transportation) has been radically reduced; thus the number of us having the opportunity directly to attend to the production of desired characteristics in living things through breeding has correspondingly lessened. Many of us have never even looked out the back door of our supermarket. Second, we have fewer reasons to follow the comings and goings of aristocrats. The old distinctions have small impact on our being; no one cares who the true prince is. Third, particularly in the United States, we have made it almost an article of faith that marriage (usually) and progeny shall follow exclusively upon romantic attraction between the sexes. At least that is the dream. And finally we have been taught to believe that environment is the overwhelmingly important factor in determining who shall and who shall not lead a life of meaning and felicity. The effects of heredity are something that we blissfully ignore—or, more to the point, of which we remain ignorant. The result is that among the public at large the only people with first-hand knowledge and continuing concern for genetic effects are those who grow food for consumption or animals for display; that is, farmers, ranchers, and eccentrics.

The last of these points we shall examine in more detail later. For the present it is enough to note that our disaffection on the matter of heritable differences, especially in behavior and especially in humans, seen in the last 50 years, represents an unusual interlude in history. Otherwise we have always been interested, sometimes to excess and with grievous consequences. Ironically, however, we have only recently understood the actual means of genetic transmission and what is transmitted. For millennia the first rule of the breeder was: like produces like, while the second rule was; like does not necessarily produce like. It took Mendel and those who followed him to show how that this could be so.

The intellectual frustrations of some under the previous conditions of ignorance is well represented in a series of plaintive questions by Montaigne:

> Tis to be believed that I derived this infirmity from my father, for he died wonderfully tormented with a great stone in his bladder; he was sensible of his disease till the sixty-seventh years of his age; and before that had never felt any menace or symptoms of it, either in his reins, sides, or any other part, and had lived, till then, in a happy, vigorous state of health, little subject to infirmities, and he continued seven years after, in this disease, dragging on a very painful end of life. I was born above five and twenty years before his disease seized him, and in the time of his most flourishing and healthful state of body, his third child in order of birth: where could this propension to this malady lie lurking all that while? And he being so far from the infirmity, how could that small part of his substance wherewith he made me carry away so great an impression for its share? and how so concealed, that till five and forty years after, I did not begin to be sensible of it? being the only one to this hour, amongst so many brothers and sisters, and all by one mother, that was ever troubled with it. He that can satisfy me on this point, I will believe him in as many other miracles as he pleases. (Quoted in McClearn & DeFries, 1973, pp. 5–6)

All societies regulate to some extent who mates with whom. There is in this a recognition, vague or explicit, of kinship relations with various families, clans, or classes, that are thought to have the capacity to prescribe or proscribe the production of offspring with a member of another group. However, it was not until after the publication of *The Origin of Species* in 1859 that the question of the improvement of the human species through evolutionary techniques was raised at large. Before that almost everyone, it seems, had opinions on the issue of mating, but the theoretical means whereby the desired ends should be achieved were not at hand. Belief in the special creation of species was of course a hindrance in this respect. But by 1869 that true believer, Francis Galton, had taken up the question and had little doubt as to its outcome:

> I propose to show in this book [*Hereditary Genius*] that a man's natural abilities are derived by inheritance, under exactly the same limitations as are form and physical features of the whole organic world. Consequently, it is easy, notwithstanding those limitations, to obtain by care-

ful selection a permanent breed of dogs or horses gifted with peculiar powers of running, or of doing anything else, so it would be quite practicable to produce a highly-gifted race of men by judicious marriages during several generations. (p. 1)

In 1883 Galton coined the term *eugenics* to apply to the improvement of man through breeding. In 1907 he helped found the Eugenics Society in Great Britain, and in 1911 the Eugenics Laboratory, to promote research upon the effects of heredity upon human characteristics.

In the U.S. the movement was not far behind. In 1913 the Eugenics Research Association was established. Similar societies quickly sprang up in France, Germany, Italy, Norway, Russia, China, and Japan.

This was fast development, but, given the intellectual preoccupations of the day, hardly surprising. The eugenics movement was conceived and nurtured through the years of survival of the fittest, of concern for loss of racial purity, degeneration through mixture, and the collapse of culture-bearing civilizations. And, much like the first and second stages of Anglo-Saxon racism, it began in hope and ended in fear. As practiced, it was essentially intended to reverse, or at least delay, the decay within.

Galton emphasized the positive aspects of the approach. Why not, he thought, establish fellowships that would enable the "well bred" to marry young and thus produce greater numbers of children? We would simply get more of the desirable. But whereas some of Galton's positivism remained, eugenics quickly turned the issue around and began to ask instead about the elimination of the undesirable.

These were disquieting times. The rapid growth of cities during industrialization had created huge slums in which were amassed the poor, the diseased, and the deranged. The impression thus arose that mental capacity and moral character were genetically linked to social class. Curiously, an increase in philanthropy and appropriations for public health multiplied the impression. More and more instances of diseased and defective families came to the attention of physicians and social workers, resulting in a larger proportion of cases detected relative that and existing. Overall, it seemed as if the slums were quite literally the breeding grounds of degeneracy.

Some famous research studies contributed to this conviction. In one report (1877), R. L. Dugdale (a member of the executive committee of the New York prison association), in an inspection of the county jails found six prisoners in one country to be related. Intrigued, he undertook a survey of the family and traced the lineage back to six sisters, whom he called the Jukes. One had left the country, but the descendants of the remaining five presented a notable picture of criminality, immorality, feeblemindedness, and pauperism. Dugdale was chiefly a social reformer, and was cautious in naming heredity or environment as the cause, but made it clear that with this family there *was* a problem.

In 1911 the real name of the Jukes was found in Dugdale's manuscripts. G. H. Estabrook (1916) traced the family over the 40 years that elapsed since Dugdale's report, and summarized:

> For the past 130 years they have increased from 5 sisters to a family which numbers 2,094 people, of whom 1,258 were living in 1915. One half of the Jukes were and are feebleminded, mentally incapable of responding normally to the expectations of society, brought up under faulty environmental conditions which they consider normal, satisfied with the fulfillment of natural passions and desires, and with no ambitions or ideals in life.

In 1912 H. H. Goddard published an account that seemed clearly to show the inheritance of degeneracy. Two branches of the Kallikak family (a pseudonym) showed radically different behavior. One was very much like the Jukes, whereas the other was normal mentally and upstanding morally. They were both begun by the same man, Martin, who while a soldier in the Revolutionary War had impregnated a feebleminded girl whom he met in a tavern. She named her son Martin Kallikak, Jr., and it was among this branch that the problems arose. Meanwhile, Martin Senior returned home and married, beginning the other branch of the family. The feeblemindedness, to which the problems were traced, appeared obviously to be heritable and were attributable to the condition of the tavern girl.

Reports such as these provided sustenance for the general fear of decay. Crime, delinquency, vagrancy, pauperism, prostitution—all were also suspected of genetic control. And, in the United States the immigration of large numbers from peasant countries in central and southern Europe added alarmingly to the milieu. Because of their language difficulties and their rustic habits, these people seemed to provide evidence for the notion that the standard of American intelligence was being lowered. The economic deceleration at the end of the century was interpreted as the beginning of a national decline, ascribable to the disappearance of "the American type."

Eugenicists seized the moment and translated their zeal into political action. They sought the isolation, in order to prevent reproduction, of the feebleminded (and were instrumental in achieving humane treatment for the retarded). They also advocated sterilization and were successful in having appropriate laws enacted in 15 states. The first was in Indiana in 1907, which made mandatory the sterilization of confirmed criminals, idiots, imbeciles, and rapists in state institutions when recommended by relevant medical authority. Such laws were not unenforced. In California between 1909 and 1929 there were more than

6000 sterilizations of the insane and feebleminded. Similar programs were established in Denmark, Finland, Germany, Sweden, Norway, Iceland, Switzerland, and Mexico.

The movement was perhaps at its zenith in 1924 with the passage of the Immigration Restriction Act (granting high priority to those from northern and western Europe) and following a favorable decision from the U.S. Supreme Court by Oliver Wendell Holmes. The case involved Carrie Buck, a feebleminded girl who had already borne one illegitimate child, and who was chosen for the first sterilization to be performed in Virginia. Said Holmes:

> We have seen more than once that the public welfare may call upon the best citizens for their lives. It would be strange if it could not call upon those who already sap the strength of the State for these lesser sacrifices, often not felt to be such by those concerned, in order to prevent our being swamped with incompetence. It is better for all the world if, instead of waiting to execute degenerate offspring for crime, or to let them starve for their imbecility, society can prevent those who are manifestly unfit from continuing their kind. The principle that sustains compulsory vaccination is broad enough to cover cutting the Fallopian tubes . . . Three generations of imbeciles are enough. (Quoted in McClearn & DeFries, 1973, pp. 267–268)

Ironically, however, continuing research into genetics contributed to the decline and ultimately to the loss of credibility of the eugenics movement. In the beginning, such scientific proposals as Weismann's germ plasm theory and de Vries's mutation theory served the eugenicists. Here was evidence suggesting that the germ plasm was not modified by the environment—that in fact it occasionally made sudden large changes on its own—and therefore it could not be an environmental impress that accounted for familial or racial difficulties. Mendel's discoveries seemed to solidify the position. For some time many geneticists studying humans supported eugenics on this basis. But as more was learned about the principles of genetics, particularly about genes in populations, it was found that the total gene pool of a species acts as a conservative and quite stable system. The genetical affairs of humanity were thus not in the degenerating state earlier supposed; they were not becoming worse at a noticeable rate. It also followed that the probability of improving our genetic condition through differential reproduction was quite low, and thus the remedies of the earlier eugenicists would be considerably less effective than they had imagined.

The real problem lay with deleterious rare genes that are recessive—that is, with those whose effects are seen only when two of the genes are present together in the same organism, as in the case of Mendel's dwarf peas. In these cases there cannot be selective breeding using carriers of one such gene, because its effects are not seen in a single dose. Selection must be only against those with a double dose. This is a slow process. If, for instance, such a gene were found in a population with a frequency of 0.01, it would require approximately 100 generations to reduce the frequency to 0.005, or by one-half. At five generations per century, the time required for substantial improvement thus becomes rather longer than the eugenicists had envisaged. Also, it was discovered that many genetical conditions were not nearly so simple as originally believed. For these and other reasons, many geneticists began to abandon the movement. And without their corrective support, "By the 1920's American eugenics degenerated into a mixture of pseudo-science, Bible-belt religion, extreme reactionary politics, and racism, so that the very term became repulsive to geneticists" (Lerner, 1968, p. 269).

A second, emotionally devastating, reason for the demise of eugenics derived from the practices of the Nazis, which were revealed in full only after the end of the war. Few could begin to conceive of the madness of those directing what must be admitted to have been negative "eugenics" at its most grotesque. (What a long way from Galton's fellowships!) Such events understandably generated a strong reaction against eugenics programs of all sorts, effectively putting them to an end.

This is hardly to say that geneticists have stopped their interest in human problems or for that matter have ceased to dream of improving the species. Enormous advances have been made in genetic counseling, for instance, whereby those contemplating having children may be apprised of the probability of producing a child who is severely disadvantaged. Tests for carriers of defective recessive genes have been and continue to be developed and are of great service to the counselor.

Looking beyond, the American Nobel geneticist Herbert Muller wrote profusely on some possible steps to improve the genetic quality of mankind. Artificial insemination is used for women whose husbands produce too few functional spermatoza, but the donors are unselected. Why not, suggested Muller, use donors with outstanding desirable characteristics? At a more advanced level, why not use the preserved sperm of men long since dead who have passed the historical test of greatness? What woman, Muller asked in 1935, would refuse to bear the son of Lenin? In 1959, after his disenchantment with communism, he nominated Einstein, Pasteur, Descartes, Leonardo, and Lincoln. (The changes in nominees illustrates a part of the difficulty.) Muller may have been right nevertheless. Still, the picture of thousands of distinguished gentlemen making regular visits to have their sperm hygienically collected makes one wonder at the lengths to which one would be expected to go in the interests of science and posterity.

Of course, it need not end there. The eggs of outstanding women could be flushed from their reproductive tracts, preserved, and later fertilized by spermatoza from the known distinguished men. (We need not speculate upon them here, but the potential combinations are marvelous to suppose.)

However, it should be understood that even though such medical techniques are at hand, the program itself would still be very much a shot in the dark. The gravest difficulty is that we know nothing whatsoever about the relationship between particular genes and greatness. There are clearly hereditary elements in talents and aptitudes of all kinds, but direct connections between genes and notable contributions to society remain to be traced. In addition, the spermatoza of a male individual are not genetically identical, as the eggs of a female are not. The heredity combination that made the person admirable may not be reproduced in any of his or her germ cells. Even were it available, in other words, a mating such as Einstein × Madame Curie might generate much mediocrity—and could produce some real losers.

Each person of imagination has his or her own vision of utopia, and—particularly if a scientist or inventor—this individual is likely to see his or her innovations as offering humanity a cure. So it has been with many of the artifacts of modern society. The internal combustion engine, electricity, atomic energy, transistors, vaccines, organ transplants, and urban renewal all were intended to reduce labor or enhance the health and dignity of the human condition.

But in many respects technology is a plastic Jesus, and one may wonder whether the species is any happier for being physically better off. There is, however, no choice but to seek further, for we are in a fast chute with no handles on the sides. And we shall, unfortunately, know less before we know more. Perhaps, as remarked by Dobzhansky (1962), from whom the discussion on Muller is taken, the only fair way to criticize a utopia is to offer a substitute. I know better than to try. It can only be remarked with safety that although it is desirable that there be technological visionaries and that we know their thoughts, it should be remembered at all times that the single program or device guaranteed to perfect or even to save us has yet to be invented. Nor is it likely to be.

Social and Biological Laws: A Lingering Divorce

"I have received in a Manchester newspaper rather a good squib, showing that I have proved 'might is right,' and therefore that Napoleon is right, and every cheating tradesman is also right." So wrote Darwin to Sir Charles Lyell, following publication of *The Origin of Species*. He had every reason to be disheartened, if somewhat amused, at the article, for it only presaged a long series of misinterpretations of the nonbiological implications of his work. Philosophers at all levels would seek an ethic from natural selection with an avidity that revealed both their disenchantment with the old and their myopia regarding the new. We have described some of the results. The fires of quick romance so often grow quickly dim; what is embraced in haste is seldom what it seems.

There was nothing in *Origin* or in Darwin's later books that made them an apology for rabid competition or force. Darwin firmly believed that a moral sense had evolved in man, that we are now and had evolved from social beings, and that our sociality had been of enormous importance in our survival. The human moral sense, indeed, was seen as an inevitable outgrowth of our social instincts. Sensitivity to group feelings and family affections ranked equally with intelligent self-interest as the biological foundations of moral behavior. Nor did Darwin personally espouse a strong eugenic cause, fearing it was treasonous to the "noblest part of our nature." The most he hoped for was that those hereditarily less well endowed would "not marry so freely as the sound." Those who could not spare their children the horrors of abject poverty, he thought, should refrain from reproducing. Thus, it was only a portion of Darwin's science (not to mention his humaneness) that was read early from his works. The temper of the times and the dispositions of certain men obscured the whole. The rest had always been there, awaiting only new eyes, as noted by Hofstadter (1955): "If there were, in Darwin's writings, texts for rugged individualists and ruthless imperialists, those who stood for social solidarity and fraternity could, however, match them text for text with some to spare" (p. 91).

Some had objected to the early interpretation at once. Thomas Huxley in 1893 said that people in society are undoubtedly subject to the "cosmic process" of evolution, which includes the struggle for existence and the selective propagation of the fit. However, he saw that humans had evolved to the extent that they can alter the environment so as to change the conditions of evolution itself. A raw struggle for existence need not occur under circumstances in which, through the intervention of some external agent, room and sustenance is left for all. The external agent would of course be human ethics. But, noted Huxley, although evolution had produced in man a moral sense, moral credos are themselves not derivable from the principles of evolution; thus "cosmic evolution may teach us how the good and evil tendencies of man have come about; but, in itself, it is incompetent to furnish any better reason why what we call good is preferable to what we call evil than what we had before" (Dobzhansky, 1962, p. 341.)

Still, Huxley failed to see the synthesis implicit in Darwin. He compared ethics to gardening, proposing that as horticulture is a branch of the evolutionary process because of its elimination of the struggle for existence, so morality defies evolution through the imposition of human ideals, benefiting some and restricting others in ways that would not otherwise be. Indeed, said Huxley, the more advanced a society becomes, the more it prevents the struggle for existence among its members. To adhere to natural selection in the crudest sense would destroy the bonds holding it together.

Huxley was in his analysis far more sagacious than was Spencer, who would in effect have left the garden to the weeds on the premise that if they could get it they deserved it. There was nevertheless an obvious and unsatisfying cleft in Huxley's position, for it put ethics in a state of limbo vis-à-vis nature. Ethics, though derived from a natural process, was not contained in it; in fact it was even contradictory to it. The task of man was then not acquiescence to nature, but opposition to it, a "constant struggle to maintain and improve, in opposition to the State of Nature, the State of Art of an organized polity" (Hofstadter, 1955, p. 96).

Clearly, the evolutionary process had here been far too narrowly defined. The "struggle for existence" was to Darwin a metaphor, suggesting the broad range of actions necessary for the continued life of an organism or a species, without undue emphasis upon the occasions when there was direct confrontation. Certain animals, for instance, in order to stay alive in the winter, may "struggle" to do so, but the only competition involved may be in building warm burrows and storing food. In severe cold those who built and stored well should survive and leave progeny; those who did not may die. It is rarely a matter of one animal throwing another out into the snow.

The competitive element in many cases therefore is not between one individual or population and the next (although such instances obviously occur) but is related to the capacity to make form and behavior suitable to the totality of circumstance. That is, much of competition lies in the differing capacities of different individuals to adapt to changes in the environment. Furthermore, there is evidence for the evolution of cooperative behavior among many social animals (see, e.g., Allee, 1938; Wilson, 1975).

The "natural" in natural selection thus did not refer, as some believed, to brutal force, but was merely the antonym of "artificial," as in artificial selection as practiced by the breeder of plants and animals. (Recall that Darwin was much impressed by the changes that were wrought by artificial selection and sought a similar principle that could apply in nature without man's intervention. Natural selection is thus an appropriate term and logically carries no surplus meaning.)

The social Darwinists who found nature red in tooth and claw had therefore "set up a theory . . . without bothering to include in it more of Darwin than his name" (Dobzhansky, 1962, p. 133). The overly gladiatorial interpretation of the struggle for existence seems more appealing to some intellects, ranging no doubt from the heroic in outlook to the ruthless, than is the view that evolution, rather than being a process of personal victories, is the tediously long, unseen accumulation and elimination of genes. "Social Darwinsm," noted Dobzhansky, "really never had sound biological roots" (p. 341). It was not really Darwinism.

It may be wondered why Darwin did not himself put an end to the misuse of his ideas. One reason was that he was unprepared for the philosophical assault in which his theory would be used. He had argued on the biological issue of the origin of species, and, except for the religiously based notion of special creation, had not actually much concerned himself with its potentially radical ramifications. (Although he did consider ethics at some length in *The Descent of Man and Selection in Relation to Sex* (1871), published 11 years before his death). He believed, as remarked, in the evolution of a moral sense, but on the matter of an evolutionary ethics he was hesitant and ill at ease. He did not know, even if others claimed to, what were the implications of natural selection for relationships among people. He was not, in fact, particularly concerned with humans.

As Eiseley (1956) astutely remarks, "No man afflicted with a weak stomach and insomnia has any business investigating his own kind." Darwin indeed had fled London to work in peace at Down, where "he was not bedeviled by metaphysicians, by talk of ethics, morals, or the nature of religion." He did not wish to exclude humanity from his system, but was content to consider it merely as "a part of that vast, sprawling, endlessly ramifying ferment called 'life'." The rest of it he left to the philosophers.

Too, much of the furor came after Darwin died. In 1882 he was 73 and had long been ill. Then the one-time clerical student, by now bereft of faith in paradise, succumbed to a weakening heart. Of his final thoughts we know nothing—only that when the end came he remarked with simple dignity: "I am not in the least afraid of death." No one who believed so much in life would be.

Darwin was not, however, universally misinterpreted. Peter Kropotkin in 1902 published *Mutual Aid*, wherein he developed the thesis that cooperation among animals of the same species was a major factor in evolution. He filled his book with observations taken on organisms of all sorts, from insects to man. The important point, though, was that Huxley's dualism was rejected. Socialness was as much a part of nature as individualism; hence the basis for an ethics of cooperation was firmly rooted in biology, not requiring, as Huxley had asserted, ceaseless opposition to it. Darwin, Kropotkin claimed, had unequivocally recognized the social and moral nature of human beings, and was thus innocent of the brutish conclusions drawn in his name. Unfortunately, Kropotkin's arguments were unheeded, partly because of the thrust of the times and partly because of the uncritical nature of the evidence that he, a dilettante in biology, provided in the attempt to make his case. Cooperation and "altruism" have however, continued to excite the interests of many scientists (e.g., Allee, 1938; Wilson, 1975) to this day. We shall discuss the evolution of ethics in some detail later.

The final legacy of the Social Darwinist movement, both to society and to science, was the exacerbation of a malignancy that we are only now in part beginning to excise. The direct personal costs of that legacy are immeasurable. The scientific costs (and therefore indirectly those to society) may be measured in one sense in terms of scientific progress, and by this assessment also society has suffered. Reaction against the apparent alignment of biology and Spencerian sociology, imperialism, racism, and politics for the rich created a cleavage between the behavioral and biological sciences that for a short time retarded them both. Lasting damage, however, was done on the behavioral side. Biology went its own way and produced momentous discoveries. The behavioral sciences fell victim to environmentalism.

In the past, naturalists had always considered behavior to be as much their province as were fossils, rocks, trees, or flowers. Lamarck, Huxley, Darwin, even Lyell, treated the actions of organisms as they would any other class of attributes, in fact *along with* the other classes of attributes. Behavior was expected to be as subject to natural law as was bone formation; there was no reason for it to be considered separately. Comte and Spencer, the founders of modern sociology, both believed that their discipline should be based upon biology. Sociological texts of the time, indeed, included long disquisitions upon all manner of biological material.

The difficulty, which eventually led to the schism, lay in the early sociologists' use of analogy. Spencer likened human society to an organism, and argued that what was good for the individual organism was, ipso facto, good for society. This of course included the beneficent effects of evolution, read by Spencer as severe individual competition and survival of the fittest until the final equilibration was reached.

This interpretation, we have seen, was rejected by Lester Ward, the first and most formidable of Spencer's critics. Ward rightly objected to Spencer's simple analogy and claimed for sociology a unique position in dealing with a novel level of organization. But, unhappily, in so doing he began a trend in the social sciences that ended in the rejection of the relevance of biology for understanding behavior.

Spencer's approach, it must be remembered, offered only the status quo and time to a political and economic system begging for reform.

Thus there were more than scientific or philosophical reasons for wishing to discredit it. With Spencer's influence declining and that of the reformers in the ascent it is unremarkable that the new zealots should carry the day and that all of the old associations, even those that were valuable, should be cast out. Thus between 1890 and 1915 there were sweeping changes in the social sciences, some based on positions not altogether consonant with original postulations, as noted by Hofstadter (1955):

> The most important changes in sociological method was its estrangement from biology, and the tendency to place social studies on a psychological foundation. Spencer had not long completed *Principles of Sociology* (1897) when the tide began to turn powerfully against him, and repudiations of his method were so drastic as to be unmindful of the qualifications Spencer himself had made. (pp. 157–158)

Said sociologist Albion Small, "Spencer's principles . . . are supposed principles of biology extended to cover social relations. But the decisive factors in social relations are understood by present sociologists to be psychical, not biological" (Hofstadter, 1955, p. 158).

Perhaps the most devastating criticism came from economist Thorstein Veblen in his *Theory of the Leisure Class* (1899). It is of interest here because Veblen turned around the argument, derived from Spencer and promulgated by William Sumner, that the giants of business were the most evolutionarily fit and therefore were deserving of whatever they could acquire. Such a position is tenable only from the individual-competitive point of view. But it is not true in general, and particularly among humans, that the most acquisitive are necessarily the most productive. In fact, from the societal viewpoint such individuals may be seen as destructive. The business class, argued Veblen, far from being the evolutionary elite, was in fact essentially predatory on its own species, both in outlook and in habits. Veblen again and again attacked the proposition that a person's productivity in society could be equated with his capacity for acquisition and the fitness of his character with his monetary status. Indeed, the personal attributes of the ideal pecuniary man he described "in terms ordinarily reserved for moral delinquents" (Hofstadter, 1955, p. 152).

Where money could be regarded as a reward for social service, Veblen distinguished between the beneficent competitiveness of *industry* and its productive function as an expression of *workmanship*, and the treacherous character of *business* as an expression of *salesmanship* and deception. Whereas a rivalry in productive service was desirable in maintaining industrial efficiency, business rivalry was a social toxin because it had become mainly a contest between seller and buyer, involving a large measure of fraud and exploitation. The economic ideas of Spencer and Sumner held only when business and industry were one, but now, argued Veblen, business had become dominant over industry; salesmanship had become dominant over workmanship.

The business manipulators not only were predatory and nonproductive, but also represented not evolutionary progress, as Spencer and Sumner believed, but evolutionary regression. They were society's barbarians, substituting for simple physical aggression the ruthless and ignoble practices most likely to lead to the accumulation of wealth. The result of their continued and ever more powerful existence was the conservation and increase of the barbaric temperament.

Veblen's position well expresses the general disenchantment reformers felt for the interpretations of "Social Darwinism." The progressive era had begun, and with it came a consciousness of society as a whole rather than as an aggregate of individuals. However, although the new approach was ethically, politically, and economically valuable, it unhappily continued an unnecessary error. Biology was, in large measure, read out of the behavioral sciences, at first on the grounds that it had been ethically misleading and later by the assertion that it had nothing to offer to the scientific understanding of behavior. Environmentalism came to permeate the study of behavior to the extent that biological factors were considered irrelevant. Behavior was thought to be a self-contained system, to be explained only in terms of behavioral laws.

Leslie White, an eminent cultural anthropologist, was one who stated the case squarely:

> From a biological standpoint, the differences among men appear to be insignificant indeed when compared to their similarities. From the standpoint of human behavior, too, all evidence points to an utter insignificance of biological factors as compared with culture in any consideration of behavior variations. As a matter of fact, it cannot be shown that any variation of human behavior is due to variation of a biological nature. (Dobzhansky, 1962, p. 54)

White is not only out of date now; he was out of date then (1949). There has always been evidence that individual differences in behavior are in part biologically determined. But one sees what one is prepared to see, and clearly the years just following World War II were not vintage ones for viewing genetic influences on human behavior.

Of more importance than mere scientific blindness (with which, alas, we are all gifted), however, is an element that runs like a dead vein through the body of all these matters: fear. There is fear that biology may enter into ethical considerations too little and fear that it may enter too much. We have been substantially in the grip of the latter for many years. Even so astute a mind as Hofstadter's, which clearly saw the differences between Darwin and Social Darwinism, could not relinquish it:

> Whatever the course of social philosophy in the future . . . a few conclusions are now accepted by most humanists: that the life of man in society, while it is incidentally a biological fact, has characteristics that are not reducible to biology and must be explained in the distinctive terms of a cultural analysis; that the physical well-being of men is a result of their social organization and not vice versa; that social improvement is a product of advances in technology and social organization, not of breeding or selective elimination; that judgments as to the value of competition between men or enterprises or nations must be based upon social and not allegedly biological consequences; and, finally, that there is nothing in nature or a naturalistic philosophy of life to make impossible the acceptance of moral sanctions that can be employed for the common good. (p. 204)

One would not want to argue with all of this (although portions are certainly debatable) (e.g., Dobzhansky 1962), and Hofstadter's reaction is understandable given his own evident predelictions and the sort of material he was reviewing. But his statement is nevertheless defensive, as if to let biology in is to risk putting humanism out. It is also representative of the *least* prejudiced of professional social thought over the past several decades.

To some, biology is forever, but many kinds of environmental manipulations are acceptable, presumably because any resulting damage is believed always to be reparable. The intelligence behind this escapes me. To argue, for instance, as some psychologists have, that in guiding your children you should never display anger or spank them but rather should show your displeasure by withholding love is, to my mind, to argue for a program with potentially profound effects on the basis of the flimsiest scientific evidence. Who knows how many maladjusted, manipulative little folks have been nurtured in such a cultural gazebo? Yet advice of this order was given with apparent impunity because it suited the prevailing tide of political and scientific sentiment.

The above is not an apologia for biology; it does not need one. Rather, it is intended to reduce bias on both sides. Neither environment nor biology alone can claim behavior. This, however, is the subject for extended discussion later. For the present, the object has been to expose some relevant history and to develop the case for the importance of scientific conclusions for conventional wisdom and, hence, for political and social action. This is merely one instance. What we think of one another is greatly influenced by what we believe ourselves to be. Yet what we are is a question that has had many answers—and will have more still.

Chapter 2
Environment and Behavior:
Retrospect

"Psychology," noted one of the first professional psychologists, "has a short history but a long past." It is obviously true. It is also something we would do well to remember, regardless of our familiarity with the current *apparat* of behavior modification, encounter groups, primal screams, and the happy proposition that you are OK and I am OK. Science is, as much as any other human endeavor, a culture, and, like all cultures, it is by turns fruitful or barren, expansive or constrictive, of the horizon of knowledge. It is both a safeguard and a trap. Progress of course lies in the enhancement of the fruitful and expansive phases. This usually requires essential reformulations.

The behavioral sciences have freely partaken of reformulations—but, alas, often with more enthusiasm than acumen. This is not to say that changes have not been forthcoming: only that they have not been particularly efficacious. We are little closer, in terms of scientific laws, to the substance of human thought and action than were our forebears of many centuries.

Particularly in psychology there is so much old in the "new" that it is impossible to know what is in fact new without knowing what is old. The reasons for this are many, but perhaps the most obvious is that the *problems* of psychology are old. We have been observing ourselves in our present evolutionary condition for some 50,000 years and have, where possible, taken pains to record and systematize what we saw. The result is that each generation has produced a set of compelling, basic questions about the determinants of behavior and bequeathed it to the next generation. The difficulty is that it is always essentially the same set. The questions get "answered" from time to time in one theoretical perspective or another, but these perspectives fail and fall away and the basic issues are never actually resolved. Psychology has thus been living with its questions for a long, long time. To know a psychological problem in the present is to know it only in part.

Having old problems leads to frustration and sometimes to simplistic solutions. It also leads to a degree of outside ridicule. Scientific psychology is often accused of being a failure. In terms of results, it is. But all of the behavioral sciences are comparative failures. None of them approaches physics, chemistry, or parts of biology in generality and precision of theory or even in sheer number of provable facts.

The charge of failure usually is countered by the claim that psychology's lack of performance is due to its youth. Newton, after all, published the *Principia* in 1687, whereas Wilhelm Wundt, probably the first official experimental psychologist, did not establish his laboratory until 1879, some 200 years later. Such an argument, superficially, appears not only plausible, but promising: Give the new science time and it will develop an equivalent of physics, with general laws, high-level theory, explanation, and precise prediction. However, as noted by philosopher Michael Scriven (1964), the argument is invalid for the very reason that it does falsely distinguish psychology's short history from its long past. As "science" the rational study of behavior may be only 100 years old, but as a human enterprise it is as old as humanity itself.

The more revealing observation is not that psychology is one of the recent sciences, but that it was not the very first. (Actually it was, in the limited sense that awareness of psychological causality preceded that of physical causality. The perception that human needs and desires appeared related to human activities prompted the investiture of all nature with more powerful but manlike overlords, having much the same desires and needs. For millenia our fortune was seen as the perogative of such personified beings. Health and soft rain were their reward, drought and disease their punishment. People thus applied the knowledge obtained from insight into their own acts to physical phenomena, and attempted to manipulate the physical universe in the same way in which they manipulated other humans; that is, they attempted to apply to physical phenomena the same psychological techniques used successfully for influencing others: "persuasion, sup-plication, bribery, and intimidation" (Adler & McGill, 1963, p. vi). Demonology and magic, extrapolations of the perception of psycho-logical causality, constituted the first cosmology.)

But demonology is not science, and it is to be wondered why a science of behavioral phenomena was so long in coming. We could always have used one, for our most urgent problems have always been behavioral. Even assuming that as presciences physics and psychology were equals in the Renaissance period, when Western civilization be-gan again self-consciously to approach the world through reason, psy-chology would seem to have had the advantage. There were thousands of years of human self-observation with which psychologists could be-gin, whereas few physicists had had experience with bodies moving in a vacuum. It seems we would have understood the workings of our minds before we understood the motions of the planets. Yet it was physics, first in celestial and then in atomic mechanics, that triumphed. In contrast, some argue, doubtless with overstatement, that psychol-ogy, when stripped of its facade, the litter of its statistics, the rhetoric of its promoters, and the restrictions of its laboratory demonstrations, is in very much the same state it was.

If this is only partially true we still must ask why. One reason is that those who would explain psychological phenomena have from the beginning suffered not from a paucity of data, but from a surfeit. No human has ever been a planet or an atom, for instance, so physicists were spared direct personal experience *as data*. Yet everyone behaves, and all but a few of us perceive what we take to be the cause of our behavior and that of others. In psychology one is always burdened, however subtly, with being the subject.

Relatedly, psychologists are burdened by others being subjects; reporting, formulating, systematizing subjects. There are so many ob-servations, so close at hand, that one cannot know how to grasp them. Moreover, many of the observations are usable; that is, they are valid, if contradictory, reflections derived from thousands of years of living. The scientific psychologist has thus inherited a great body of folk wisdom that is not easily dismissed. It might be thought that such an accumulation serves his ends, but, paradoxically, it does not for in it have been drawn away the first easy pickings upon which a new science thrives. Prescientific psychology, in effect, combed through human be-havior, gathering the roses and leaving the thorny residue.

This is another point made by Scriven (1964):

> . . . we find, if we view the study of human behavior as an enterprise to produce systematic organized information, that a colossal quantity . . . has already been long since snapped up and incorporated in our

ordinary language. It has either been incorporated in the very meaning of the concepts themselves or become a stock part of the body of truisms which we all know about human behavior. Psychology as a science must begin beyond that level. . . . We are faced therefore with a task which Galileo never faced, which Newton never faced. The fact that they did not face this task made it possible for their subjects to undergo the kind of fundamental change and unification which those individuals contributed to them. (p. 167)

When it declared itself a science psychology thus did so in the shadow of a mountain of folk knowledge which appeared ready to be mined. But 100 years later some find the mountain as impenetrable (and as enigmatic) as ever, and much of psychology only seems to have been marching across its surface. The compelling, basic questions remain, and little in the way of sound psychological explanation has been forthcoming. This is, as noted, common to the behavioral sciences; it is not exclusively a psychological dilemma. Psychologists' only real fault has been undue ambition—and that is, in fact must be (in all but a species of saints) forgivable.

However, it appears less and less supportable. Behavior may be inexplicable at the behavioral level, and therefore psychology may be forced to relinquish its goal of becoming an independent science of behavioral laws. The fact is that many of psychology's basic questions appear to require answers from other disciplines. The result that its territory is constantly being annexed. The neurosciences, genetics, and evolutionary biology in particular are seeking more and more to explain behavioral phenomena. Explanations, when they come from these sources, will not be psychological, but will be biological, leaving psychology in the place of its ancient mentor, philosophy, with most of its subject matter usurped.

Psychology's failure, if that is the word, cannot then be ascribed to its youth. Rather, it may be ascribed both to the intransigence of its material and to unlucky choices as to how that material might be pursued. What appears now to have been one unfortunate influence is the sentiment in much of this century that behavioral laws could be formulated without reference to the specific organisms to which they were to apply. Environmental determinism was the result, something with an understandable emotional appeal, following as it did the prior excesses of eugenics and racial slander, but ultimately leading psychology the science into intractable space. We shall shortly discuss the events that may be expected to result in a change in this condition. For the present, however, a glimpse at how psychology found its way into its present situation may be salutary.

THE PRESCIENTIFIC PHASE

If there are devils, there must be exorcists. This role in philosophy was played first in Western thought in the sixth and fifth centuries B.C. when Thales, Anaximander, Empedocles, Anaxagoras, Leucippus, and Democritus attempted to replace an ageless demonic cosmology derived from the extrapolation of the psychological to the physical experience with a cosmology that was rational and materialistic. At first glance there may seem little difference between belief in the Olympic gods and the espousal of the proposition that the universe is created from differing proportions of earth, air, fire, and water, but it was nevertheless revolutionary. Leucippus ended, indeed, with a very modern physical theory, requiring no soul substance whatsoever; all matter consisted of atoms in motion, in the void.

However, the relevance here lies not in the particular philosophic system derived, but in a change of concept, at first implied, and later explicit. A different way of looking at the world, a reformulation, had occurred, from the religious to the scientific. Thus in Cornford's (1957) view,

> Ionian science supersedes theology, and goes its own way, without drawing any fresh supply of inspiration from religion. Science, with its practical impulse, is like magic in attempting direct control over the world, whereas religion interposes between desire and its end and an uncontrollable and unknowable factor—the will of a personal God. The perpetual, if unconscious, aim of science is to avoid the circuit through the unknown, and to substitute for religious representation, involving this arbitrary factor, a closed system ruled throughout by necessity. (pp. 158–159)

Hippocrates, called the father of medicine because he was the first to leave a body of medical writings and because he is said to represent the ideal physician by his intelligence, honesty, and moderation, was among the famous beneficiaries of the new approach. A contemporary of Socrates, his greatness was acknowledged by both Plato and Aristotle. (Indeed, Plato has Socrates, in the *Symposium,* agree with Hippocrates' view that the body can only be treated as a whole. Considering that medicines for specific diseases were unknown at the time, the position seems not unreasonable.)

For our purposes, however, Hippocrates is of interest as a representative of the spirit of scientific inquiry, continuing in the tradition of the Ionian philosophers. In his remarks on the "sacred disease" (evidently epilepsy), for instance, he vigorously rejected sacrifices to the gods and declared that medicine must seek natural causes for all maladies. His attack upon the magic and the supernatural in medicine is sometimes regarded as the "charter of science."

Hippocrates was, however, among the last to subscribe to the naive realism of the older Greek cosmologies, untroubled by epistemological and psychological considerations. With the arrival of the Sophists, and in reaction to them, came a concern with the nature of knowledge and how it is acquired which would seldom be relinquished. The Sophists argued that we can never know truth, or, if we could know it, that we could not communicate it to others. They therefore taught their students to forswear the search for truth and develop instead the art of rhetoric, of persuasion dependent upon ornamentation of language, appealing to the emotions rather than through logical argument to the intellect. Protagoras put the position squarely when he announced that man is the measure of all things, thereby not only encapsulating Sophist thought but epistemologically prefiguring Hobbes, Locke, Berkeley, Hume, and Kant.

The decline of Athenian democracy accelerated the trend toward self-scrutiny. For Socrates and Plato it was ethics rather than cosmology that was of central concern; speculation on the constituents of the good man and the good life replaced speculation on the constituents of the physical universe. Before one can describe the path to what ought to be, however, one must describe what is. There can be no theory of ethics without a theory of humanity. Human nature, its origin, predilections, and prescription thus became the abiding interest of secular philosophy.

Socrates and Plato opposed the Sophists not with regard to the formal skills that they taught but with regard to the ethical consequences of their teaching. Sophistry is only a means of gaining power over others. On the other hand, Socrates, through his student, argued that it is ultimate virtue and the manner of obtaining it that should be our concern (thereby fixing upon a contrast that still survives to embarrass us: that between the just and the merely successful person). And for Plato the just life was both desirable and possible.

To understand how to obtain it, however, one must speculate upon human nature. Plato did, devising the first comprehensive personality theory on record. Anticipating later theorists, he divided the mind (soul) into three parts: reason, will, and desire. But most importantly for consideration here, he proposed that dominance of one of these three traits over the other was inborn. There were thus three "natural" types of person.

For those in whom reason was dominant great wisdom was obtainable. In them desire was regulated and will was directed toward rational ends. They were capable of all the Greek virtues, the highest among which was justice, occurring when the other traits were properly subordinated to reason. Further, it was these individuals who were to direct the utopian state: Since they could govern themselves, they were capable of governing others. They were the natural aristocrats, who, in the local State described by Plato in *The Republic,* were to have sole authority (Adler & McGill, 1963, p. 18).

Those in whom will was dominant were courageous but not wise, and were inclined to violence. They would require the guiding hand of the just individuals, who alone knew due proportion. In *The Republic* they are destined for the army and the constabulary.

The great majority were so constituted that desire, including the desire for goods and luxuries, was dominant. They were capable of little justice or courage but could learn to obey their rulers and contain their desires when their lives were regulated in a just and orderly society. They would be the workers and artisans.

Aristotle was Plato's finest student, but opposed him on fundamental issues, particularly on the doctrine of ideal forms. Plato was too much a mathematical rationalist whose views bore little relation to reality, according to his more naturalist successor. Rather, Aristotle saw ethics as a practical endeavor, designed to help attain virtue through training. The term *ethics* derives from a Greek word that originally meant the "accustomed place" or "abode" of animals and then was transferred to humans to mean "habit" or "disposition." Aristotle's use of it signifies his interest in training and habituation in the attainment of virtue and his predominant concern with the study of human character and conduct rather than in abstract formal ethical principles (Adler & Cain, 1962). Of more immediate concern, Aristotle, in stressing habit and training, likened the mind to a tabula rasa, a blank tablet upon which experience would carve out the defining features of behavior. This is in sharp distinction to Plato's inborn human types and brings us quickly to an issue that we shall examine throughout—namely, the differential scientific and social perspectives deriving from an emphasis upon the innate or the experiential origins of behavior. Opinion on the question varies to extremes, as one would guess; but in Western thought the difference is first comprehensively expressed in the views of these two most famous philosophers, one of whom (appropriately, it seems) was the teacher of the other.

There is, of course, more to choose. It is sometimes said that every philosopher is either a Platonist or an Aristotelian, leaning toward the inborn and the rationalist-ideal or toward inculcation and sense experience. Certainly much of later philosophy can be understood in terms of this distinction. But we need not follow it further here. It is enough to know that the thought of these men dominated the Western mind up through the Renaissance, and that a fair amount of what mental activity there was in the Middle Ages (such as that of Aquinas) was concerned with the reconciliation of Aristotle's concept with a Christian world view.

One of Aristotle's ideas, duly so reconciled, indeed survived until Darwin's time: It was Aristotle who was ultimately responsible for the Grand Ascending Chain of Perfection. His theory of the soul ordered nature hierarchically from inorganic material to man, the latter possessing the most and best capacities. From there it was but a short theological migration to species fixity. It was not, however, a move that Aristotle intended. Rather, he was much impressed with intermediate forms in nature, with continuity and gradations among species and subspecies. The concept of continuity, we have seen, is necessary for the theory of evolution. Aristotle did not propose such a theory (although there is some opinion that had he lived longer he might have), but his observational and classificatory abilities were praised more than 2000 years later by one who did: "Linnaeus and Cuvier have been by two gods," wrote Charles Darwin, "but they were mere schoolboys to old Aristotle" (Adler & McGill, 1963, p. 47).

Major progress in scientific and philosophical thought was, for various reasons, thereafter deferred until the seventeenth century. Philosophy continued in Greek hands through the period of the Cynics and early Stoics, then passed to the Romans, the most famous of whom were Seneca, Epictetus, and the emperor Marcus Aurelius for the later Stoic period, and thence to the Christians. Among the former there remained a professed admiration for natural law, the better to live in harmony with it.

Such a goal can hardly be faulted, provided one does not act too soon, being prematurely certain of just what those laws are. Epictetus's remarks, for instance, "I must die. Must I then die lamenting? I must be put in chains. Must I also lament? I must go into exile. Does any man then hinder me from going with smiles and cheerfulness and contentment?" (Adler & Cain, 1962, p. 76) reflect in some respects an admirably courageous personality, but one must wonder about the relationship in nature from which the laws that prompted them derive. For both the Cynics, who advocated a life of pure virtue, involving complete simplicity and indifference to external affairs, including even those of family, and the Stoics, then, the result, in spite of their commitment to the search for knowledge, was intellectual and emotional detachment. The intention, not unshared by much of practical Christianity, was chiefly to cope.

The Renaissance changed all that. As people again began to look to themselves rather than to the Divinity and his host of interpreters for guidance, basic human questions reemerged and new answers were sought. The first act was to sweep away the constrictions of Middle Age doctrines—to formulate a different view. It was to this task that Francis Bacon and René Descartes set themselves.

The genesis of modern science and philosophy is usually assigned to the early seventeenth century. Kepler in astronomy (1618), Galileo in physics (1638), Harvey in physiology (1628), and Descartes in mathematics (1637) demonstrated the power of rational thought for all future generations and thus reset Western man upon the road of intellectual conquest begun by the Ionian Greeks and so long in detour. The initial step in philosophy, however, was taken by one unacquainted with these events, but who, gifted with prescience, assured us of their coming.

Francis Bacon seems an unlikely candidate for such a deed, if in fact, almost all of his life "was spent in seeking preferment and position at the Court" (Adler & Wolff, 1960, p. 136) under Elizabeth I and later James I, and, further, without lasting success. (Nevertheless some of what is best in history has been unlikely, and it suits posterity only to give thanks.) Bacon found the theologized science of his day blind and futile, and he declared it his intention to restore the "original commerce" between humans and nature. To this end he envisaged a huge masterwork that could accomplish the restoration of all knowledge. It was, however, to be an altered restoration. The Greeks, and following them the pre-Renaissance philosophers, looked upon the acquisition of knowledge as an end in itself, in service of serene contemplation. Bacon's concern was power, the power that learning and knowledge gives us over nature. Science in his terms was to exist as the benefactor of humanity and was therefore held to answer for its productivity. (In Bacon we thus find the harbinger of what might seem a uniquely nineteenth- and twentieth-century theme: "progress." Science increasingly has been viewed as a source of societal well-being, to the extent in recent times that it underwent an almost religious transformation—thereby, ironically, acquiring the exemptions of the mystical. Much of this has recently been put to right, or is tending to be, although there remain few other receptacles for that seemingly endless human product, faith.)

It was Bacon's intention to reengender a respect for learning, in particular a respect for inquiry, for discovery. To do so he had first to challenge the pedants, an effort that apparently set him at odds with perhaps the greatest universal genius of whom we know, Darwin's dear old Aristotle. But it actually was not the clear, swift Greek mind now so highly admired that guarded the retrogressive bastions; rather, it was Aristotle made over and stultified at the hands of the pedagogues. So enormous had been his influence in ensuing ages that his conclusions, or some of them, had in the Christian synthesis become dogma. The same had happened to a lesser extent in the sciences. What was could not be unless it had been forespoken by the master.

The difference between a world caught in a given order and one seeking further to inquire cannot be overstressed. Bacon's service was aggressively to state the case for a new freedom, and in so doing to open the gates of modern thought. He succeeded well, although compared with the task he had proposed for himself his actual contribution would have been small. He desire was to publish a huge volume, *The Great Insaturation*, in which he would "try the whole thing anew upon a better plan, and to commence a total reconstruction of sciences, arts, and all human knowledge, raised upon the proper foundations" (Adler & Wolff, 1960, p. 138). Actually he completed only the second part of that work, and it is upon this, the *Novum Organum*, which his fame justly rests.

The *Organum* to which Bacon refers is Aristotle's *Logic*, the six treatises placed at the head of Aristotle's works being collectively designated the *Organon*, or in Latin, *Organum*. Bacon's objection to Aristotle, however, was not logical in character, but methodological. Aristotle had dealt with the value of deduction, of moving from the general to the particular, as the method of acquiring knowledge. This was evident in his famous syllogisms.

Bacon's "New Logic" proposed a different approach, that of induction. Scientific facts, he held, should be drawn from nature bit by bit until they coalesce into a larger law. We infer, then, from the many particulars to the general. Bacon's quarrel with deduction had to do with productivity; syllogistic logic was of little use in interpreting nature, but rather, was the drawing out of a few obvious truths to great lengths. "The creations of the mind and the hand," he wrote, "appear very numerous if we judge by books and manufacturers; but all that variety consists of an excessive refinement, and of deductions from a few well-known matters" (Adler & Wolff, 1960, p. 139).

René Descartes, born some 35 years after Bacon (in 1596), took the next step in scientific philosophy. Like Bacon, he rejected the wisdom of the "ancients," but he did not follow Bacon in the method whereby truth was to be obtained. Bacon was in modern terms an empiricist, one who believes that an ever greater accumulation of facts will necessarily lead to general principles; that scientific laws more or less naturally fall out of massed and shaken hordes of data. Descartes likewise valued experience, for he found it to be at variance with the intellectual pronouncements of the time, and it taught him "to believe nothing too certainly of which I had only been convinced by example and custom" (Adler & McKeon, 1963, p. 157).

However, the resemblance stopped there. Descartes was a rationalist, trusting above all the power of reason to arrive at fundamental truth. So much did he trust reason, indeed, that he made a single reasoned judgment the absolute foundation of all knowledge. Resolved to doubt all except that which was utterly incontrovertible, all data from the senses, all capacities of the mind, he arrived at last at a proposition which he believed indubitable: *cogito ergo sum* (*I think, therefore I am*). This was of course his famous first principle; that he could think meant that he had to exist, for obviously what did not exist could not think.

From this first principle Descartes drew an assertion, namely that the truth of our thoughts lay in their clarity and distinctness. What appeals directly and clearly to the mind he adopted as certain. One certain thought was that doubting as a method of arriving at knowledge, though adequate to prove the existence of the doubter, was actually the product of an imperfect being. He saw that it is better to know in the first place than be forced to make conclusions through doubt. But realizing that he was imperfect, he had necessarily to ask from where the idea of perfection could have come. Certainly an imperfect being was incapable of the idea of perfection; therefore, the idea must have come from a perfect being who placed it in his mind.

Who might that be? It was God. God must exist, since there was nothing perfect in the real world and yet there was the idea of perfection. From this conclusion Descartes derived the doctrine for which he is best known in psychology, that of *innate* or *inborn ideas*. Since God had implanted the idea of His perfect self in the mind of humans, He was of course capable of implanting other ideas. Thus, the axioms of geometry, concepts of space, of time, and of motion were all innate, being nonderivable, that is being dubitable, from the actions of reason and sense experience alone.

Descartes's conclusion was a powerful one (which, we shall see, was quickly opposed by what came to be the British experiential school). He represents for us a loose line of descent from Plato in his concern with the perfect, his doctrine of the inborn, and his heavy reliance upon reason. Curiously, he is also considered the father of French materialism, because he held that although the mind is free, the body is a physical automaton. This radical notion led him to espouse that for which he is probably best known in philosophy, dualism: the existence of two sorts of substances, the extended substance of the body and the unextended substance of the soul. The two were held to affect one another in humans as a particular place in the brain, the pineal gland. Animals, having no souls, were complete automata.

In 1588, some eight years before Descartes was born and 62 years before his passing, the Spanish Armada approached England. The wife of the vicar of Westport thereupon prematurely delivered a son. His mother, wrote Thomas Hobbes later, "bore twins, myself and fear" (Adler & McGill, 1963, p. 110). The peculiarity of his birth may have affected his disposition, as he claimed, but it did not prevent him from energetically pursuing a sometimes dangerous career up to the age of 91.

Hobbes was provided for by a wealthy uncle, and by the age of 14 had a remarkable command of Greek and Latin. He continued at Oxford for the next five years, where he studied physics and logic. In 1608 he read and was greatly impressed with, Kepler's *New Aetiological Astronomy*, one of the pioneer works of the science of the seventeenth century. He thereafter determined, as had others before him, to renovate the system of knowledge according to new rules.

With Francis Bacon, Hobbes shared a distrust of past wisdom and a conviction that knowledge is power. But whereas Bacon saw induction as the sure path, Hobbes at the age of 40 had discovered Euclid's *Elements* and become enamored of the prospect of constructing a deductive system based upon scientific principles, which would extend even to ethics and politics. Like others, he envisaged a grand network founded upon a few basic laws. Unlike others, however, his conceptions were governed almost exclusively by the physics of Galileo; he would treat matter, people, and human society (the state) in terms of the laws of motion.

For our purposes, Hobbes's importance lies in his espousal of an experiential epistemology. He is the first in a long line of what has come to be called the school of British empiricists, whose influence upon the development of psychology was enormous, and which is indirectly responsible for much of modern learning theory. Hobbes proposed that all one can know is derived from sense experience. The motions of the mind, consisting of sense data, thoughts, and ordered trains of thoughts, originated in the motions transmitted from external objects to the senses. Images we retain exist as they do because particles stay in motion; they fade because fewer and fewer particles so remain.

Trains of thoughts, or imaginations, could be unguided, as in dreams, or guided. When guided, they were seen as connected by *association*—similarity, contiguity, and so on—or by persistent desire or purpose. The concept of association of thoughts or ideas was proposed as far back as Aristotle, and Hobbes does not pay it a great deal of attention. Nevertheless, we find at this point the two central principles of the British school: *experientialism* and *associationism*. These would soon be joined into the proposition that what is in the mind is derived solely from sense data and connections among them. Those who followed the tradition only elaborated, refined, and codified this proposition until it slipped together with German sense physiology more than 200 years later to form the first experimental psychology.

The experiential tradition is opposed by what has been called the *nativist*, the view that the association of sense data cannot alone account for behavioral phenomena. Rather, the nativist argues that an organism comes to the experiential situation with predispositions that shape meaning from sense data in ways that are independent of prior experience and may be unique to an individual as well as species-specific. The mind is not, in this approach, a tabula rasa upon which anything may be written with equal felicity, nor indeed is it a blank page, as John Locke proposed; rather, it is at inception an instrument, by whatever metaphor, with many spaces already occupied, and is therefore particularly receptive to certain kinds of input, chiefly at certain times during its development.

Both the experiential and nativist traditions are ancient, the former loosely tied, on record at least, to Aristotle through Hobbes, and the latter to Plato through Descartes. They are logical extremes when polarized, when one exclusive of the other is considered to be the basis of all behavior. Such extreme positions have, as one could guess, been espoused, or appear to have been, in the long history of humanity's concern with its own behavior, which is to say throughout human history. They need not occupy us except as coordinates in the topography of that history. However, without such coordinates the contours of an acceptable solution are less clear, and we must therefore continue briefly to explore the lines of their descent.

Hobbes was, as mentioned, one of the first to build a psychology upon the physical concepts of the seventeenth century. He began, in fact, what became a restless and unrelenting determination among those concerned with social and psychological events to use the physical sciences as a model. Time and again if one is to comprehend events inside psychology one must look outside it, as did the psychologists themselves, for guidance and understanding. With care and such an outside eye, one can see straight to the twentieth century.

John Locke, George Berkeley, and David Hume are usually considered to be the cardinal British empiricists. Although Hobbes began it, the latter three are known especially for their experiential epistemology, and through them experientialism gained preeminence over the rationalism of Descartes, Spinoza, and Leibniz on the continent. (It was, however, short-lived, for only a few years elapsed before German idealism as propounded by Kant and Hegel superseded British experientialism. Such are the winds of the intellect. The matter is of course yet to be settled.)

Locke was born in 1632. England was at the time rife with civil war and religious persecution, and his father, a strict but kindly attorney, was often away in the army of Parliament. The impression on the boy told: "From the time that I knew anything," he wrote at 28, "I found myself in a storm which has continued to this time" (Adler & McGill, 1963, p. 134). He was given a classical education, which he found unrewarding, but soon turned with avidity to the study of Descartes, Gassendi, and Hobbes. A man of deep religious feeling, he was at first drawn to the clergy but found he could not accept the required articles of faith. He turned instead to medicine and chemistry, forming friendships with Boyle and Sydenham (and later Newton). He was greatly taken by their empirical methods, a fact that doubtless influenced his later experiential theory of knowledge.

His *Essay Concerning Human Understanding* was a long time in coming (1690) and began as the result of an early discussion between Locke and his friends from which he concluded that before one could answer certain (religious) questions one would have first to deal with the means whereby knowledge is acquired. He accordingly set himself the task "to examine our own abilities, and see what *objects* and understandings were, or were not, fitted to deal with" (Adler & Cain, 1962, p. 186).

The primary materials of knowledge, he concluded, are "ideas," and the sources of ideas are "sensation" and "reflection." Sensations, sense data, are the single largest source. Reflection consists of operations of the mind upon the ideas we have. Although the ability to reason was seen as innate, Locke opposed Descartes on the proposition that anything else could be. The mind, he wrote, is originally as "white paper, void of all characters, without any ideas" (Adler & Cain, 1962, p. 189). Even such exalted notions as perfection, God, or the primary geometrical axioms could be explained through the accretion of sense data and the mind's use of them. Locke thus elaborated and extended Hobbe's position, while sharpening the contrast with the nativist doctrine by his direct attack upon Descartes.

Berkeley is known primarily for carrying experientialism to its limit. Locke had been concerned with how we know the world, how we know what is "out there." This leads to some obvious technical difficulties, never fully resolved. To Berkeley the solution was simple, and seemed to have pleased him: Forget about transactions between "things" and "the mind"; all we know or can ever know is what is in our minds *about* things; all we can know are our own perceptions. *Esse est percipi: perception is the reality.* Matter does not generate mind, as Hobbes and Locke contended; rather, mind as it were, generates matter. (Leading one wit to sum up the philosophical scene of the day with "No matter? Never mind!")

Berkeley's proposal is, of course, inescapable. Finally, we are indeed dealing with our own perceptions; it is the image of the tree (person, building) and not the thing itself that is in our heads. But it cannot be left at that for it would be scientific suicide (and result in the loneliness of solipsism) to abolish the social nature of science as collective thought. We cannot rest content, that is, with the fact that the basic data of all knowledge are our own perceptions and sensations; we must *act as if* there is a physical reality. (The tactic of *acting as if* in the face of uncertainty is often useful. In this case it is a necessity.) Berkeley's suggestion, therefore, while incapable of disproof, may be relegated to the logical limbo especially reserved for a reducto ad absurdum.

Hume occasionally wrote in Berkeley's arch-subjective fashion, but for the most part took the existence of material objects as given. Much more important to him was the conviction that confusion, ignorance, and error could be eliminated by consistent adherence to empiricism. He distinguished between sense perceptions (impressions) and ideas, the former being much more lively than the latter. New ideas come about through the compounding, transporting, or augmenting of impressions (explaining how we could conceive, for instance, of a mermaid). Philosophical disputes could be settled, and confused affairs set aright, by referring to the impression from which an idea presumably derived. If there were none, the term was questionable.

Hume is most famous generally for his analysis of cause and effect, something that seems obvious and that we take for granted. But he demonstrated that we cannot directly *know* cause-effect connections; we can only perceive what appears to be a necessary relationship. He further argued that there is no sense impression to act as a basis for the idea "necessary"; hence, the concept of a necessary relationship between cause and effect is an illusion that exists in the mind and not in nature. Scientific laws therefore do not inhere in natural events; rather, they are an imposition of the mind of man.

Such a skeptical position could, as well as Berkeley's extreme subjectivism, be scientifically devastating. If necessity is only in the mind, what of the quest for order that lies at the foundation of the scientific enterprise? How can one do physics (or psychology for that matter), without knowing what is beyond oneself? The answer, once again, is to *act as if* there were laws of nature that can be found and are not merely the fantasies of creative and wishful minds. The advance of science is proof enough that the technique works (or we all are in a sometimes blissful, often nightmarish dream). Necessity, that is, may be an illusion, but if so it is a shared and useful one.

Another man destined for the church but who turned to medicine was David Hartley. He figures little among the giants of philosophic psychology except, as Boring notes, as a "founder." Hartley

> . . . took Locke's little-used title for a chapter, "the association of ideas," made it the name of a fundamental law, reiterated it, wrote a psychology around it, and thus created a formal doctrine with a definite name, so that a school could repeat the phrase after him for a century and thus implicitly constitute him as its founder. Whoever discovered "association," there is not the least doubt that Hartley prepared it for its *ism*. (Boring, 1950, p. 194)

With Hartley in 1749 British experientialism had thus developed into British associationism. Associationism, in turn, was to culminate some 100 years later with the Mills. In James Mill, association as a principle of mechanical compounding reached its zenith; he felt the experientialist's need for making sense data primary and the associationist's need for postulating many different elements. As the physicists had discovered larger things to be composed of smaller, so Mill analogously saw complex ideas and thoughts as composed of the smaller sense data.

James's son, John Stuart, modified his father's doctrines of association and mental compounding. Instead of a mental mechanical physics he substituted a mental chemistry, and it was an important step. For James, thoughts and ideas were complicated mixtures derived ultimately from the senses, but with the individual elements, though perhaps disguised, still intact. John essentially applied the maxim that the whole is not simply the sum of its parts; in the combination of elements some original properties may emerge that are not predictable from the properties of the elements themselves.

The chemical analogy is fruitful and also offers a penetrating insight into the reasons for the difficulties of learning theory some 100 years later. We might think ideally, for instance, that if one knew all there is to know about hydrogen and all there is to know about oxygen, one would without further consideration understand all there is to know about water. But it is not so. To maintain that it is could be called the compounding error, and it must be obvious that such a view is a matter of faith. We never know for any set of elements all that is required for precise prediction of the behavior of a compound composed of them. We must study the compound directly, in addition to the known or supposed properties of the elements. Were this not true there would be no reason for other sciences at all; one would need only physics and a large computer.

The compounding error, derived directly from associationism, appears early in American theories of learning, and accounts for many of the problems it has encountered. It was assumed that basic laws of learning were the same for all organisms and that complex behavior could be analytically decomposed into its simpler elements, which, when understood, would explain it fully.

John Stuart Mill brings us to 1865. But it should not be thought that British experiential associationism dominated the intellectual world at this time. On the contrary, from East Prussia all across the philosophical landscape had spread the ponderous, Teutonic figure of Immanuel Kant. Whether the Mills thought so or not, Kant had revolutionized philosophy with the *Critique of Pure Reason* in 1781, and it would never be the same again. It was David Hume who roused

Kant from his "dogmatic slumbers." And Hume's skepticism of our capacity to know causality, to perceive necessary relationships, was in fact the galvanizing agent. Hume had argued that causality was an illusion, and Kant, alarmed at the prospect that certainty in science, metaphysics, and even religion, was unobtainable, reacted with a synthesis that was intended to provide certainty.

His solution was to postulate the ground of necessity in the knower. It is the individual perceiving the world who has an indwelling, pre-experiential capacity to impose a causal order upon natural events. We know, he argued, even before experience, that everything must have a cause and that the effect will always follow. Such capacities he called *a priori*, and they figured strongly in his epistemology. Our subjective faculties thus guarantee the validity of causal laws, and also the rigor of pure reason, such as seen in arithmetic and geometry.

Kant was not, however, merely a rationalist, on the order of some of his predecessors (he had been a disciple of Leibniz). The rationalists (Descartes, Spinoza, Leibniz) had spoken for reason—reason independent of the vagaries of perception—as the chief source of knowledge. The experientialists (Hobbes, Locke, Berkeley, Hume, and their successors) had presented sense data (and the operations of the mind upon them) as the main contributors. Kant argued that both reason and perception are circumscribed and neither is alone to account for knowledge. It is true that we cannot grasp reality, "the thing in itself" directly with the mind; we have only "the phenomenon" with which to deal. But knowledge does not come from a more or less passive distribution of the nearly infinite little pellets of sensations put in through the senses; rather, there is in the mind an innate capacity to impose order upon the world. The mind operates upon sense data, but has necessary, innate capacities also.

Kant's emphasis upon the innate capabilities of the mind puts him, more by contrast with the British experientialists than anything else, on the nativist side of the nature-experience issue. He thus represents one view in the old and continuing conflict on the question of what is inborn and what is experientially derived in the behavior of organisms.

A SCIENCE OF THE MIND

The difference between scientific and nonscientific knowledge lies not in its intrinsic validity but in the manner in which it was obtained. That is, whether or not a piece of information is correct in the final configuration of things need have no relevance to its origin, but, lacking certainty now, we are forced to load our prejudices with reference to the means by which it was acquired. The scientific method does not guarantee truth; it only assesses its probability. What is believable is what is properly obtained.

Scientific psychology therefore began when the methods of science were generally and systematically applied to psychological subject matter. What passed for psychology before was not thereby rendered invalid, it was merely nonscientific—or, as we have used the term, prescientific. Classification and description are usually the first phases of scientific endeavor, and in this respect much of "prescientific" psychology might actually be called scientific, many people in all ages having spent much of their lives observing and commenting upon human behavior. But the distinguishing feature of modern science, beginning with physics in the seventeenth century, is the experiment, the deliberate manipulation of presumed causes and the attempt to understand their presumed effects. We may therefore consider the inception of scientific psychology to be the point at which experimental methods were applied. This takes us to Germany.

The ground there was especially well prepared, but in a peculiar way. The Germans had come late to science, the Renaissance having begun in Italy and spread northwest and north. At the time of Galileo and Descartes the new science, emphasizing mechanics and astronomy, was confined to Italy and France. England promptly joined, but the English contribution was restricted chiefly to a few individuals, among whom Newton was the foremost. In the eighteenth century France led all nations in devotion to the new pastime, the German interest having begun with the foundation of the Berlin Akademie in 1700.

This order of events is important only because it seemed to leave certain sciences, for a short period, to certain nations. The French and English, having been participants, were impressed with the physics of Galileo and Newton and the mathematical-deductive approach, which leads to grand generalization. To the Germans, therefore, fell the even more recent science of biology, still in the classificatory stage and not ready to yield to mathematicodeductive principles. Consequently, according to Boring, "It was left for the Germans, who have always had great faith that sufficient pains and care will yield progress, to take up biology and promote it" (p. 19).

Germany was thus the first home of biology, and the willingness of German scientists still to classify and describe, as they were required to, led directly to the genesis of psychology under their care. Through their interest in phenomenology there developed a flourishing sense physiology, which, when joined later with British associationism, led to the taxonomy of consciousness, the first science of the mind.

Although many in one way or another preceded him, Wilhelm Wundt is recognized as the first person who without reservation could be called an experimental psychologist; he was both a "founder" in Boring's sense and senior among those to follow. Wundt studied at Heidelberg, forced by economic necessity to train in medicine rather than the sciences. The effect was seen as quite important:

> So Wundt, like Lotze and Helmholtz before him, went into medicine because of the necessity of earning a living. It is thus, in a sense, because young men have to be self-supporting and because the medical faculties of the German universities gave a truly academic training, which could nevertheless be made profitable later in the practice of a physician, that modern psychology began as physiological psychology. (Boring, 1950, p. 317)

Wundt's "physiological psychology" was not, however, anything like what would be so named today. He was—strangely enough, considering his training—a dualist, believing that mind and body cannot be compared. (Boring credits Kant's influence upon all of German science for this.) When Wundt turned to the mind, he therefore turned to it without reference to physiological functions. Psychology was to deal with the stuff of the mind, which, obviously, was consciousness.

The problems for psychology as outlined by Wundt were: (1) the analysis of conscious processes into elements, (2) the determination of the manner of connection of the elements, and (3) the determination of the laws of connection. The method of psychology was determined by the definition of its subject matter. If psychology were to deal with consciousness and describe the relationships among its elements, then self-observation or introspection was the only means at its disposal, for the experiencer was the only one who can describe his or her experience. Psychology was thus to be

> . . . introspective, sensationistic, elementistic, and associationistic. It was *introspective* because consciousness was its subject matter. Consciousness was then the *raison d'etre* of psychology. It was *sensationistic* because sensation shows what the nature of conciousness is. It was *elementistic,* because the whole conception at the start was of a mental chemistry, and it seemed as if sensations, images, and feelings might well be the elements which make up those compounds that are stuff of psychology. And it was *associationistic* because association is the very principle of compounding, and because the British school had shown how you can get perception and meaning out of the association of parts. (Boring, 1950, p. 385)

The influence of the long line of British empiricists and associationists, from Hobbes to Hartley, upon the first (and, as will be evident, the later) experimental psychology should now be clear. We shall encounter it again and again.

Such, in any case, was the new psychology, consisting of consciousness, the elements of consciousness, and the laws of association by which the elements combined. It was, however, almost exclusively German psychology. Could it be made international, and in the process what would survive? Of Wundt's formal psychology, of the taxonomy of consciousness, almost nothing remains. However, the foundations of that psychology—the sensationism, the elementism, and the associationism—were retained and are in fact the foundations of behaviorism.

Wundt's psychology was replaced initially, however, by functionalism. The United States became the new home of psychology, since Americans were more receptive to functionalist views. To understand the reasons for this, it must be remembered that Wundt's laboratory was established in 1879. Twenty years prior to that Darwin had published *The Origin of Species*, and its effects on all aspects of thought, particularly in the United States, had been enormous. Later, Galton's influence came to bear, bringing with it a concern with the inheritance of mental capacities, especially intelligence, and the means of its measurement, mental tests. The defining feature of American interest in mental processes therefore became functionalism, an emphasis upon the role played by a character or trait in a population and its meaning for the organism or species. This was in opposition to Wundt's structuralism, which emphasized the analysis of the content of the normal adult mind. With Darwin, Galton, and J. McCattell stressing the importance of individual differences, then, American psychologists became interested in *minds* as contrasted with *the mind.*

The process was apparently a subtle one. It was William James who introduced scientific psychology to America with his recognition of the new experimental approach. Soon, with the avid guidance of Stanley Hall, a wave of laboratory founding swept the land. E. B. Tichner, an Englishman who had studied under Wundt, was imported to Cornell in the early 1890s to administer the German tradition. But James found himself opposed to Wundt's elementism, and proposed instead his famous stream of consciousness. There was no question but that James believed that analysis was necessary to the scientific method; however, he also believed that the analytic description of mind should not be taken to mean that the real mind is a mere congeries of particles. To do so would be to lose the whole to the artifact of method.

James's interest, however, did not last. Having begun by introducing psychology to America, he gradually abandoned it. In his principal text he had both supported and condemned the German view. He supported it by presenting with great care many of its experimental findings. He condemned it often in the interpretation of the results. His book, published in 1890, was an instant success. But, begun as a manual of the new scientific psychology, he wrote of it when finished that it only proved "there is no such thing as a science of psychology"

(Boring, 1920, p. 511). Thereupon, he returned to his first love, philosophy, developing with Dewey the philosophical school called pragmatism.

It was Dewey, more than any one, who formally developed and proclaimed the functionalist doctrine and set it in opposition to structuralism. He favored change, and looked for the value of things not as they are but as he thought they could be. As a democratic evolutionist he noted two basic and related propositions: First, struggle is necessary for progress, and second, use and functional practicality are the factors differentiating what survives and what does not. What is useful—that is, pragmatic—survives and evolves; what is not becomes vestigial or extinct. This he applied to all affairs, including those of his own professon. "Philosophy," he wrote, "recovers itself when it ceases to be a device for dealing with the problems of philosophers and becomes a method, cultivated by philosophers, for dealing with the problems of men" (Boring, 1950, p. 553).

Thus at the end of the nineteenth century there were two opposed approaches to the study, even the definition, or psychology. The one had its origins in British associationism and German sense physiology; the other had taken the same concepts and superimposed upon them derivations from Darwin's theory of evolution. Both approaches, however, dealt with consciousness in humans; the difference was a matter of how it should be treated. Both would shortly be forsworn.

ANIMAL PSYCHOLOGY

Darwin was to affect American psychology in a number of ways. Not only did he provide its defining characteristic through his *Origin of Species* in 1859, he also laid the foundations for animal psychology with his publication of *The Expression of the Emotions in Man and Animals* in 1872. It was a pregnant event, for it would at length provide the basis for a revolution in the new science. Consciousness would be abandoned as the basic psychological datum in favor of behavior. This would occur because of the impossibility of introspection in animal minds, which would never have been a problem were there no animal psychology. This we now briefly examine.

In *The Origin of Species* Darwin had demonstrated the evolutionary relatedness of all life, and, although he wrote of no group in particular, it was clear that his conclusions could apply to humans. However, so that the reception of his ideas would not be overly prejudiced and to avoid a discourse upon human provenance without evidence he considered sufficient, he deliberately refrained from mentioning humans, except "in order that no honourable man should accuse me of concealing my views, to add that by this work 'light would be thrown on the origin of man and his history'" (Darwin, 1882, p. 49). He did, of course, believe in human evolution. When he found that many naturalists accepted the doctrine of evolution of species, it seemed to him advisable to publish a special treatise. The result was *The Descent of Man*, published in 1871.

The Expression of the Emotions in Men and Animals* was intended to be a chapter in *Descent*, but as usual Darwin found that his notes had become so voluminous as to require a separate work. It was published the next year in 1872. *Expression* began many things, among them the modern science of ethology, although it was some time coming. Most important at the time, however, in presenting the thesis that the expressions of humans could be traced to evolutionarily antecedent animal origins, Darwin established the general credibility of the study of animal behavior. So long as species were considered to have been created separately (presumably for separate purposes), there was little use in studying one in order to discover something about another. But if man and animals were evolutionarily related, there might be great use indeed. The animal mind, dismissed by Descartes and simply not discussed by many others, became something of interest in understanding the evolution of consciousness.

First to seize publicly upon these implications was Darwin's friend, John Rommanes, who in 1882 published a work on animal intelligence. Rommanes foresaw the development of a psychology of comparison among different forms that would be similar to the study of comparative anatomy. However, his book did not treat directly the problem of mental contiguity between humans and animals, but rather was a compendium of nonexperimental observations on the high level of what was thought to be animal insight.

In the United States, however, E. L. Thorndike had undertaken a series of actual experiments. The precepts under which they were conducted would charter the field, eventually gaining hegemony over the entire science, for decades. Thorndike's early work was directed toward uncovering the nature of animal learning, and he published his results in *Animal Intelligence* in 1898. Later he applied the concepts and principles so derived to humans.

However, his results interest us less than one of the assumptions upon which his work was based, for it was this, in somewhat modified and often covert form, that would be prerequisite to the development of American learning theory. Thorndike was convinced that laws of learning were equally applicable to all individuals regardless of species; that basic learning laws were applicable across species—to man, cat, or chicken, alike. This supposition would come to underlie so much of subsequent work that an entire genre of research and theory is uninterpretable without it.

It was not that Thorndike denied the existence of innate or "instinctual" differences; on the contrary, he was quite aware of them. He worked with a fairly wide variety of animals, and well knew that cats did not peck, chickens hiss and claw, or dogs do any of these when confined in one of his puzzle boxes. But differences in such preexperiential reaction were of little moment. It was the *modification* of behavior that interested him. It was the effect of experience upon the organism; or rather, it was what appeared to be the common effect of the same experience (or at least circumstances) upon a number of different organisms with which he was impressed. Thus:

> On the basis of his studies of animal intelligence Thorndike was led to the formulation of his basic laws of learning. The entire behavioral repetoire of the organism may be conceived as a complex system of connections between situations and responses, i.e., between the effects of stimulation on the one hand, and impulses to action on the other. Each organism is innately equipped with a fund of such connections which are modified through experience. The laws of learning state the conditions under which connections are strengthened or weakened. (Postman, 1964, p. 336)

In this we see again the sensationism, elementism, and associationism of the empiricists and structuralists. These, along with Thorndike's emphasis upon the modification of behavior and his assumption regarding species differences would be incorporated directly into behaviorism by Watson and passed on intact to his successors.

The assumption that differences among organisms are irrelevant to laws of learning was made presumably because those laws were thought to express something so deep and primary as to be necessary for existence. There was, in this system, a single "type" of learning, for which one could write a function rule relating stimulation and action that would apply to all organisms, regardless of differences among them. Implicit also are two derivative assumptions, which came equally to characterize learning theory in the decades following: (1) If species differences are irrelevant, it is permissible, even desirable, for investigators to use one convenient laboratory animal; if what applies to it is expected to apply to others, then the pertinent findings can be extrapolated later. This, in part at least, accounts for the ubiquity up through the 1950s of that most famous psychological specimen, the white rat. (2) If the species matters little, perhaps the particular behavior studied matters little also. Might it be possible to use simple behaviors convenient for study, on the assumption that function rules written on them will apply to complex behaviors inconvenient for study? Yes. Once the importance of differences is eliminated, it is but a short shift to the proposition that one can examine now what is most accessible, leaving to the future the prospect of minor reconciliations.

This becomes doubly significant for understanding not only the past and therefore recent present of learning theory, but also of psychology in general, because for a period learning theory and psychology were practically synonymous. Learning theory, it was hoped, would make psychology a behavioral physics. Its promise and prestige were great; it occupied the high center of the science and earnestly sent disciples to the lesser stations (social, clinical, and so on), where conversions were made. And, while it lasted, it did not receive, it only gave. Stratagem flowed out, rarely in, until there was a learning approach to problems in all psychological fields—until, indeed, all of psychology was reconceived as a problem in learning.

The intention here is by no means to disparage Thorndike. At the time his approach was as reasonable as any other and was certainly worth attempting. Perhaps, as Boring would have it, it was the zeitgeist, the spirit of the times, which would gather up connectionism, reflex arcs, stimulus and response, conditioning, a lack of respect for organismic differences, and an abiding faith in the decomposition of the mentally complex into the simple, and shake them until learning theory fell out.

Nevertheless, much of this was in one form or another present in Thorndike. He firmly believed, for instance, that his principles of animal learning applied equally to humans:

> These simple, semi-mechanical phenomena . . . which animal learning discloses are the fundamentals of human learning also. They are, of course, much complicated in the more advanced stages of human learning, such as the acquisition of skill with the violin, or of knowledge of the calculus, or of inventiveness in engineering. But it is impossible to understand the subtler and more planful learning of cultivated men without clear ideas of the forces which make learning possible in its first form or directly connecting some gross bodily response with a situation immediately present to the senses. (Postman, 1964, p. 339)

Further, the principles governing such simple learning will "still be the main and perhaps the only facts" required to explain complex human learning.

Although Thorndike recognized innate differences among organisms, like all investigators his experiments were guided by his convictions. He sought evidence to support his hypotheses. It was thus that he put animals into his puzzle boxes, enclosures from which they could escape "by some simple act, such as pulling at a loop or cord, pressing a lever, or stepping on a platform" (Postman, 1964, p. 332), and it was by observing their behavior that he concluded that learning occurred by trial and error. He was initially certain that animals do not solve problems by reason or inference, and, of course, he found no such manifestations. Rather, he saw in operation two "laws": effect and exercise.

The characteristic course of learning was trial and error, with accidental success the pivotal event. When the animal faced a problem, its initial responses were determined by its innate dispositions and by habits acquired in similar situations. Responses that were successful led to the strengthening of connections between the situation and those successful responses; responses that were not successful led to the weakening of the connections between the situation and the unsuccessful responses. Further, the more frequently a successful response occurred, the stronger became its connection with the situation. Reward and frequency—effect and exercise—were the keys to the modification of behavior.

Thorndike was later actually to apply his principles to humans and find that in major respects they did not work. Nevertheless, he never changed his conception of learning as the strengthening of connections, between, in the new language, stimuli and responses. Both "stimulus" and "response" were defined broadly and flexibly (ever a sore and harassing difficulty for precise psychological theory). By stimulus (S) he referred to events varying widely in scope and complexity, ranging from specific sensory events to the general features of a learning situation. Similarly, response (R) referred to various reactions, from specific muscular movement to an integrated series of actions. In any given instance, stimulus and response were to be defined in the terms that yielded stable relationships and served the interest of the investigator.

Connections between stimuli and responses constituted the sum total of the organism's behavior, and changes in the strengths of connections determined the modification of behavior:

> Any given person is what he will think and feel and do in various circumstances. He is the probabilities that each of the R's that he can produce will be evoked by each the S's that can evoke anything from him. He is the total of his S-R probabilities. It is by adding to these S-R connections, and by changing the probability of one or another of them up or down, that the environment changes him. (Postman, 1964, p. 350)

Without the mentalistic terms "think" and "feel," this is the model of modern behaviorism.

A paradoxical composite thus emerges from the juxtaposition of Darwin's *Expression of the Emotions* and Thorndike's *Animal Intelligence*. The theory of evolution is of course a theory of descent. It requires contiguity among organismic forms; if they are ancestrally related, there should be at least some similarities among them. This was the fact emphasized in *Expression;* it put humans and animals on a continuum, claiming that the distinctions inspired by the Grand Chain and special creation were false. In one respect, therefore, Thorndike's premise that learning laws are the same in all organisms could be seen merely as a logical ramification of Darwin's principle of organismic contiguity. If all organisms are related, must there not be something behavioral, at some level, common to them, such that the complex higher is an extended compound of the simple lower? Initially, it seems not a poor choice.

But it ignores Darwin's other principle, that of organismic diversity. For evolution to proceed there must be differences among organisms sufficient for selection to occur. Evolution thus requires a somewhat delicate balance between similarity and contrast: too much similarity and there can be no selection among organisms; too much contrast and too few forms survive environmental changes. In presuming that the same fundamental learning laws apply regardless of phylogenic differences Thorndike therefore failed to heed the limit on continquity imposed by the necessity for variety, and, ironically, in so doing developed a science that was antievolutionary, or at best nescient, in this regard.

America thus became the home of animal psychology almost from the beginning. The chief reason was that American psychology was functional and could find a place for the study of animals whereas structural psychology could not, because its methods were entirely introspective. This is not to say that the matter of animal consciousness was never discussed. It most surely was. Psychology, to both functionalists and structuralists, was still the science of consciousness, and an animal psychology had perforce to deal with it. But not for long.

CLASSICAL BEHAVIORISM

Psychology remained a science of the mind for roughly 40 years, if we date its inception in the 1870s and the beginning of its end somewhere just past 1910. Following the latter period it changed the data upon which it was to be based, thereby changing its methodology, its purpose, and its entire concept as a science. The agent of this revolution was John B. Watson, who founded the new school in 1913 with a paper entitled "Psychology as the Behaviorist Views It." The operative word here is of course "behaviorist," for Watson argued that it was not consciousness but behavior that psychologists should study. He thus redefined the discipline, seemingly for all time.

Watson, like other historically important personages, however, did not spring fully armed de novo upon the world. Behaviorism too has a short history but a long past. Generally it may be said that a number of tendencies toward objectivism in the approach to the subject matter of psychology prefigured the behavioristic position. Even Descartes, for instance, with his reliance upon reason and his doctrine of innate ideas, offered animals as automata, complete physical machines devoid of distinguishing mental characteristics. The materialist La Mettrie had no difficulty thereafter in extending Descartes's doctrine from animal to man.

Pierre Cabanis followed La Mettrie with a much more specific formulation. (He was also another one of those interesting contributors: A failure at formal schools, he was largely self-educated, finally settling down to the study of medicine at the age of 21. Then came the French Revolution, and Cabanis became Mirabeau's personal physician and friend. Medically his important ideas began to emerge when he was asked in 1795 to determine whether those treated to the guillotine were conscious after having been beheaded. He concluded that they were not.) Cabanis agreed with La Mettrie that the brain is the seat of consciousness, and contributed papers on the subject. With the heritage of Descartes, Condillac, La Mettrie, and Cabanis, the French are properly considered the progenitors of physiological psychology— not in the sense that Wundt used the phrase, which, as discussed, had nothing to do with physiology, but in the sense of being based upon what physiological facts there were. The French materialist position was thus one of those that prepared the way for Watson.

Other, more specific trends, some direct derivatives of the objective tradition, aided behaviorism as well. For one, the development of animal psychology lent itself quickly to an emphasis upon behavior rather than consciousness. At first, in keeping with the structural framework, an attempt was made to infer animal consciousness from behavior. Watson had himself produced a doctoral thesis on the psychical development of the white rat in 1903, and in 1907 was writing upon the (again inferred) "kinaesthetic and organic sensations" of rats in mazes. But in this very process of inference lay both the problem and its proposed solution. The rule of functional animal psychology was first to observe behavior and then to use those observations to guess at the animal's consciousness. Watson merely shortened the circuit, arguing that, since the primary data are the observations of behavior, the science should be made on them. This would serve two purposes: First, behavior could be observed by more than one person, whereas consciousness could not: thus, the data come closer to fulfilling the requirement that they be shareable and that their very existence, not to mention their qualities, be agreed upon. Second, a science of behavior was expected to be more parsimonious; since with animals we cannot communicate directly about consciousness, and since with humans we must rely upon their verbal report, which is itself behavior, it is the behavior that is actually used, and inferences to consciousness became an unnecessary, even harmful, protraction.

Animal psychology contributed in an equally important way to what behaviorism would become through the nascent learning theory that had already developed. It was, after all, in 1898 that Thorndike presented an essentially stimulus-response analysis of animal learning, something that could be well and easily tailored to a behaviorist position. His emphasis upon the universality of simple learning laws and upon explaining complex behavior in terms of those simple S-R connections would also find its way into, if it had not determined, Watson's scientific cosmology.

Finally, there was the influence of the Russian school, consisting chiefly of the work and opinion of I. M. Sechenov and I. P. Pavlov. Sechenov argued that all thinking and intelligence depend upon stimulation, and that "all acts of conscious and unconscious life are reflexes" (Boring, 1950, p. 635). Although this appears to be a rather drastic view, it came, in modified form, to be the foundation both of early behaviorism and learning theory, an emphasis upon stimulation from without ("the environment" or "the stimulus") or from internal organs and the reduction of all acts to many reflexlike connections being their distinguishing features.

Everyone is familiar with Pavlov's work. In 1888 with Heidenbain he discovered the secretory nerves to the pancreas (beginning the research for which he received the Nobel Prize in 1904). In 1890 he began his operations to bring ducts to the surface of the body so that digestive secretions could be easily observed. It was then that he discovered that the digestive juices began to flow when the animal anticipated food. From that observation grew the concept of the conditioned reflex (as opposed to the unconditioned reflex, which is more or less built in).

The conditioned reflex was a most invigorating discovery and eventually became the fulcrum of Watson's psychology, for it suggested a sort of transferrence of connections from one stimulus-response to another stimulus and the same response. Pairing a bell and food soon led to flowing juices upon presentation of the bell alone. Pavlov's technique seemed to provide a physical measure for processes hitherto called psychical; indeed, he at first spoke of "psychical secretions," and only later of "conditioned reflexes" (Boring, 1950, p. 637). It is notable that it was acquaintance with Sechenov's work and that of Thorndike that encouraged Pavlov to continue in research of this kind.

If we consider this background, Watson's proposals seem less revolutionary. (But in the larger context most revolutionary proposals do. The history of the moment consists of only those events to which most people are then attentive; it requires the future's hand upon the present to uncover the myriad contrasting attitudes that are always waiting to converge into a new view.) The effect was, nevertheless, to redefine the purpose and method of the science. No longer was psychology to be a taxonomy of consciousness; it was now to be an analysis of behavior. It was no longer, thus, a science of the mind, but a science of overt acts.

In itself Watson's initial redefinition was both liberating and limiting: It offered a way around the dilemma of animal consciousness, yet it still allowed for the treatment of consciousness in humans through their verbal reports; at the same time, complex mental events were bound to be of less immediate concern since there description is difficult, often contradictory, and their bodily representations unreliable. We were therefore left with a rather simplistic approach to what is in fact psychology's only raison d'être: understanding the human condition. (Protagoras was right: Man *is* the measure of all things. What is our choice?)

In his favor, Watson found that 40 or more years of structural psychology had produced no interesting body of knowledge, and that whatever facts the structuralists had managed to come by were objective in character anyway (reaction times, learning of lists of material, and so forth). "Borrowing the spirit of German polemical writing, the psychological journals were filled with protocols, long wordy arguments, disagreements and the inevitable invective. Psychology claimed to be a science but it sounded like philosophy and a somewhat quarrelsome philosophy at that" (p. 642). Watson's reforms were clearly due and worth trying.

It is not, however, what the new study of behavior might have been that commands our interest, but what it was. The behavioral method quickly became a philosophy when a number of attendant urges and assumptions were gathered under one rubric. We have noted a number of these in an informal way previously. Watson generally represents a continuation of British sensationism, elementism, and associationism, coupled with and influenced by Russian reflexology and the assumptions underlying the American study of animal learning. More specifically, five features have been listed by Koch (1964) as defining behaviorism in the classical (1912–1930) period: (1) *Objectivism*. Measurement techniques were to be objective. "Only such observations were to be considered admissible as can be made by independent observers upon the same object or event—exactly as in physics or chemistry." (2) *S-R orientation*. All lawful psychological statements were to be expressed in terms of stimulus and some measurable (glandular or muscular) act. (3) *Peripheralism*. The congeries of approaches adopted by Watson, including the primacy of S-R connections and of the conditioned reflex, led him to emphasize immediately (on the surface) observable processes as responses, even to the extent of positing that thought must have some peripheral, muscular basis, namely, subvocal speech (which will appear here again). Thus, "Watson's program necessitated that he consider how phenomena traditionally classed as 'mental' might be treated in objective S-R terms. Most of his positive systematic ideas are thus attempts to show that processes formerly conceived of as determined primarily by the brain could be better understood if allocated mainly to receptors, effectors, and their most direct nerve connections." (4) *Emphasis on learning and on some form of S-R association as the basic laws of learning.* The older psychology had been primarily sensory and perceptual. An S-R emphasis, however, is prone to make learning the central problem, and learning was in addition beginning to appear amenable to the S-R approach, as shown by Thorndike.

The story of how Watson came to fix upon conditioned reflex principles as the basic learning laws presents an interesting insight into the inception of an approach that guided a whole field in science for generations. Koch notes that he was told by Karl Lashley that Watson spent much of the summer of 1916 attempting to obtain photographs of subvocal speech movements. His hope was to present pictures of the physical basis of thinking in his presidential address to the American Psychological Association. But by two weeks before the address he still was unable to find such evidence, so he rapidly shifted tack. Lashley, who was then a student in his laboratory, had been doing work on human salivary and motor conditioning. For his address Watson presented this research, along with a vigorous recommendation of the use of conditioning methods (only slightly known outside of Russia at the time). Thereafter, he assigned more and more importance not only to the utility of conditioning *methods* (i.e., for the study of animal sensory acuity), but to the value of conditioning *principles* for the explanation of behavior. By 1924, he was prepared to phrase *all* problems of learning in terms of "conditioning" (Koch, 1964).

(5) *Environmentalism.* Stimulation, muscular or glandular response, peripheralism, conditioning—all emphasize the environment as the source of control. It is little wonder then that Watson came to discount any innate properties in the organism, for they, he believed, quickly entered into the conditioning process and were themselves conditioned. We are therefore little more than what we have been conditioned to be.

All the features listed by Koch are evident in Watson's *Behaviorism,* published originally in 1924 and revised in 1930. The S-R orientation is stated clearly: "The rule, or measuring rod, which the behaviorists puts in front of him always is: Can I measure this bit of behavior I see in terms of stimulus and response?" (p. 6). Peripheralism is recognized and specific:

> Because he places emphasis on the facts of adjustment of the whole organism rather than the working parts of the body, the behaviorist is often accused of not making a place in his scheme for the nervous system. For the behaviorist the nervous system is . . . a specialized body mechanism that enables its possessor to react more quickly and in a more integrated way with muscles or glands when acted upon by a given stimulus than would be the case if no nervous system were present. (p. 49)

This is reflexism at its limit, and of course presents the central nervous system in a role quite the opposite from what we know it to be, as some sort of simple transducer of stimulus energy rather than as an initiator and modulator of responses.

The emphasis upon learning is apparent in a chapter entitled "Are There Any Human Instincts?" It was Watson's position that whatever might be seen as innate or inborn properties of the organism could be explained in terms of conditioning processes begun at an early age. He admitted genetic differences in structure, but believed them to be immediately overridden by learned responses. Thus:

> There are then for us no instincts—we no longer need the term in psychology. Everything we have been in the habit of calling an "instinct" today is a result largely of training—belongs to man's *learned behavior*. As a corollary from this we draw the conclusion that there can be no such thing as an inheritance of *capacity, talent, temperament, mental constitution, and characteristics*. These things again depend on training that goes on mainly in the cradle. (p. 94)

Finally, there is the somewhat gratuitous offer: "Give me a dozen healthy infants, well-formed, and my own specified world to bring them up in and I'll guarantee to take any one at random and train him to become any type of specialist I might select—doctor, lawyer, artist, merchant-chief and, yes, even beggar-man and thief, regardless of his talents, penchants, tendencies, abilities, vocations, and race of his ancestors" (p. 104). (Clearly this is a challenge that could be made with courageous immunity, for it would never be accepted. Still, even the bare proposition might be lost one, say, a Mozart, who wrote his first piano concerto at the age of four.)

Environmentalism, of course, permeates Watson's *Behaviorism,* as it must, given his other commitments:

> In the case of man, all healthy individuals . . . start out equal. Quite similar words appear in our far-famed Declaration of Independence. The signers of that document were nearer right than one might expect, considering their dense ignorance of psychology. They would have been

strictly accurate had the clause *"at birth"* been inserted after the word equal. It is what happens to individuals after birth that makes one a hewer of wood and a drawer of water, another a diplomat, a thief, a successful business man or a far-famed scientist. (p. 270)

The political sentiments expressed above are both admirable and necessary. People must be equal before the law. We may go further and argue that the opportunity to succeed in life should be made as nearly equal as it can be, or at least, realistically, meet some minimum standard. But it is a deadfall trap to confuse equality in politics or social relations with equality in ability, personality, or biology. The real differences among us are neither obscure nor fear-inducing; they are commonplace, and, more, they offer the only texture upon which we as individuals may find definition.

Further, we have all in the ordinary course of events had to face our strengths, weaknesses, talents, penchants, inabilities, and so on, and most of us have done so with relative goodwill. To promulgate the notion that such differences are made and unmade by whim in the cradle is not only to obfuscate by fiat a genuine developmental issue but to destroy that goodwill. (One need believe neither in behavioral destiny nor in the feudal system to feel that both society at large and we as individuals are better off when each of us concentrates on exercising his or her special constellation of assets rather than pining for others that, but for a too-tight blanket, he or she might have possessed.)

Behaviorism as espoused by J. B. Watson was then an S-R, peripheral, environmental psychology wherein the most complex acts and mental events were held to have been developed from and were reducible to conditioned reflexes. As such, it was experiential extremism, leaving no room for the effects of genetic or biological differences, which, according to the thesis, were immediately covered over and rendered nugatory by the conditioning process.

Because of the strength of his views in this regard Watson has become something of an obligatory stopover on the list of malefactors composed by those of almost any other persuasion. This is in part unfortunate. There is a genuine legacy in the study of behavior and the removal of the science from the exclusive control of those who would have dealt with mind alone. But it is also deserved, in the first place because the study of behavior grew without adequate scrutiny into a many-appendaged *Behaviorism* under Watson's aegis and, second, because Watson publicly stated in the name of psychology, and apparently believed, some of the most atrocious things in the modern history of the discipline.

A few of his remarks have already been noted. Others are easily found. For instance, in speaking of the "birth equipment of the human young," Watson observes that "Defaecation can also be conditioned at a very early age. One of the methods of course is to introduce a glycerine or soap suppository at the time the infant is placed on the chamber. After considerable repetition of this routine, contact with the chamber will be sufficient to call out this response" (p. 122). He seems serious that it should be done.

At another point in his book Watson relates an incident in which he was invited over the weekend to give a man some practical psychological advice. The passage is a little funny because of the behavioristic language in which it is occasionally couched. For example, "I said to him 'Sling your arms and legs out a bit and do your daily dozen and take a tepid bath. This will set you up.' This verbal stimulus led to the act" (p. 45). Or again: "Previous commonsense observation had given me, as a behaviorist, a store of data to predict that with his temperament his day might go very wrong indeed, considering the start he had. This situation called out from me the overt verbal response: 'You'll have to watch your step all day . . .' " (p. 46).

Less amusing, even somewhat chilling in its bouncy confidence, however, is Watson's conclusion:

> Without asking this man to introspect or psychologize or psycho-analyze himself, I could detect his weak spots, his strong points, where he went wrong with his children, where he went wrong with his wife. There can be little doubt that the behaviorist, by training him both in principles and in particulars, could almost remake this very intelligent individual in a few weeks time. (p. 46)

Although they sound as if they were uttered by a cross between Sherlock Holmes and Dr. Strangelove, these are nevertheless Watson's words, and presumably they reflected his beliefs with a fair degree of accuracy. Such ebullience and tendentiousness has carried through to the most popular behaviorist movement of this day.

NEOBEHAVIORISM: THE AGE OF THEORY

The behaviorism of John B. Watson, though sometimes seeming to be, was not unique. Rather, behaviorology, as mentioned, had been in the air since before the turn of the century. The instant acceptance of the American version derived in some measure from its contentiousness and from the ability of its founder to popularize his view. Nevertheless, as has happened to other doctrines, it no more stood fully revealed under the midday sun of critical judgment than the ardor of its admirers paled, many finding second thoughts about embracing a creature with so many warts.

Behaviorism was in need of repair in the late 1920s, in spite of Watson's claims. With its near past realized and much of its present called to question, some wondered where the contribution actually lay. It is not unusual. Behaviorism had in fact suffered the fate of many movements that achieve too much too soon: there had to be an accounting for all the stir once its goals were, at least partially, in hand. It was in this spirit that Woodworth could remark in 1931: "Exactly where the bogey of subjectivism hides himself nowadays I cannot guess. It seems to me I have not seen him rear his head in psychological circles for lo these thirty years or more. But he must be somewhere about, for do we not see the behaviorists charging all over the field and jousting at him?" (Turner, 1965, p. 12).

It really did not matter. The basic tenets of behaviorism were preserved and psychologist's attention diverted from their close examination by four proposals, at least three of which, by the time they reached psychology, were in question in their own disciplines. Nevertheless, they merged with classical behaviorism and for a brief, quixotic period dominated psychology in the forms of what Koch (1964) has called neobehaviorism.

The character of neobehaviorism was, it should be clear from the beginning, quite different from that of its antecedent. Watson's approach had been philosophically unsophisticated, if forceful and straightforward. In contrast, neobehaviorism was philosophically over-sophisticated. It derived its initial momentum chiefly from logical philosophy and was therefore greatly impressed with philosophical propriety, accepting the proffered hegemony of the logician in establishing the rules for sound scientific theory. There was therefore little of the simplistic effervescence of the classical period; rather, neobehaviorism sought refinement in the complex (and, as it happened, incorrect) dictates of the philosophy of science. It was period of grand composition, concerned with comprehensive theory and the criteria for its design, and also thereby a period of great optimism. Psychology had the means at last, so it was thought, to be the equal of the physical sciences. The effort was as we know premature. To understand the psychology of today, however, we must examine neobehaviorism and its consequences in some detail.

Perhaps the earliest and most pervasive influence of the four mentioned above was logical positivism. "Positivism" is a term that comes from August Comte and carried for him and subsequently others the meaning of the immediately observable, the given, the preinferential, and hence the basic, the secure, the undebatable. Positive, preinferential data are, if they exist, of the utmost importance in any science, for they would provide the foundation upon which an inductive-deductive superstructure could immutably rest. Indeed, much of science and philosophy have been given to a search for such facts. There has not, however, been an abundance of agreement on their having been found.

For Comte the basic data were social. He believed introspection of the single private consciousness to be impossible. Therefore there could be no psychology of the individual, but only social science. There could be no "me," but "us," since people could be understood only in relation to their fellows. Physicist Ernst Mach was also a positivist because he argued that his science should return to basic data, but in his case the data were immediate sense perceptions. Comte and Mach then represent two varieties of positivism that contrast strongly in their implications.

Logical positivism represents a third. If Comtean positivism pressed the social and Machian the sense-experiential basis of knowing, the new positivism emphasized the logical. It was, or was supposed to be, a window on the world through logic. This alone is hardly atypical in philosophy. The distinctive feature of logical positivism was that it chose to do so by means of prescriptions and proscriptions on science. The movement attempted to provide the criteria not only for the success of the sciences but for their unification through logically treating that single common scientific commodity, *language*.

The new positivism was, not strangely, a reaction against the relativism encountered in science at the beginning of the twentieth century. It should be recalled that at this time long-accepted notions of certainty were being successfully challenged. Ideas of space and time that served as the matrix of Newtonian mechanics, even the applicability of mechanical description, with its deterministic premises, had become suspect. The fundamental faith that science puts one in touch with reality was being questioned. Mach and Einstein corrected and reinterpreted classical physics, depriving it of timeless truth. Others suggested that all the revered scientific verities had a time-bound, conventional character. Thus, "scientific truth was found not to be absolute, enduring, and ontologically basic, but rather to be contingent upon the frames of reference, the constructions, the conceptual premises that give to the scientific argument its unique possibility. Science lost its ontological status" (Turner, 1965, p. 105).

The reach for certainty promoted by the logical positivists in the face of relativism was not, as might be supposed, toward additional fact, but toward method. If facts were indeed time-bound and changing, then the new faith would be fixed not upon them but upon their theoretical connections. Facts may come and go, thought the positivists, but good theory must remain the same. It was thus that they turned to the logic of scientific method, to the construction of scientific theory, and finally to the unification of the sciences for a fresh grasp at the ultimate.

It should not be imagined, however, that the positivists relinquished fact. It was quite the opposite. The failure of empirical certainty led rather to a complete renunciation of metaphysics and, somewhat paradoxically, to a revived respect for, indeed a reaffirmation of, scientic empiricism. In this case it would not be empiricism alone, but empiricism in conjunction with logic, that would produce the new assurance—the assurance of method.

Logical positivism had its roots not only in Comte and Mach, but also in the logical atomism of Wittgenstein and Russell, from which it drew its concern with language. It was Wittgenstein who developed the metaphor of language as picturing—that is, as a representation of reality. As such, it could of course be more or less accurate. The goal of logical atomism was to say perfectly what can be said about the world and to say nothing else. Thus, what we can say about the world is one set of things and the actual condition of the world is another, but, given an ideal language (picture), the logical structure of appropriately derived propositions would reflect that actual condition. Language as a picture of the world and logic as operations upon language could, it was thought, map the topography of experience.

To be thus used, however, language must be restricted in form to workable units; hence the *atomism*, which refers to the use of simple declarative statements, such as "the barn in blue." Assessment of the logical operations upon sets of atomic statements then allows one to judge the truth status of the whole, and hence, so it was intended, to gain insight into the meaningfulness of our descriptions of reality. What sorts of operations? The same sorts as in arithmetic.

Russell and Whitehead, among other things, demonstrated that even the most complex manipulations and groupings of atomic statements could be reduced to a few logical connections, namely: "or," "and," "not," and "if-then." We thus have four ways of forming propositions; by disjunction, conjunction, negation, and implication. And, according to the thesis of extensionality, the meaning of any complex proposition is a function of the truth status of the atomic propositions of which it is constituted.

The manner of evaluating complex propositions need not concern us here. The relevance for us in logical atomism lies only in its role as a predecessor of logical positivism and the consequent intrusion into psychology of (at least the semblance of) some considerations derived therefrom. It is enough to know of the attempt the atomists made to draw truth from the logical (mathematical) treatment of the language we must use both to describe and to communicate.

Or nearly enough. The atomists provided an additional idea that would germinate into a central concern of positivism: the concept of meaninglessness in language. This was not the meaninglessness that is easily discovered, obvious gibberish, but meaninglessness uncovered by truth functional analysis. Thus, when we accurately decribe the world we express a sentence or group of sentences that are in fact true. But not all sentences need be true; some may be false, some contradictory. They are meaningful nevertheless if their truth value can be established. Even false or contradictory statements are informative if they are known to be such. Meaninglessness occurs when sentences are uninformative; that is, when there is no possible means of establishing their truth status.

This concept was seized upon by the positivists and used with great force in their system. Ordinary language, they concluded, is unreliable, its verbal pliability seducing us into false conclusions and false disputes. Rudolph Carnap in particular offered a veritable catalog of so-called pseudosentences based upon syntactical confusions. Reform of the language that scientists use therefore became the chief positivist program, which would, they hoped, avoid pseudoproblems in the future and lead to a unified science through the development of terms that were clearly translatable from one discipline to another. The positivist movement was, at last, mainly syntactical, its cardinal tenet being that the methods of science would improve if the problems of science were more clearly understood, and that this could in turn be achieved by a reformation of the language of science. The goal was therefore to discover the propositions that were meaningful and to cast away forever those that were not.

However, the effort foundered almost before it was launched. Being committed to empiricism as well as to logic, the positivists altered the criterion of meaning from that dependent upon the "mental calculus" of the atomists. Rather, the meaning of a statement, they asserted, is the method of its verification. If one does not know the conditions under which a statement is to be held true or false, one does not know the statement which was made.

While this did and does make great common sense, it was quickly found not to satisfy the rigorous explicative demands put upon it. The verification principle excluded much in science of what the positivists themselves agreed was significant. There were two types of verification, one syntactical, in which case a proposition could be reduced to another set of terms or propositions, and another empirical, in which case verification was achieved through reduction to the objects of experience. (The positivists were divided on the verifiability issue from the beginning; schlick, the founder and leader favored the empirical and Neurath and Carnap the linguistic.) Whichever view one took, however, there remained a large number of propositions that could be verified neither syntactically nor empirically but were considered meaningful by scientists.

Universal propositions and general laws, for instance, can never be verified by a finite set of observations. Single observations may exemplify the law, but no finite number of them can cover the infinite range of possible empirical instances. Also, some statements express hypothetical constructions that are essential to scientific inference yet are beyond the reach of direct verification. Most of the hypothesized entities of nuclear physics fall into this category, for there is no direct experiential evidence of their existence. We know them by their effects. Historical statements fare no better, which leaves archaeology, paleontology, and history in questionable positions. Predictions of future events are equally impossible to verify by syntactic or empirical reduction.

The exclusions do not end here, but they are enough for our purposes. It becomes obvious that science as we know it would hardly be practicable if restricted by positivistic rules. In fact those rules represented more than anything the attempted bureaucratization of what is essentially a free forum of thought. Method, it became clear, is not the spirit of science but merely its chosen mode of expression.

But this was not immediately salient, and positivisim persisted, chiefly in the form of the syntactical approach. Carnap in particular developed what came to be known as "physicalism," which, when it penetrated to psychology, seemed to present as its goal a system whereby the behavioral sciences were to be able to reduce their terms to the language of physics, and thereby gain status and meaning both separately and as part of a unified discipline. But "physicalism" so interpreted was not what Carnap had in mind. The "physical language" to which all others were to be reduced was not, in fact, the language of physical objects but rather a set of protocol sentences having no ontological status whatsoever. Meaning was both determined and defined as convention, explicited entirely within the formation rules of the syntactical system. Meaning for the terms of psychology came about in their reduction to the protocol sentences. But it was all language. As noted by Turner (1965): "Language, its sentences and transformation rules, provides the content as well as the structure of our knowledge. Meaning comes under the purview of syntax" (p. 134).

This does, indeed, as Turner also notes, make "for a very puzzling doctrine" (p. 134), but it went further, and here had its major impact upon psychology. Psychological *theories* were to be constructed using terms that could be logically reduced to the positivistic protocols. In that way the terms would be both meaningful and part of a grander program.

For sheer deliciousness such an offer to a creature like psychology could rarely have been surpassed. Two haunting goals appeared immanent in the positivistic schema: sound, comprehensive theory, and a union with the basic sciences. But the schema was not what it seemed. For one, rather than providing a relationship to physics, which its choice of terms seemed to indicate, logical positivism provided only linguistic connections to the ontologically groundless protocol sentences. How psychologists somehow did not know this remains mysterious, except that avidity is notoriously weaksighted. According to Koch (1964), psychology was enthralled by the apparent authority of logical positivism—not its content. "What seems to have been imparted to the typical psychologist might be characterized as an ocean of awe surrounding a few islands of sloganized information. . . ." (p. 11).

Thus, while the positivists were falling out among themselves over the verification principle and beginning to devour their own philosophical remains, psychology accepted the whole corpus as if it were not only intact, but inviolate. Curiously, it then turned away from developments in philosophy and, using as the basis this doctrine which was in fast disintegration, attempted to construct a logical foundation for behaviorism.

How could this be? We cannot easily see, except that the behaviorists saw in the sophisticated, incomprehensible positivistic canon an opportunity to rescue their commitments. Such is the only explanation one could reasonably believe for psychology's adopting positivism in the first place, or for its continuing with a long-defunct philosophy in the second. That it was so held is hardly in question. Scriven could write in 1964: "I spend my life going around campuses and finding in each new psychology department a new burst of colossal enthusiasm; the leading lights of the graduate student body turn out to be enthusiastic, tough-minded positivists circa 1920" (p. 81).

The acceptance of positivism may also be understood because a different sort of appeal. It was glamorous. It emanated from sparkling men joined in a society with a memorable name, the Vienna Circle, which self-consciously and publicly announced its goal as the reformation and affiliation of all science. It was bound to attract attention, and evidently did not mind: "From its inception in Vienna (positivism) has been a brash movement, conscious of its own destiny and garlanded with manifestos and programs to enlighten the critical public. Its spokesmen have been direct, articulate, and poignant" (Turner, 1965, p. 104). Further, "The scattering members of the original circle [fleeing the Nazis] found students, confederates, and university chairs awaiting them in England, Scandinavia, and America. Their way had been prepared, for logical positivism was little more than a programmatic exploitation of ideas that were well entrenched in American pragmatism and English empiricism" (p. 104). Perhaps psychology could be forgiven its infatuation.

Overlapping the positivist movement and espoused by many in it, was a second, short-lived attempt at methodological certainty. It is always a temptation, given the success of comprehensive scientific systems such as that devised by Newton, to attempt to reconstruct their development, discern their necessary features, and prescribe those features for new systems. The project seems simple enough: find what has worked in the past and prescribe it for the future. A number of philosophers have, as one would guess, succumbed to the temptation. And the project, as one would also guess, has been found not to be simple.

Nevertheless there was at the height of positivistic influence a corollary interest in the "rational reconstruction" of science that, taken together with the reduction-unification program, appeared to offer, or at least was often represented as offering a coherent design for progressive science. The reconstruction then promulgated was the *hypotheticodeductive method*.

As described by Hans Reichenbach,

> The story of Newton is one of the most striking illustrations of the method of modern science. Observational data are the starting point of scientific method; but they do not exhaust it. They are supplemented by a mathematical explanation, which goes far beyond a statement of what has been observed; the explanation is then subjected to mathematical derivations which make explicit various implications contained in it, and these implications are tested by observations. It is these observations to which is left the "yes" or "no," and thus far the method is empirical. But what the observations confirm as true is much more than what is directly said by them. They vouch for an abstact mathematical explanation, that is, for a theory from which the observable facts are mathematically deducible. (1963, p. 102)

This is very concise and may be expected to have value in a highly advanced discipline. However, its value in other disciplines is restricted just by the fact that it is so concise: Too much is required to be known to quite put it into use beforehand, and when enough is known, it is actually not needed afterward.

Any complex historical reconstruction seems to, perhaps must, suffer this flaw: It cannot include all that actually occurred, or it would be as large as the event itself and serve no purpose. We therefore are given an abstract, which may be more or less representative. With the delicacy, dedication, and luck involved in science there is obviously great room for abstractive error.

Kaplan (1964) well recognizes this difficulty (as have others). He distinguishes between logic-in-use, consisting of some sort of cognitive style involved in ongoing science at a given time, and reconstructed logic, which we later impose to explain the accomplishments of that science. Pertinently, he notes that a logic-in-use may be superior to its own reconstruction. Thus, Newton and his successors made use of the calculus in spite of telling logical criticisms made by Berkeley that were not answered for 200 years.

Conversely, Kaplan observes that reconsruction may itself become, or at least influence, a logic-in-use. Aristotle's syllogistic reconstruction certainly influenced the logic of successive millenia, so much as to cause Locke to remark that "God has not been so sparing to men to make them barely two-legged creatures and left it to Aristotle to make them rational" (Kaplan, 1964, p. 9). The point, of course, is that reconstructed logic can become preemptive, and this is precisely what happened in psychology with respect to the hypotheticodeductive method. For a number of years knowledgeable psychologists believed that great progress would follow, could we but emulate the methods of theory construction and validation (thought to have been) used so well by classical physics. They thus devoted their efforts and enlisted those of others in the quest.

This might seem little more than an acceptable—if, we know now, misguided—determination except for a tendency, which seems inevitably to accompany such programs, to operate more than necessary by fiat. A problem with programmatic, self-conscious choice often lies in what fails to be considered. As Kaplan again astutely remarks,

> The greater danger in confusing the logic-in-use with a particular reconstructed logic, and especially a highly idealized one, is that thereby the autonomy of the science is subtly subverted. The normative force of the logic has the effect, not of necessarily improving the logic-in-use, but only of bringing it into closer conformity with the imposed reconstruction. (p. 11)

That is indeed what transpired. For a period some of the central forces of experimental pyschology sought to subject the science to the rigor of the hypotheticodeductive method. But the variables of psychology are not those of early physics; psychology is not, nor has it been, usefully mathematicized, so naturally the attempt was abortive. This failed to prevent all sorts of formulas, mostly restatements of verbal positions upon which mathematical operations could not be performed, from consuming journals, books, and the heavy air of graduate classrooms. Nor, in reason, should one expect otherwise. Once the endeavor had entered such a path there was no way out until its inadequacy lay exposed.

It must be noted that, to the surprise of some, mathematics is a comparatively easy way to do science. Indeed, there are more formulas and logical systems around that can profitably be used. But, as Bertrand Russell found to his great dismay (for he, too, sought certainty), mathematical propositions and arguments are tautological, true only by the conventions of logic, but empty of content. They bear no relation to the real world (although under certain circumstances they may describe and predict real events). Thus the important work of a science lies not in its mathematicization but in the process of isolating and describing relevant variables.

The strange thing in light of the times when these movements developed is that its their basic insufficiencies were known long before. Wrote Norman Campbell in 1921:

> If we attempted to describe science as a purely logical study in which propositions are deduced one from the other in a direct line of descent from simple ultimate assumptions to complex final conclusions . . . all scientific arguments would appear "circular," that is to say, they would assume what they intend to prove. But the result that follows from our discussion is not that science is fallacious, because it does not adhere to the strict rules of classical logic, but that these rules are not the only means of arriving at important truths. And it is essential to notice this result; for since logic was the first branch of pure learning to be reduced in order to be brought to something like its present position, there has been a tendency in other branches—and especially in discussions of science—to assume that, if they have any value and if they do really arrive at truth, it can only be because they conform to logical order and can be expressed by logical formulas. The assumption is quite unjustifiable. Science is true, whatever anyone may say; it has, for certain minds, if not for all, the intellectual value which is the ultimate test of truth. If a study can have this value and yet violate the rules of logic, the conclusion to be drawn is that those rules, and not science, are deficient. (p. 47)

Nevertheless, a large part of psychology did, from the middle 1930s to the middle 1940s, successfully immerse itself in mathematical logic, or at least in mathematical imagery. That could, in itself, be considered a routine divertissement, except for the side effects and the residue, which we but slowly surrender.

The third principle contributing to the renovation of behaviorism was also a reconstruction, also came from physics, and also was produced by the relativism of the early twentieth century. P. W. Bridgman was moved by the alteration in outlook with regard to the absoluteness of physical concepts that Einstein had wrought, feeling in effect that his discipline had duped itself for years and would not otherwise have been so unprepared for the attitudinal changes required by Einstein's theories. He proposed that if physicists remained alert to the nature of their dispositions toward concepts this would not happen again.

The answer to change is permanence. In one respect this is of course obvious: having been shocked by change once, one should be constantly on the watch for it. Bridgman's proposal, however, does poignantly illustrate once again the need for certainty underlying the professed acceptance of continual variation in the formulations of the era. It is characteristic of the positivists as well. Herbert Feigl, perhaps their ablest spokesman, could write: "It is a sign of one's maturity to be able to live with an unfinished world view" (Feigl & Brodbeck, 1953, p. 13). And yet the thrust of the entire genre, with its emphasis upon mathematics and logic, was toward the irrefutable. To be sure, the positivists were acutely aware of uncertainty; however, Feigl notwithstanding, it generated in these men not acceptance of the eternally dubitable but a search for another form of the indubitable. Uncertainty produced in them only a new hope for certainty. So it does, one must suspect, in us all.

For Bridgman, what was enduring in the relativity of Einstein was that he showed what the concepts useful in physics are. Physical concepts had previously been defined in terms of their properties. Bridgman gives as an example the definition of absolute time, which, according to Newton, "from its own nature flows without regard to anything external, and by another name is called Duration." Bridgman asked how one could possibly measure such time, and, failing measurement, how such a concept could be useful in an empirical enterprise. He then turned to other concepts in physics and discovered that those that have found empirical justification are those that are measurable, and that the actual definition of such concepts are the operations by which they are measured. Einstein, concluded Bridgman, showed what the useful concepts in physics should be by showing that they are operationally defined.

As one example, he took "length." What do we mean by length? "To find the length of an object, we have to perform certain physical operations. The concept of length is therefore fixed when the operations by which length is measured are fixed; that is, the concept of length involves as much as and nothing more than the set of operations by which length is determined. In general, we mean by any concept nothing more than a set of operations" (Bridgman, 1953, p. 36).

The "nothing more" would be operationism's undoing. For some time, however, definition by the operations of measurement seemed messianic. Psychology's concepts would be defined operationally and thereby be ready to take their place in the hypotheticodeductive, syntacticoreductive schema. Problems, such as what precisely is meant by "hungry" in sentences like "Hungry rats were placed on an elevated maze," would be simply solved by stating instead "Twenty-four-hour food-deprived rats. . . ." We would define "hungry" by what we did to measure it or in this case produce it. The iterations are endless.

Altogether, a cognitive blanket of great warmth and promise may by virtue of these events seem suddenly to have settled over psychology, bringing with it the assurance of logic, the prestige of mathematics, the friendship of philosophy, and the blessings of physics. All one need do was turn the key in the lock.

It was too good to be true. Logical positivism, as we have noted, collapsed upon itself, and the hypotheticodeductive method was shown to be only a questionable reconstruction. Operationism, in its turn, performed the secondary function of aiding different investigators to acquire a more specific means of communication but failed in its primary function as a means of definition. The problem, despite defenses, remained unresolved. The strict equation of a concept with a set of operations multiplies the number of concepts beyond use and design, for if we alter even slightly the operations by which we measure something we would by definition have measured something different. This quickly becomes unmanageable and is obviously self-defeating. The most reliable concepts in science are in fact those that meet numerous criteria from various approaches—that is, those that can be measured by different sorts of operations. We are most confident of things that prove out in various ways.

Bridgman, it should be mentioned, was not responsible for the legerdemain worked by psychology in his name. He was in fact later (1945) to write that he felt he had "created a Frankenstein which has certainly gotten away from one. I abhor the word *operationalism* or *operationism*, which seems to imply a dogma, or at least a thesis of some kind. The thing I have envisaged is too simple to be so dignified by so pretentious a name . . ." (Turner, 1965, p. 320). It was too late and it was not his fault in any case, but in the meanwhile psychology's hopes for self-sufficiency had been raised to an intolerable level.

The fourth step in the revival of behaviorism came from within. Edward Tolman in his purposive behaviorism went beyond Watson's reflexology (molecular behavior) to what he considered to be the meaningful whole (molar behavior). The person you see walking down the street carrying the brown sack is not merely putting one foot in front of the other, argued Tolman; he or she is going home from the grocery store. But more relevant to our present considerations, Tolman introduced the concept of the intervening variable, which was then taken up by Clark Hull and his contemporaries and successors; that is, by virtually all of the major figures in learning, in what Koch has called the Age of Theory.

To Watson's S-R formulation, Tolman added incidently a repository for antecedent conditions, A. Thus we might write: S, A-R, or, more commonly, $B = f(S, A)$; that is, behavior is a function of the current situation and antecedent circumstance. This opened the door to a concern for organismic variables of a certain sort, namely, those due to prior experience. Between the S, the A, and the B, however, lay the intervening variable. Such variables, as originally proposed, were not to be real, but rather logical concepts with "firm anchorage" via language to empirical relations; that is, they were not to contain any words not reducible to empirical laws between S and R.

The intervening variable was, thus, essentially a creature of the other two classes of variables, stimulus (situation, antecedent) and response (behavior). It grew to be more, as we will discuss briefly later. But at the time it served chiefly to bind the psychologist's yearning for sound theory to some form of the positivist, hypotheticodeductive, operationist structure.

First, it offered objectivism at the level of theory, since the intervening variables as explanatory concepts were to derive from the two observable classes of variables (S and R). Second, it offered order to the behavioral house: Psychology would deal with three classes of variables, and its business would be to derive the interconnecting functions. Third, such functions could be stated as postulates, thereby taking their place in the hypotheticodeductive system. The accompanying mathematization could be sought in quantitatively specified intervening variable functions. Fourth, the requirement that intervening variables be explicitly linked to observables could be equated with the tactic of the operational definition. Finally, values for the intervening variables were to be generated empirically from "defining experiments." The intention here was to design a series of experiments, the empirical variables of which could then represent the value of the intervening variables in future work (Koch, 1964).

There appeared, then, to be a nearly infallible system for the production of theory, a cohesive whole wrought from logicomathematical philosophy, physics, and from psychology. The system seemed to offer a decision procedure that was self-corrective, a formula for science.

But it fared badly, and with hindsight's graceful ease we can see why. With regard to mathematics and logic, Scriven (1964), for instance, notes that the crucial point is not whether a precise law governing an ideal case can be produced but how precisely the ideal case can be related to actual cases. Symbolic logic has not helped. Indeed it became a minor branch of mathematics of some intrinsic interest but of little value for solving typical logical problems.

This was something of a letdown. Psychology had invested much, discovering little in the way of fact and humbling itself before method, only to awake and find itself in the ruins of a shrine long overgrown and out of date. Such are the perils of a borrowing science.

What of its own product? The intervening variable suffered the same fate. A distinction was made by others (e.g., MacCorquodale and Meehl, 1948) between intervening variables, which originally were to have no meaning beyond the empirical relations of the other variable classes, and so-called hypothetical constructs, which had "surplus meaning" giving them an ontological status. Whereas Tolman had intended the former, Hull and his collaborators had subtly and without notification substituted the latter, under the same name. A controversey ensued, naturally, over the value of each sort of variable, but this too, in spite of whatever terms may currently remain, proved futile. MacCorquodale and Meehl's distinction was helpful in showing the differences between Tolman and Hull (and the drift toward reification to which we are inevitably prone), but it did not produce two workable classes of mediating variables. Tolman's original concept has in fact been found wanting in formal properties and lacking in everyday use.

Turner (1965), for instance, notes that were an intervening variable merely a logical construction from observables it could not "by virtue of its status in the calculus, generate more interesting experimental hypotheses than would arise within the radical atheoretical stimulus-response psychology itself. That is, there is nothing inherent within the intervening variable . . . that would lead to hypotheses which could not be derived from a set of laws eliminating [it], as it were, by using only primitive terms in the observation language" (p. 260). There is thus no formal advantage to the restricted definition.

Turner also makes the point, so necessary to the understanding of these affairs, that scientific terms always have certain psychological properties even when we wish them not to. We *will* furnish them with personal meaning: "Regardless of the aura of empirical respectability with which intervening variables have been endowed by people such as Tolman, Hull, and Spence, it is doubtful that anyone holds to a strict reductive definition of an intervening variable. There is always some presumptive factor beyond the empirical ingredients of pure logical construction" (p. 260).

This, whether or not it may seem to be, is usually to the good. Reichenbach once asked Einstein what led him to the theory of relativity, and Einstein answered that it was because he was so strongly convinced of the harmony of the universe. Could Newton have been any less convinced or, for that matter, could Darwin? Great searches are made on conviction, the conviction that something—an entity, a principle—is in some sense "there." It seems impossible to search in quite the same way for a purely logical construction. Science, certain humanists to the contrary notwithstanding, is very personal. It is curious that it could seem otherwise to gifted men, even for a short time.

Theories that utilize only intervening variables, that insist upon the strict operational reduction of their terms, are thus, in Turner's words, heuristically sterile. Tolman himself later reevaluated the status of intervening variables and indicated serious doubt as to whether they could be applied objectively. His general conception of the significance of intervening variables indeed changed dramatically: He later came to regard them as "merely an aid to thinking," no longer determinately linked to empirically stipulated independent and dependent variables. With regard to the "defining experiments," he later also doubted their transsituational generality. We are very much back, then, where we started. There are no guarantees.

However, the journey was not without cost. During the Age of Theory almost the whole science was convulsed by an effort to adhere to some form of the "new view." Perhaps the best known single effort centered around the work of Clark Hull, who proposed to construct a hypotheticodeductive, quantitative theory of mammalian behavior, and who, in a series of offerings in the period between 1943 and 1952 produced four books (*Principles of Behavior* (1943), *Essentials of Behavior* (1951), *The Mathematico-Deductive Theory of Rote Learning* (1940), and *A Behavior System* (1952)—the last of which was in galley form at the time of his death)—and a number of articles. His influence, although not uncontested, was great. Was it possible that psychology would, at a stroke, be transformed into the psychomechanics of its dreams?

Alas, no. The exercise, however gallant, was presumptuous, serving chiefly to illustrate the seductive power of rules and reconstructed logic. Valid data were too few, postulates too unsteady, and mathematization too hasty to support such an elaborate superstructure.

Interestingly, the proposal, as have others in the behavioristic mold, presumed to treat all of mammalian behavior without having examined much beyond that of laboratory rats. Again one finds buried deep within and continuing through behaviorism's heritage the cardinal assumption that the basic (principally learning) laws are observable in almost any organism and applicable in form to all. Hull's work foundered upon this assumption, we suggest, as well as those assumptions already noted. But he at least recognized the problem and attempted to deal with it. In others it is often hidden, if altogether in force.

Hull's formal system broke with his health. His disciples did not perpetuate his views, except in miniature, and the grand effort he inaugurated simply came to ground of its own impossible weight. And the dust settled quickly. Looking over *A Behavior System*, published less than 30 years ago in what must have been great pride and optimism as well as great sorrow, one almost palpably senses it, as a relic, as surely interred, unearthed, and obscure as the calculations of some ancient astrologer. The same is true of Tolman's major work, as well as Watson's. But particularly Hull's work epitomizes the tragedy of a wrong dream lost: how much it must have meant, how little difference it finally made.

Psychology's next phase was one Koch calls neo-neobehaviorism, the essence of which was a broadening, if sometimes bewildering retrenchment. The great designs were abandoned and far more was admitted into the behavioral purview, particularly human-related problems. "Higher mental processes" once again became legitimate subject matter, as did the physiological bases of behavior. However, through it all the true behaviorist maintained the same orienting prescriptions: what was readmitted was translated into behavioristic parlance—to the extent, for instance, that little S's and R's were postulated for the central nervous system. If this misapplication of peripheral concepts warrants the enthusiastic rejection it received from neurophysiologists, it is at the same time indicative of what people do who are trained in a fading dogma. Kaplan (1964) again puts it most succinctly: "The price of training," he observes, "is always a certain 'trained incapacity'" (p. 29). It goes further. "I think of behaviorism," notes Scriven (1964) ". . . as something which will thus be with us for thirty or fifty years" (p. 181). That is, the "trained incapacity" tends to last, and is but sorely bent from its purpose.

How, then, is change engendered? Max Planck, creator of quantum theory, had one answer: "A new scientific truth does not triumph by convincing its opponents and making them see the light, but rather because its opponents die, and a new generation grows up that is familiar with it" (Feuer, 1974, p. 87). Although Planck's funereal view is more than one would like to accept, it is nevertheless true that scientific opinions often seem to be dolorously aligned with the ages of the scientists, and one does get the impression that fresh ideas await acceptance in near-archaeological order. The point is: Dogma sticks. It often takes not only time but new minds to get it unstuck.

Psychology may be said, loosely, to have been undergoing such a process for the past 20 years. It began quietly with the collapse of the Age of Theory, the initial thought being that the approach was right and the theorists had merely gone too far. But as contradictory data from other sciences—particularly neurophysiology, anthropology, genetics, ethology, and pharmacology—as well as psychology's own evidence of doctrinaire behaviorism's intrinsic futility, accumulated, it became obvious that the approach would have to be abandoned. Chiefly, the role of the organism, its biology and its genetic history, had been ignored to the degree that the formulation of general laws was impossible. This at least seems clear at present.

Where will it end? We shall speculate upon that later. For the moment one should note simply that we are at present witnessing a double act: first, the unraveling of the final threads of the knot that bound us to neobevariorism, and second, the reintroduction of biological features of the organism into psychology's subject matter. The two are of course related: Without behaviorism the biological features would have never been kept at such a distance from the center of psychological concern, and there might not have been an internal collapse. Without a collapse there would have been no outside contradiction and no need for the rapid assimilation of new fact. Without behaviorism, thus, psychology could have been a far more open science between 1920 and 1960, could have integrated smoothly, or at least lived side by side with the incredible advances of the biological disciplines and would not now be experiencing what is in effect a culture shock. The same is true of other environmental disciplines, especially sociology, which finds its old doctrines unworkable and is currently in a state of crisis and biological redress. One can only regard the crisis as having been inevitable and the redress as none too soon.

RADICAL BEHAVIORISM: A RETURN TO SIMPLICITY

If neobehaviorism was a self-conscious exercise in formality, both in the construction of theory and the logical rules of that construction, radical behaviorism is a self-conscious exercise in simplicity. The movement is not new; indeed, over 40 years ago its chief proponent, B. F. Skinner, took much the same position he holds now. Rather, it waited, as it were, through the neobehavioristic period, until its time arrived. When psychology, weary and dismayed with the failure and complications of the Age of Theory, looked around, radical behaviorism was already in place, as unique and serene as ever.

Skinner is, in many respects, a direct decendant of Watson, circumventing neobehaviorism entirely. Like Watson, he is much more of a public figure than were any of the neobehaviorists, and, like classical behaviorism, radical behaviorism appears straightforward and is forceful and promissory. Judging from the popularity of Skinnerian techniques, particularly those falling under the rubric of "behavior modification," a fair amount of that promise was accepted.

However, the apparent plainness of radical behaviorism is misleading. Although it consists of a set of simple assertions, these are mixed with philosophical notions, technique, and ramifications in such a way as to generate questions, the answers to which are most complex. This is not unusual. The imposition of simple assertions, if they are not exactly right, generally creates contradictions that the system cannot tolerate. It is then either reformed or abandoned. Neither has been the case with radical behaviorism. It remains undeterred, unchanged from its original formulation, it could be said, to a fault. It is thus all the more difficult to understand.

The basic assertion of radical behaviorism is that an adequate analysis of behavior consists in finding correlational relationships between behavior and features of the environment. A secondary assertion is that theoretical structures are not helpful in this regard; the relations are to be entirely empirical. The basic assertion in turn depends upon an underlying philosophic assumption—namely, that organisms respond to the selected features of the environment with the exclusivity and to the extent necessary to make the relationships between behavior and such selected environmental features sufficiently predictive.

It follows then from the basic assertion and the underlying assumption that behavioral analysis can (in fact it has been argued that it should) be unconcerned with mediating links between environment and behavior, be they Tolmanian intervening variables, hypothetical constructs with surplus meaning, or variables brought straight from physiology. Since all behavior is considered to be determined by the environment, internal conditions that might otherwise be considered to motivate behavior are similarly of little direct interest.

The relationships to be sought are in the form of operant responses and the schedules of delivery of their reinforcers. An operant response is one that occurs independently of the manipulation of a known relevant stimulus. A reinforcer is some feature of the environment. A positive reinforcer is said to strengthen any behavior that produces it; a negative reinforcer is said to strengthen any behavior that terminates it. "Strengthen" refers to the probability that the response will be emitted again. The relationships between the delivery of reinforcers and the probabilities of operant responses are called contingencies of reinforcement. The program of the radical behaviorist consists of determining these contingencies of reinforcement. The aim of the program is to predict and control responses through manipulation of their contingencies of reinforcement.

This is superficially a simple doctrine indeed. In foreclosing on theory it avoids the usual tests of self-consistency and external relevance to which theoretical constructs are heir. It also avoids the difficulties produced by methodological issues in theory construction. Radical behaviorism was, for instance, not caught in the positivist–deductionist–operationist–intervening-variable box into which neobehaviorism fell. (It is, of course, caught in a box of its own.)

Closely tied to, in fact inextricably mixed with, the assertions and assumptions of radical behaviorism is its technology. The technology consists chiefly of a means for manipulating contingencies of reinforcement and recording the responses thereto. Perhaps the most famous technical device is the Skinner box, which has some standard manipulandum (a bar for rats to press, for example) on the inside, and a device to record the distribution of responses over time on the outside. The organism is placed in the box, and the schedule most efficacious in modifying its behavior vis-à-vis the manipulandum is determined. This, once again, seems a simple procedure. However, its beneficial use involves assumptions, unstated by the radical behaviorist, that require study in detail.

Finally, certain ramifications of the radical behaviorist program for everyday human behavior have been announced unambiguously by Skinner in a number of publications. Once again, however, overlying the simple claims are to be found questions of substantial complexity.

It should be obvious that there is huge freedom in a purely empirical position, particularly if it is held in concurrence with an inventive technique and in an area where data are few. One can, then, for a while, merely gather facts and leave the worry of their connections with other levels of scientific discourse to the future. One can even argue against theory itself as being premature and unnecessary.

This, as indicated, has happened. Neobehavioristic theory was found wanting, and upon the demise of the big systems the field was increasingly left open to an approach that seemed to offer the promise of a return to plainer themes. The proposals are, however, perhaps too plain, and it is little wonder that "radical" is the term associated, by Skinner himself, with this approach. Not Hull, not Tolman, not even Watson (who was in fact quite interested in physiological mediators) went so far. All in one way or another took the internal part of the organism into account. Radical behaviorism, however, would deal only with the "empty organism," or, more precisely, would leave psychology empty of organismic variables. This seems, a priori, unnecessarily preemptive, and one must wonder at the basis of radical behaviorism's appeal.

What generates the appeal that it does have? First, and most important, is a viable concept intimately associated with a technology. The concept is that of the operant response, and it makes a great deal of common sense. Obviously, all else equal, if we are rewarded, or reinforced, for doing something, however strange it may seem, the likelihood that we will do it again is greater than it would be otherwise.

Without the technology, however, the operant response and its reinforcer would merely be one more tandem on the psychological scene, filling out the general conditioning paradigm derived from the Russian reflexogists, from Thorndike, and from Watson. But radical behaviorism provides in addition something which to some appears genuinely mesmeric—namely, a machine: the Skinner box.

The Skinner box is without question an ingenious way to gather data of a given kind. More to the point, it is also a prepackaged technology. This need not necessarily be harmful, except that prepackaged technologies tend to generate a substantial amount of trained incapacity. A movement, that is, can become so engrossed in the minutiae of its apparatus that it loses all meaningful contact with the larger issues with which it intended to deal. Or it can become so engrossed in its technology that the usual order is reversed: Rather than the technology deriving from the science and the philosophy underlying it, the philosophy and the science derive from the technology.

The effect, in any case, is that the science is controlled by its machines; the commitment is to the apparatus and what it can do rather than to the problem (which usually means that the problem is restricted and/or redefined to fit the apparatus). Kaplan (1964) describes this as the *law of the instrument*, which he rather humorously formulates as follows: "Give a small boy a hammer, and he will find that everything he encounters needs pounding" (p. 28). Specific to the present situation, such a law would read: "Give a group of scientists a machine, and many will find justification in work which consists of using the machine in whatever way possible." A derivative corollary is that those same people will also engage in competition with regard to modifying the fine points of the machine.

However, the appeal of radical behaviorism lies not only in its concept and its device: A second important feature contributing to its continuing existence is its advocacy. It is somewhat startling to find that what is scientifically such a non-theoretical and self-contained program is publicly tendentious and proselytic. One would suspect that if the real business of a discipline were only the accumulation of data, it would be less inclined to debate and more content to let data decide the issues than might some other disciplines of a more theoretical mode, anxious to make causal connections among varieties of fact. But radical behaviorism, beyond any other current movement, persistently appears to wish to sell itself as much as it wishes to uncover its functions.

Skinner's program of radical behaviorism obviously consists of more than recording the responses of organisms under certain circumstances, appearing at symposia, and writing utopian novels such as *Walden Two* (1948) and behavioral-political tracts such as *Beyond Freedom and Dignity* (1971), as well as the usual explanatory vehicles, such as *About Behaviorism* (1974). Its essential concern is the *modification* of behavior. This was Thorndike's and Watson's interest, and so too is it Skinner's. However, whereas Thorndike and Watson attempted to explain the modification in terms of theoretical connections between stimuli and responses, Skinner finds such attempts not only superfluous but damaging. His preoccupation is the probability that an operant response will be controlled by a reinforcer, that is, in the contingencies of reinforcement.

The modification of behavior through manipulation of contingencies of reinforcement may appear, and is presented by Skinner, to be a recent development. Outside of the terminology and the click of the machine, however, it is ancient. Furthermore, it is and has been well used. Any parent who has stopped chastising a child when his or her behavior changed knows what negative reinforcement is (as well as does the child); similarly, any parent who has rewarded an offspring for doing well knows what positive reinforcement is, and, again, so does the offspring. I choose these situations because they come as close to the conditions of control required by the manipulation of reinforcement in humans as we ordinarily have. Other situations are more loosely relevant; indeed, at the limit one could turn it all around and ask which conditions are not? Anthropologists long ago recognized that a portion of cultural difference could be accounted for by differential reward. Sociologists also recognized this. There is little of behavior, in fact, that cannot be said to be in some way under reinforcement's sway.

This suggests that a certain kind of learning plays a role in the organism's responses to its environment. But it is not news. If it seemed to be it is because Watson had limited learning to the change in connection between a stimulus and an unconditioned ("natural") response to that between a new stimulus and the same response. Operant conditioning—wherein some stimulus is not first deliberately used to elicit an unconditioned response and wherein the response must appear first and then be reinforced—may have appeared next to the Watsonian version to be quite fresh and liberal.

Like much of behavioral thought, however, the notion of reinforcement is buried in folklore and common wisdom, perhaps to the extent that it is unrecognizable as such. It is nevertheless surprising to find Skinner write in 1974: "Reflex phenomena, conditioned and unconditioned, have, of course been known for centuries, but it is only recently that . . . contingencies of reinforcement have been investigated" (pp. 38–39). If this means that contingencies of reinforcement have been studied in a modern way only in modern times, then one must certainly concede his point. However, little altogether has been studied in a modern way prior to modern times. On the other hand, if Skinner means that nothing like the notion of contingencies of reinforcement had been expressed until recently, as he appears to, then we must either conclude that his remark is poorly phrased or find a very long hiatus in history.

The concept of the operant response and its reinforcer is novel only in the context of late nineteenth- and early twentieth-century reflexology, and beyond those confines it is and always has been a useful part of folk knowledge. It is this background, rather than its recent elaboration, that reflects its viability. It is a simple fact that reinforcement works. There have been historical successes in training in all manner of things other than the successes that Skinner and his disciples claim.

But, and this is a most urgent consideration, reinforcement has never worked perfectly, nor has it ever come close to working perfectly. In fact, except under the most limited of circumstances, we have no evidence that it even works well. What we find is that although some individuals under some circumstances are susceptible to certain kinds of reinforcement, others are less so, or are even refractory. The "functions" do not apply equally, or nearly equally, to all organisms or to all situations. How reinforceable a response is, what the most appropriate reinforcer is, and indeed, whether or not the response is reinforceable at all vary from organism to organism and from condition to condition to such a degree as to make the concept, without further specification, unserviceable. It is within "further specification" that the complexities arise.

Before an attempt at reinforcement can even begin, one must designate and control a very large number of things. The species of organism is one thing that is essential to know. A bale of hay is probably more reinforcing to an elephant or a horse than to, say, a cat. The age of the organism may also have something to do with the likelihood of reinforcement. So does gender.

One must also know something about the internal state of the organism of a given species, age, and gender. Satiated rats respond less well, if at all, to food reinforcers; in fact, the common practice is to deprive them of food so that they will respond. Similarly, a male rat castrated at birth will find small reward in an estrus female, no matter how nubile. In addition, particular manipulanda are more likely to be successful with certain organisms than with others (rats get bars to press and pigeons get disks to peck, and not the other way around). There is a whole class of variables about which the investigator must have knowledge and/or exert control before it is possible to think much about reinforcing responses. (If this is in doubt, imagine the difficulties to be encountered by the reinforcement specialist seeking to modify behavior but knowing no more about the organism than that it is alive.)

Species, age, gender, and general internal state by no means exhaust the variables with which the operant experimenter must deal. Even with these controlled he or she must face a large subcategory that often goes under the descriptive but rather vague name of "individual differences." That is, even among organisms of the same species, age, gender, and (presumed) general internal state vis-à-vis the reinforcer there will still be a distribution of response likelihood. The organisms do not all do the same thing, at the same time, in the same way. Further, the more complex the organism, the more potent is the factor of individual differences. The human situation is of course the most complex of all.

Individual differences do not disappear even when environmental histories are controlled. Brother or sister rats, for example, raised together in the same pen by the same mother, may respond differently. There is at least one good reason for this, in that unless special animals that share all of their genes are used, there will be genetic differences among the animals. Genetic considerations have not, however, played a large part in the experiments of radical behaviorists.

This brings us to the circumstances under which operant conditioning works best and simultaneously returns us to technology. The Skinner box (or some facsimile) is not only a device for recording the responses of an organism to schedules of reinforcement, but also a device for depriving the organism of anything to do but to respond to those schedules. It is a case, as noted long ago by Cronbach (1957) of essentially "decorticat[ing] your subject by . . . giving him an environment so meaningless that his unique responses disappear." It is to this end that the whole radical behaviorist program is turned, because only when and unless the features that are unique to the organism are suppressed do the reinforcement schedules appear to exert the control necessary for study of their functions. Therefore the radical behaviorist must either make some potent assumptions or endure some powerful contradictions in order to implement his or her program. We have touched upon some of the assumptions in discussing Thorndike and Watson, but there are others. For instance, there are very few responses that the Skinner box is equipped to record. One must assume that these responses are scientifically meaningful. Skinner answers that "A behavioral process is none the less real for being exhibited in an artificial setting. What an organism does is a fact about that organism regardless of the conditions under which it does it" (1966).

It must be granted that an organism does in reality do what it does. But this argument disregards what in other sciences has proved to be a most valuable consideration: Some observations are more important than others. And it is to the important observations that creative scientists address themselves. Indeed, choosing the "right" observations to study may be the most difficult part of the whole process.

The point is that since *all* behaviors are real, their reality is not a differentiating issue. Some kinds of responses are likely to be more informative than others, however. To guess which ones will be, the experimenter must have sufficient knowledge of the organism's repertoire in relation to the problem being studied. This cannot be done merely by looking at the results taken from a Skinner box, unless of course one wishes to generalize to the behavior of organisms in such boxes, something that Skinner himself manifestly eschews. Even if one wished to study operant conditioning, one would have to demonstrate that the responses made in the restricted environment were important in the normal operant repertoire of the organism and hence useful to extrapolate to less narrow situations. This has not been accomplished. The functions generated by the radical behaviorist are derived from organisms placed in unusual and limited circumstances, often in a deprived state, and almost always under conditions from which they cannot escape.

Such a situation eliminates a substantial part of the foundation of scientific progress. Restriction of the environment and the response repertoire of various species until by admission and design the differences disappear shows only how in a restricted way organisms can be *made* to behave, not how they ordinarily behave. But it is upon the *exploitation* of differences, not in overwhelming them with limitations, that science thrives.

What, then, accounts for the long commitment to the radical behaviorist program? One must suspect that the law of the instrument (Kaplan, 1964) plays a substantial role. That the commitment exists cannot be questioned, and it exists to such an extent as to jar the intellect even at this late date. What, for example, can Skinner (1966) mean when he writes that "The provenance of learned behavior has been thoroughly analyzed"? Is this only a case of ambitious phrasing? Unfortunately, no. Skinner appears to believe he has solved the problem of learning. If so, it would certainly be imposing. It might be, indeed, the single greatest human accomplishment. Would that it were true.

The "provenance of learned behavior" encompasses far more than reinforcement-response functions, far more than that upon which we can profitably speculate using data combined from all sources, and far more, we must surmise, than that upon which we ever could profitably speculate. The problem of learning is not solved. Or, if it is, all but Skinner are deluded.

The real distinction is in all likelihood more pedestrian, revolving around the level of explanation one is willing to accept. Skinner proposes that explanation is to be found in the correlation between a reinforcement schedule and the probability of responses. He categorically refuses to recognize anything that is not either a part of the environment or a part of the behavior. For others, some intervening sequence is necessary, and learning is not explained until that sequence is understood.

Unhappily whereas much has been written on explanation (e.g., see Turner, 1965, chap. 10; Kaplan, 1964, chap. 9; Feigl & Brodbeck, 1953, chap. 4), no one has yet explained it, for in truth an event is explained when we are satisfied that we understand, which leaves us mostly going around with words. The only escape is to fall back on what has been successful before—on what has been required in the past by way of explanation—for a concept or principle to have been demonstrated to the degree that all now accept its validity. By this criterion, Skinner's notion of explanation is inadequate because correlation has in the past provided dangerous proof. We may not, as Hume would have it, know what causation is, but the empirics of success suggest strongly that it is more than the gross sort of relationship that radical behaviorism proposes. If we cannot accept this notion of explanation, we in turn have grave difficulty accepting this notion of science.

The lack of concern for explanation extends even to the central concept of the Skinnerian program: reinforcement. Skinner's interest is only in establishing the conditions under which reinforcement operates. Why or how it operates does not appear to trouble him: "I am *not* interested in explaining why something *is* reinforcing. I identify a reinforcing stimulus only in terms of its observed effect. . . . I do not know why and I do not care . . ." (1964, p. 104). This is another powerful philosophical proscription. It is true that a science is not immediately required to provide an explanation for every phenomenon it finds of use (for example, mechanical physics has employed the concept of gravity while being unable to explain it), but it is rare for scientists to be unconcerned with such an explanation. Physicists have never ceased work upon understanding gravitation simply because it could be well measured and successfully applied at the gross mechanical level. If only because it works so well it *demands* explanation.

Despite the controversy that swirls constantly about it, radical behaviorism appears to go serenely on, tending its own garden as Skinner says, reposed in philosophic security and the comfort of its schedules of reinforcement. It was never guilty, as we have noted, of the systematic excesses of the Age of Theory, and it does, again, have a viable concept and a technology. However, it buys its protection by risking little: It is so restricted in concept that it is difficult to criticize on its own grounds, and one is forced to take exception chiefly not to what it does but to what it does not do.

How, for example, can fault be found with construction of theory in an approach that is avowedly atheoretical? Not easily. One must then call for the development of supporting theory, which may be useful to progress but can be rejected from a purely empirical standpoint. How can fault be found with correlation as explanation when explanation is itself a matter of personal satisfaction? Not well. One must look to achievements of the past, which may be rejected with the argument that the present approach has yet to be sufficiently tested. And how, similarly, can fault be found with the limited conditions under which the reinforcement schedules are generated when the schedules can be reasonably predictive under those conditions? One can only debate the possibility of extrapolation to the normal world and point out cases of failure.

Radical behaviorists are thus, for the moment at least, safe from formal criticism and face only the more general question of whether their work is useful. That they have demonstrated a phenomenon and can go on demonstrating it ad infinitum must be conceded (although one can certainly take issue with the proposal that the phenomenon is new); the debate centers upon its possible value. Skinner has little doubt: "In the behavioristic view," he writes in the last lines of *About Behaviorism*, "man can now control his own destiny because he knows what must be done and how to do it" (1974, p. 251). Others, for better or worse, believe some confusion remains. The "what must be done" seems more a political-moral question with many answers, whereas the "how to do it" appears to have no answer at all, unless radical behaviorists are privy to special information. Failing that, one can on the contrary only conclude that an informative investigation of the what and how of human behavior is in its infancy at best and is therefore somewhat removed from the definitive state ascribed to it.

All of this is not to say that the reinforced response is not an important part of an individual's adaptation to its environment. We have acknowledged that it surely is. Rather, it is to say that the Skinnerian strategy has no means of fully investigating such responses. At the very least, species, genetic, developmental, and other organismic variables must be manipulated simultaneously with reinforcement schedules. The goal, that is, should not be to reduce variation until all organisms behave alike, but to use it systematically so as to uncover the differences that do exist among organisms. The choice of which responses to examine, in addition, must not be made on the basis of convenience or what can be reliably measured by an automatic device. Relevant behavior must be observed. To provide meaningful analyses the operant investigator must take the risk of leaving his philosophically secure domain; he must shed his simplistic assumptions and relinquish his machine.

The strategy that a discipline adopts is most pressing to its success. One cannot investigate everything at once, and so a few important phenomena are chosen for study in such a way as to contribute to increased understanding. However, any such undertaking requires assumptions, and the success of the venture often depends upon their validity. The assumptions in turn usually derive from a guess about the way the world is. Einstein first believed that there was harmony in the universe. He made assumptions and then a revolution. Darwin did much the same. Thus strategy usually derives from some world view, from suppositions consonant with that view, and, if possible, with appropriate techniques. Radical behaviorism, in this analysis, has devised not a strategy that is merely expedient and momentary, but one that reflects the belief of its practitioners that life is in fact the way it would be if the world view it presupposes were accurate.

This may seem a unique assertion, and it is, to be sure, without documentation. Nevertheless, both the changelessness and advocacy of radical behaviorism suggest that the assertion has merit. Skinner's views appear to have been formulated more than 40 years ago and, in spite of major advances in adjoining fields, have remained essentially the same. His doctoral dissertation was a defense of the correlational strategy, ignoring mediators, a thesis he currently espouses. In addition, he was and has been willing to see those views translated into act and put to use, not merely by the animal trainer but by the author of human social policy. No one else, given the transitory state of our understanding, would have the confidence to write:

> [man] can remedy these mistakes and at the same time build a world in which he will be freer than ever before and achieve greater things. He has failed to solve his problems because he has looked in the wrong place for solutions. The extraordinary role of the environment opens the prospect of a much more successful future, in which he will be most human and humane, and in which he will manage himself skillfully because he will know himself accurately. (1974, p. 240)

What is he talking about? Reinforcement schedules.

If there is anything we need to know, it is ourselves. But much current information indicates that one-sided approaches, including those developed in the early part of this century that emphasize the extraordinary role of the environment, are insufficient. Skinner, in a curious way, has begun to acknowledge this, observing in one of his recent publications that "We must wait and see what learning processes the physiologist will eventually discover through direct observation, rather than through inferences; meanwhile, the contingencies permit a useful and important distinction" (1974, p. 67); and "A behavioral analysis acknowledges the importance of physiological research. What an organism does will eventually be seen to be due to what it is, and the physiologist will someday give us all the details" (p. 249).

These remarks lend hope that radical behaviorists are aware that the organisms whose behavior they would predict and control have brains, glands, and genes, and that these are important determiners of behavior. More relevantly, they suggest that radical behaviorists could be at least in part aware that the strategy they propose is but a temporary course, a holding pattern for the study of behavior until the "meanwhiles," "eventuallys," and "somedays" catch up; that is, until other sciences *explain*. (If true, however, this would leave the "experimental analysis of behavior," as the work of radical behaviorism is called, far from the princely and insular science it is depicted to be. Rather than being independent and definitive, it would become something—perhaps desirable, perhaps dubious—to do while the other sciences set about the proper task.)

But such notions evidently have not been incorporated into the central core of radical behaviorism, for in the same text there occur juxtapositions of passages similar to those immediately above with others reflecting the usual immodesty of the true believer. Some of the latter have already been quoted, and they (accurately, one must think) define the scope and depth of the radical behaviorist's commitment. Three pages after conceding that physiology, specializing in the investigation of the internal state of the organism, will provide the understanding of behavior, Skinner argues that if man is to be saved, "It is the environment which must be changed. A way of life which furthers the study of human behavior in its relation to that environment should be in the best possible position to solve its major problems" (p. 251). Then follows the remark that we can now control our destiny because we know what to do and how to do it.

In one respect such vacillation is welcome because it indicates something of a break in the line against the study of physiological mediators, hypothesized or not, but it is also confusing. If indeed physiology is to have the last say, might it not be prudent to find out what it is before the regimen of radical behaviorism is instituted? Is it possible to apply both propositions, the finality of radical behaviorism and the promise of physiology, to the same circumstances at the same time? Clearly not. What, then, is Skinner's position? Alas, it remains that of the radical behaviorist, circa 1931.

For instance, he has lately come to write on genetics, noting that "It is hard to understand why it is so often said that behaviorism neglects genetic endowment. A few behaviorists . . . have minimized, if not denied, a genetic contribution, and in their enthusiasm for what may be done through the environment, others have no doubt acted as if a genetic endowment were unimportant, but few would contend that behavior is 'endlessly malleable'" (1974, p. 221). This walks softly over the point. The fulcrum of such issues is rarely what is "denied" or "contended," but what is actually investigated. Behaviorists of all persuasions have as a matter of course ignored genetic variables in their research, radical behaviorists with more skill than others, and thereby, regardless of the presence or absence of specific verbal content, have removed them from effective consideration.

One cannot become too well acquainted with this sort of configuration. Relevance lies with what is done in experimentation now, not what is to be included in the oral mist of the future. It is not enough merely to claim affiliation, or perhaps a lack of affiliation, at the level of grand design. If certain variables are important, they must be incorporated at the level of ongoing research; they must become part of the normal research paradigm, a part of what is "expected" in good work. A strategy that does not actively incorporate such variables stands in mute and categorical contradiction of the verbal flourishes in its behalf.

That this has occurred is more than evident upon examination of behaviorism of all sorts, and of radical behaviorism in particular. And one cannot now negotiate into being what was never there. Beach, for example, long ago wondered how we were to interpret a 457-page book by Skinner, based exclusively upon the performance of rats in bar-pressing situations, but entitled *The Behavior of Organisms*. The title indicates the breadth of the approach Skinner believed himself to be taking; Beach's question shows how narrow that approach was in fact. Commitments such as these cannot be reconstituted decades later by statements of what was not specifically intended for exclusion.

Thus the new recognition of the "contingencies of survival" does not alter the past record of attempting to ignore, if not subdue, differences due to genes. Rather it represents an effort to substitute the contingency language of reinforcement for the genetical-evolutionary language of natural selection. Such a linguistic distinction, however, remains eminently preferable. And if there is merging to be done, surely it should favor the larger and more successful endeavor.

In spite of his recent attempt to identify with, or subsume (it is not clear which), the process of natural selection, Skinner when pressed returns to the same behavioristic ground: control. Considering natural selection and reinforcement, he writes: "Contingencies of reinforcement have the edge with respect to prediction and control. The conditions under which a person acquires behavior are relatively accessible and can often be manipulated; the conditions under which a species acquires behavior are very nearly out of reach" (1974, p. 44).

This is, first, misleading. The research of evolutionary geneticists has clearly demonstrated that alteration of the genetic bases of behaviors is well within the limits of experimental manipulation. Furthermore, it has been for some time. Second, we find once again in the concern with control the radical behaviorist's concentration upon the immediate: that which can be done at once to modify an organism's behavior, without reference to conventional understanding of it. Thus, although radical behaviorism does not formally deny that the organism has an inside, it continues to be interested exclusively in the outside and, despite disquisitions upon "phylogenic contingencies" (Skinner, 1966), *defends* its old choice on the basis that reinforcement is more quickly applied.

Unfortunately, there is simultaneously no call for the manipulation of genetic variables in the reinforcement context. The biological and operant domains, we see finally, are to remain separate. Biology is recognized; indeed, rather incredibly, the "experimental analysis of behavior" is defined as a branch of biology (1974, p. 23) but in its specialness is not to join other branches in their concern for organismic variation.

This is at first glance more than a little puzzling. But with an insistence upon control, and, once more, an enchantment with the machine, it all seems of a piece. Indeed, Skinner concludes that "Perhaps the best evidence that a science of behavior has something new to offer is the success of its technological applications . . ." (1974, p. 232). Technological application of scientific knowledge is usually the province of engineers. But then operant conditioning has in some places been rather proudly described as behavioral engineering. There is, of course, absolutely nothing wrong with engineering. The question is whether a fiducial technology should be called a science.

There remains, finally, the matter of just how good a technology it is. Instances of failure, even in the restricted conditions of the Skinner box, are commonplace. Some individuals simply refuse to be "reinforced" and may have to be removed from the experiment. Of more interest, however, is the extrapolation of the principles of reinforcement to situations outside of the confined conditions under which the principles were generated. Given Skinner's pronouncements one would think the principles were equally, or nearly equally, applicable, regardless of circumstances. Little evidence supports this view.

One series of failures was reported by Keller and Marian Breland (1961), who subjected a variety of species to operant techniques. They were successful initially: "Our first report . . . concerning our experiences in controlling animal behavior was wholly affirmative and optimistic, saying in essence that the principles derived from the laboratory could be applied to the extensive control of behavior under nonlaboratory conditions throughout a considerable segment of the phylogenetic scale." But, "Emboldened by this consistent reinforcement, we have ventured further and further from the security of the Skinner box. However, in this cavalier extrapolation we have run afoul of a persistent pattern of discomforting failures. They all represent breakdowns of conditioned operant behavior."

What sorts of breakdowns? Chickens refused to accept food reinforcement without scratching the ground, raccoons refused to stop rubbing coins together before depositing them in a container so as to receive a food reinforcement, pigs refused to stop rooting money across the floor, thus receiving no reinforcement (again food), and so on. It soon became obvious what was wrong. All the interfering behaviors were "instinctive." Chickens scratch, raccoons rub, pigs root. We all know this. So evidently did the animals, for at some point in their conditioning programs they gave up attempting to act like radical behaviorists and began acting like chickens, raccoons, and pigs. The Brelands termed this "instinctive drift," and defined it as follows: "The general principle seems to be that whenever an animal has strongly instinctive behaviors in the area of the conditioned response, after continued running the organism will drift toward the instinctive behavior to the detriment of the conditioned behavior and even to a delay or preclusion of the reinforcement."

To the Brelands these instances represented a "a clear and utter failure of conditioning theory." They were therefore led to examine and question the tacit assumptions of that theory, namely, "that the animal comes to the laboratory as a virtual *tabula rasa*, that species differences are insignificant, and that all responses are about equally conditionable to all stimuli." Radical behaviorists may not, as Skinner claims, overtly recognize such assumptions. But the Brelands appear correct in noting that they *behave* as if they do. It is certain that they are unsupportable.

Another set of instances has more recently come to light, again involving an attempted extension of operant laboratory principles to a more general environment but this time also involving the application of such principles to humans. Reppucci and Saunders (1974), in an assessment of behavior modification techniques, note that

> Although questions about behavior modification on the individual level are still to be resolved, a major new frontier of interest has opened up. This new frontier is centered around the application of behavior modification techniques to populations larger than one in natural settings. As with any emerging field of interest, this social application of behavior modification is characterized by certain anomalies. One of the most noticeable of these is the difference between what is real in the field and what is imagined. Judging from the literature, what is real in the field consists of rather small-scale attempts to change the behavior of isolated units of people within natural settings. These projects usually are attempted in a special classroom, ward, building, or, in rare instances, within an institution specially designed and funded for this purpose. At the other end of the spectrum, the standard for what is imagined in the field was set in 1948 with B. F. Skinner's publication of *Walden Two*. As is well known, in *Walden Two*, and more recently in *Beyond Freedom and Dignity* (1971), Skinner proposed that a total restructuring of every aspect of our culture, using behavioral technology as a guiding framework, is possible.

However, Reppucci and Saunders found that "there is a large gap between what is real and what is imagined about the social applica-

tion of behavior modification; . . . there are reasons for this gap, and
. . . some of these reasons fall outside the domain of behavioral technology as it is presently elaborated."

This last is a most telling consideration. The "experimental analysis
of behavior" is found not to be, when tested, scientifically equipped to
extrapolate its principles, and further, as constituted, it has no hope of
becoming so equipped:

> [T]he impression one gets from most of the behavioral literature is
> that, armed with a knowledge of learning theory and a stiff upper lip,
> the determined behavior modifier should be able to manipulate con-
> tingencies in the natural environment just as he does in the laboratory.
> This is sheer nonsense and a major weakness in a field whose imagin-
> ings lie in the direction of *Walden Two*. The issue inheres in the fact
> that the principles of behavior modification are insufficient and often
> inappropriate for understanding natural settings—their structures, goals,
> traditions, and intersetting linkages. The great social debate generated
> by publication of *Beyond Freedom and Dignity* in 1971 typically ne-
> glects the type of problems discussed in this article. Yet the implications
> are clear. We are nowhere near a technological capacity or sophistica-
> tion for Skinner's utopian notions, and Skinner's theory as currently
> elaborated is insufficient to provide for that technology. (Reppucci &
> Saunders, 1974)

It seems there is at least room for debate as to its efficacy between
the consumers and producers of the "experimental analysis of behavior." This is the likely result when the laboratory environment has been
so manipulated at the outset as to allow for the generation of behavioral principles that in an unrefined state apply universally across individuals and species. The situation is so artificial in the first place that
whatever occurs there offers little hope of generality.

The Brelands and Reppucci and Saunders (as well as others) also
provide the answer to a question that Skinner asks with regard to
primacy in exposition of the notion of contingencies or reinforcement.
In *About Behaviorism* he remarks:

> There have been many dramatic applications of operant conditioning,
> but very often what is done seems in retrospect to be little more than
> the application of common sense. Nevertheless, we have to ask why
> similar changes were not made before the advent of an experimental
> analysis. It is sometimes said that they were, and isolated instances in
> which something very much like a modern behavioral technology can be
> cited. But we may still ask why these occasional instances, scattered
> throughout the centuries, have not become standard practice. (Skinner,
> 1974, pp. 233–234)

The answer, obviously, is that they have not become standard
practice because they have not been efficacious. Reinforcement is but a

single component in the behavioral panoply. In nature, alone, it does not explain, predict, or control behavior. Nor can it.

As an addendum, it should be noted that radical behaviorism is discussed here in such detail and its chief proponent quoted at such length because at this rather critical point in the history of psychology the "experimental analysis of behavior" is still offered with serious deliberation as an alternative to a developing discipline that genuinely includes the biological perspective. It is thus necessary to examine the tenets and ramifications of radical behaviorism with an eye as to where they appear sound and where they appear vulnerable. It is especially necessary since Skinner has himself publicly made much of the applicability of the principles he espouses. The best means of assessing their viability is to study his words directly.

The same sort of application might be made to all of behaviorism. There can hardly be any quarrel with the study of behavior per se, but behaviorism in any of its forms is another matter. It has from its inception been not only rather pompous but unduly prescriptive, and for one to assess its effects, as well as to understand its origins, a brief (if seemingly involved) outline was required. Behaviorism thus occupies the principal space here because as the primal philosophy in experimental psychology it was so effective in gaining adherents. And, as noted, if we are now to understand the confusions of the present we must first understand the certainties of the past.

The word *confusions* may seem impertinent; many—doubtless most—psychologists would not describe themselves as confused at all. However, I obviously refer not to individuals, but to the discipline at large. Psychology may already be the most variegated science—and, if anything, it is becoming even more divided. There is no common approach to basic problems, even when the problems themselves can be agreed upon. Thus, although there may be opinions strongly held among behavioral scientists, there is not in the field as a whole a notable degree of order.

Behaviorism seemed to offer order. For a short period at least, psychology almost appeared "paradigmatic" in Kuhn's (1962) sense; that is, there seemed to be a general problem—learning—and a common approach to it. The failure of behaviorism following this most optimistic of times, however, gradually has left much of psychology without a philosophic house.

Finally, it should be clear that scientific strategy is of the utmost importance here, and, principally, the matter of treating organismic and environmental variation is probably the crux. Behaviorism offered environmental determinism and it did not succeed. Meanwhile, evidence for the involvement of organismic variables in the determination of behavior became massive. It is apparent that a strategy emphasizing organism-environment combination should, by the process of elimination if nothing else, be elaborated in the future. That is, of course, the purpose of this book. But accepting different concepts—and perhaps more important—putting them into practice, first requires questioning other concepts that stand in conflict. That also is the purpose of this book.

THE QUESTION OF INSTINCT

Given the foregoing, one could conclude that there has been little working concern with differences among organisms at the headwaters of American experimental psychology from the advent of Watsonism behaviorism to the most recent past. It is a safe conclusion. "Instinct" was officially summoned before the environmental inquisitor by Knight Dunlap's (1920) article "Are There Any Instincts?" and was consigned to the flames by Z. Y. Kuo, who in 1924 wrote "A Psychology Without Heredity." This was concurrent with the rise of behaviorism, and the two movements doubtless fed upon one another until a more or less uniform position was reached. If the conditioned reflex were to explain behavior, organismic propensities would have to find a minor place in psychology, or, given the psychology of at least some scientists, no place at all. We have seen that this occurred, though chiefly through choice of emphasis rather than outright denial.

The eclipse of the organism cannot, however, be put entirely at behaviorism's door, for instinct theory was at the time insufficient and with little assistance soon would have collapsed from within. The historical reasons for its vulnerability we shall briefly touch upon later. At this point it will suffice to note that instinct theory had come to be essentially an extensive list, or set of lists, of behaviors purported to derive exclusively from the organism's hereditary complement. Whatever seemed to show some degree of hereditary participation was incorporated: maternal behavior, mating behavior, fighting behavior, and so on. Almost nothing was rejected.

From a scientific viewpoint the initial difficulty with such a situation was that is was not heuristic. Certain behaviors or aspects of behavior were taken simply to be given, and there they lay. The object was enumeration rather than explanation; more precisely, there was a confusion between explanation and description. Naming a behavior "instinctive" seemed to be all that was required; that was in itself taken to be explanation.

William James believed that there were human instincts, as did his contemporaries. But the psychologist most famous in America for his espousal of the instinct concept was William McDougall, a transplanted English admirer of James. McDougall is, however, less important for what he believed than when he believed it, for his career overlapped the ascent of behaviorism. He was therefore a convenient target for the behaviorists (and an unashamed sniper in his own right).

Curiously, McDougall publicly predated Watson in choosing behavior as the subject matter of psychology, having written on the topic in 1912, but the rise of behaviorism led him to abandon the term as no longer descriptive of his own view. McDougall undertook to show that one could regard all human behavior, and therefore social behavior, as the result of instinct and its modification by experience. He also related instinct to emotion: For each primary instinct there corresponded a primary emotion; the instinct of flight, for example, corresponded to the emotion of fear, the instinct of repulsion to the emotion of disgust, the instinct of curiosity to the emotion of wonder, and so on. Other instincts included self-abasement, self-assertion, and gregariousness.

Such a simple, direct approach was sure to be popular. However, it fell into difficulty when it became evident that any list of instincts is boundless and its contents quite an independent matter for each scientist, since any behavior observed could be described as instinctive. It also fell into disrepute since McDougall, unable to divest himself of teleological notions and old beliefs about brain energy, had become something of a behavioristic jousting post in later years. Because the concept of instinct as elaborated in McDougall's time was an inadequate representation of organismic propensities, Watson thus easily carried the day.

How had things come to such a pass? Darwin, as had other naturalists before him, spoke readily of instincts. Behavior then was, as noted previously, as normal a consideration in the professional life of a natural scientist as bone length or the configuration of a shell, and a good deal of behavior seemed instinctive, that is, to come about in such a way that no schooling of the organism was required. That a spider required teaching to spin its web, that a bird required training to build its nest and incubate its eggs, or that a bitch needed inculcation in the art of bearing, cleaning, and suckling her pups seemed preposterous. This and much else in the world was simply taken to be inherent in the nature of things.

But the concept of instinct so conceived had never taken a hard experimental test. By default, it seems, it therefore came to acquire a meaning too inclusive to be useful; it suggested that complex behavior patterns appeared complete and intact with little or no environmental input. This was its burden: It was taken to be an all-or-nothing affair. As such it was forced to carry the development of a large part of behavior entirely on its own, which of course it could not do. Soon it was shown that behaviors do not develop entirely independently of the environment, and instinct became so much grist for the behaviorist mill.

Beach (1955) traces the origin of this form of the instinct concept to theology and philosophy. "From the beginning," he writes, "instinct has been defined and discussed in terms of its relation to reason and, less directly, to the human soul." Doubtless this is true. It seems doubly so because what records we have come mainly from theologians and philosophers; their works constitute the largest part of our older intellectual history.

But the concept has in addition a common appeal, particularly to those living on the land in well-defined kinship groups, and most of our time clearly has been spent in these circumstances. It is thus easy to understand how instinct could become the subject of intellectual concern. To the early Greeks, for instance, all the world was charged with *physis*, the vital force that gave life to nature and simultaneously gave identification to individual living things. From the primitive notion of *physis* developed the more sophisticated notions of the soul and of Gods (Cornford, 1957). Differentiation between humans and animals depended upon what proportions and amounts of soul they possessed. Aristotle, we have seen, developed a hierarchically ordered scale of nature based upon just such proportions and amounts, with humans at the top, having the capacities of all other organisms and in addition the nearly divine ability to reason.

The Olympic gods, in the Greek schema, had the most and best soul of all. Their powers were clearly superior to those of humans, in whose affairs they regularly, sometimes whimsically, intervened. The gods were nevertheless a part of nature, derived as humans and other creatures from the original *physis*, and therefore were *governed* by nature. The Greeks saw order, and the gods were a part of that order. The reasons for godly actions were the same as those for all living things.

As the notion of a more powerful, sometimes less capricious, single, bureaucratic diety emerged and permeated the Western world a division was for the first time formed between soul and nature, and we find the initial instance in which the concept of instinct may be said truly to be exclusively associated with theology. At this point, rather than being governed by nature, God was seen as responsible for it, and hense the author of instinct. What existed did so at God's sufferance and design; the purpose of inquiry therefore centered around His intentions. Having discovered those, one need look no further (not only because it was enough to see His will, but because He was known to take offense at continued prying).

This attitude pervaded Christian civilization rather straightforwardly until the Renaissance. In biology, all species were seen as having divinely fixed within them not only their morphological but their behavioral characters, and this applied to humans, who possessed in addition that element crucial to their damnation: free will. Animals then had neither soul nor reason; they acted purely from instinct. Humans had both soul and reason, and also instinct—which, however, was chiefly base. The task of humans was to use their reason and free will to overcome their instincts and thereby save their souls.

Darwin freed behavior (and morphology) from immediate deistic manipulation, but instinct nevertheless retained many of its former defining properties. This resulted because Darwin, in his effort to demonstrate the continuity of behavior between man and animals, had emphasized the inherited origin of instincts. The concept was thus left essentially intact. Natural selection (and Lamarckian inheritance) merely took the place of God. In *The Expression of the Emotions in Man and Animals* (1872), for instance Darwin concludes: "That the chief expressive actions, exhibited by man and the lower animals, are now innate or inherited—that is, have not been learnt by the individual—is admitted by everyone. So little has learning or imitation to do with several of them that they are from the earliest days and throughout life quite beyond our control" (p. 350). He thus accepted the identity between genotype and phenotype—between what was inherited and what appeared later in the behaving organism. (This was common at the time; biology had to wait through the rediscovery of Mendel—and Johannsen's distinction that followed (pp. 33–34)— before it came meaningfully to terms with the notion that genes, not traits, are inherited and that genotypes may produce different phenotypes depending upon the environments in which they develop.)

Thus behaviors that were seen to be inborn were seen also to derive exclusively from the germ plasm of the species. The difficulty, as mentioned, is that such a view tends to, or at least tended to, substitute evolution for God in the formula for behavior. That which was given was seen to have been given inescapably.

Darwin (1872) was aware of the participatory role of the environment in the development even of "inherited" traits; for example, "When there exists an inherited or instinctive tendency to the performance of an action . . . some degree of habit in the individual is often or generally requisite" (p. 30), and "it is remarkable that some (traits), which are certainly innate, require practice in the individual, before they are performed in a full and perfect manner . . ." (p. 351). Nevertheless, his concern remained with the inherited character of instincts, and his effort was to demonstrate that animals and man differ in degree rather than in kind. Little wonder then that the part played by "learning" does not appear prominently.

In *The Descent of Man* (1871) Darwin treated at length the evolution of morality, a subject about which he had been most circumspect in earlier works. Moral choice, he suggested, is possible because of the inheritance of social instinct and the development of extensive intellectual powers. We share the social instincts with animals, and these instincts evolved because natural selection favored those individuals most capable of living in groups. Group living, at the same time, obviously offered a selective advantage to those individuals. The instincts most active in preserving and expanding groups were thought to be mutual aid and sympathy.

Sociality, Darwin thus maintained, is natural among many species of animals, and certainly among those closest to humans. Therefore the virtues that make group life possible are not, for the most part, learned. On the contrary, our most perfect and noble actions are "done impulsively," effortlessly, because they are innate. The utilitarian stress upon the rewards and punishments associated with the past actions of the individual as the basis of moral acts was rejected. Rather, social behavior derived from instincts evolved over millions of years of communal living. Humans were not exempt. It is only their capacity to reason that made them feel so.

The *specific* practices that hold particular groups together were not, however, initially found to be appropriate for members of other groups. Murder, robbery, or deceit may be forbidden among members of one's tribe but may permitted or even encouraged against strangers. According to Darwin, with the advancement of "civilization" the "group" has grown more inclusive and eventually has come to be comprised of all humanity, indeed of all sentient beings.

Optimism aside, this has quite a modern biological ring (pages 263–266). Darwin, with his usual perspicacity, seems to have anticipated much recent ethological-anthropological thought in applying evolutionary principles to the ontogeny of group behavior and derivative moral action. He was even careful to note that no particular moral code follows from his construction, only that which enhances survival.

There are, however, certain problems, attributable not to his logic but to his premise. One, the identification of the genotypic and the phenotypic, with the implication of the absence of environmental involvement, has been noted. Another develops because of Darwin's subscription to the Lamarckian view. Lamarck, it may be recalled, was the author of the proposition that acquired (learned in the behavioral case) characters would gradually come to be inherited. Darwin admitted this to his system, and built upon it in the formulation above. "Virtuous tendencies" were seen as first acquired by habit, which, through modification of the germ plasm over a number of generations, was made heritable. It could then be subjected to natural selection.

(Persistent nonvirtuous action was, of course, also heritable. "Personal habit" thus became for society a thing of great vistas—and something potentially sinister. Perhaps one can better empathize with those who now appear quaintly to have been so concerned.)

This is in some measure familiar ground, for, as we have seen, happiness and an end to conflict were also the evolutionary products in the system of Herbert Spencer (page 39). There is, however, an enormous difference between the Darwinian and Spencerian conceptions as to the means to those ends. Whereas Spencer would have the fittest individuals survive, for Darwin it is the fittest groups, a concept that entails selection among individuals for the capacity to cooperate. Such Darwinian groups, with their evolved social instincts, would then in time be capable of extending sympathy and mutual aid to others. Spencer's individuals, on the other hand, would simply eliminate other, less competitive individuals, the world presumably becoming a safe and happy place thereby. The important difference, of course, is that Darwin makes sociability a natural product of competition, imbuing individuals with genetic propensities for behaviors suitable to the formation and maintenance of groups, and thus he opens the future to the propitious, if speculative, opportunity for world solidarity.

His formulation is, we know to our sorrow, somewhat utopian. Perhaps it would not have been made quite so freely if the evolution of social instincts were understood to depend exclusively upon natural selection rather than both natural selection and the inheritance of acquired behavior. The latter offers a more direct route to virtue than we know to be possible. (It is sure Darwin would have taken it. He liked nothing with an unhappy ending and did not even find novels to be of the first class unless there was someone in them to love (1882, p. 54).

In any case, two things are clear: (1) Darwin's temperate (and more accurate) views on the evolution of morality were lost in the sound and fury of Spencerian laissez-faire ethics and thus were never widely known. "Social Darwinism" drew little from him. (2) None of the nineteenth-century evolutionists manifestly altered the ancient concept of instinct. The idea of inherited, innate behavior fit as well into their thinking as had God-given instincts into that of the theologians and philosophers. The result we have described: Unable to fulfill the obligations that had mistakenly accrued to them, inborn propensities toward behavior were banished from experimental psychology. They would not return for decades.

In Europe, however, the traditional was not so easily discarded. (Nor was Europe ever a bastion of experimental psychology, particularly animal psychology, as it was elaborated; that was an American affair.) Rather, about the time Watson was writing the declaration of behaviorism, Julian Huxley in England and Oskar Heinroth in Germany were investigating the behavior of birds with the intention of understanding its evolution by natural selection. Heinroth in particular applied the methods of comparative anatomy to behavior patterns, emphasizing the fact that many behaviors are typical of species, genera, and even larger taxonomic groups. He also stressed that many of the behavior patterns were largely innate, being hardly modified in the environment in which the individuals were reared. This work provided the basis of a discipline quite unlike that which was to be precipitated in Baltimore.

Ethology, as it came to be called, is (as noted by its nominal father Konrad Lorenz) really the child of Charles Darwin. Darwin's ethological observations were not immediately pursued, chiefly because biologists found it enough at the time to concentrate upon the consolidation of evolutionary theory, for which they chose structural traits. Later, as it became clear that behaviors offered good taxonomic properties, evolutionary theorists turned to them more and more. The role of behavior in evolution was thus studied by zoologists and geneticists such as Huxley, Dobzhansky, and Mayr. The behavioral products of evolution were concurrently examined by ethologists such as Lorenz, Tinbergen, and Hinde.

The initial ethological statement stressed, as indicated, the inborn property of behavior, and assumed that there was very little impress from the environment. Lorenz's theory dealt with the combination of elements of behavior in a way which left an opportunity for environmental input only of certain kinds, such as "imprinting" in the young of some birds. This new, or rather reclaimed, approach,

> . . . appealed at once to a small number of zoologists, namely those who, like myself, had already chosen to observe animal behavior, and had been deeply disappointed when they looked for guidance at the psychological and physiological literature. These observers too had been struck with the relative lack of modifiability of many behavior patterns and their consequent conservative behavior in phylogeny; by the apparent "spontaneity" of much behavior, which at that time seemed incompatible with reflex theories; and by the fact that many behavior patterns could often be seen to contribute decisively to survival, and to success in reproduction." (Tinbergen, 1972, p. 133)

Heinroth and Huxley had begun publishing before 1915. Lorenz's chief theoretical papers came near the mid-1930s. However, the ethological approach had almost no impact upon American psychology until the 1960s, and only lately has it received wide attention here. The reasons why seem obvious. First, neobehaviorism was only beginning its run when Lorenz began to publish. Prior to that an even more sterilizing tariff upon instinct was exacted by Watson. Second, World War II intervened, and it was disruptive of all normal international scientific communication. (No matter how unbiased or apolitical, messages in German, suggesting that a large part of behavior might be genetically determined were, understandably at the time, unwelcome here.) It was thus years after the end of the war before American psychology was seriously willing to consider the merits of ethology. Emotional wounds had to heal, or at least be sutured, and neobehaviorism fail before Americans could relinquish their environmental preferences and readmit the concept of instinct into the scientific lexicon. Even so, with good reason, it could not be readmitted whole.

Lehrman (1953) and Beach (1955) made the essential points, Lehrman responding to Lorenzian theory in particular and Beach writing on the concept of instinct in general. Both criticized instinct theory on three counts: (1) There tended to be a confusion between naming and explaining; to call a behavior instinctual was, especially in early treatments, too often considered to be all that was required to explain it. (2) Instincts were considered to be innate, hereditary, and thus not modified by environmental agencies. This, again, left their actual development uninvestigated. (3) A class of behavior considered innate and unalterable came in short order to be contrasted with another class of behavior, that which is "learned." This further hindered research into the developmental events responsible for behaviors.

All three criticisms are obviously related. Hereditary, innate, unmodifiable behaviors seem at first to require no further explanation and to have little to do with learned behaviors, which are by definition nonhereditary, noninnate, and environmentally malleable. Investigators therefore felt free to study either instinctive or learned behaviors as they would.

This is generally what happened. For a period of time there appeared to be an almost geographical division of labor, with America fastened to the conditioned reflex in one form or another, while Europe was governed by "The Great Parliament of Instincts." In the past decade, however, to the benefit of both, the intellectual (and emotional) membranes separating the two sides have begun to dissolve, and at least a partially clear view of the foundations of a genuine science of behavior has emerged. On the one hand, the criticisms made by Beach and Lehrman have been accepted and incorporated into the thinking of a large number of ethologists; on the other hand, biological predispositions have come slowly to be recognized in the United States.

It is clear now that both genetic and environmental inputs must be considered in the development of every behavior; no behavior springs de novo from an environmentless organism, nor does any behavior develop independently of the organism's hereditary disposition. Thus, the "fixed action patterns" of the ethologists are not immune from environmental determinants, but may appear to be so only because of the rather uniform environment in which members of the species are ordinarily reared and because the investigator is unaware of the environmental determinants that do in fact exist. The development of such apparently innate characters as the highly species-characteristic songs of birds, for instance, requires that each individual have the opportunity to hear the sounds of adults of its own species. Songbirds reared alone in soundproof rooms produce a grossly abnormal version of their species song, a trait that, because of the apparent lack of variety among adults normally reared, could be taken to be totally innate (Barnett, 1963b).

At the same time, "learned" behaviors are not usefully considered to be independent of the organism's genetic composition. In the most obvious case, it is clear that there are differences in learning, at least of degree, between the lower and higher phylogenetic orders, and that these differences are relatable to differences in genes. *What* organisms are prepared to learn and perhaps *how* they learn it are as much influenced by genes as are "inborn" behaviors. Species thus differ remarkably both with regard to the range and the kind of stimuli to which they are particularly sensitive, and they differ with regard to the range and kind of responses they are capable of making. Within species, individuals differ as well as to their sensitivities and capacities, and the distribution is related to differences in genes. It is not a matter, therefore, of innate versus learned or even innate + learned, in the explication of behavior. Rather, as has been concluded by many, the organism and the environment from the beginning each contribute to the development of behavior patterns in a way that is reciprocal and complete.

It would seem then that there is ground for the recognition of both organismic and environmental sources of influence and, most important, recognition in principle of the way in which they reciprocally *combine* in determining behavior. Where does this leave the concept of instinct? Much changed, but with some essential equipment intact. We are now free to recognize inborn propensities without at the same time being forced to accept the proposition that there is a class of behaviors completely directed by heredity and entirely refractory to environmental agents. We are now free to investigate, indeed it is our obligation to investigate, the manner in which organismic and environmental variables join to produce behavior. It may be, as some have suggested, that "instinct" should be altogether forsworn as a term. But the choice of words is irrelevant. Something must be used to convey the meaning of the modern version, and whether it be "inborn propensities," "organismic programming," or innate* (or even instinct*), the asterisk being permanent, matters little so long as the basic idea remains.

Much of this has yet to be fully accepted by those with exclusive commitments to genes or the environment. Lorenz, for instance, although his position appears to be changing, remains at heart an instinctivist (without the asterisk). Other ethologists, such as Tinbergen, have moderated their view from that originally espoused by Lorenz to one which focuses upon organismic development, recognizing experiential factors. Tinbergen and Hinde represent English-speaking ethology in general. On the other side, Skinner, representing radical behaviorism, pays mild homage to phylogeny but remains locked to his environmental contingencies. A part of the remainder of American experimental psychology seems unappreciative even of the issue.

However, there has always been a viable underground, concerned with organismic variables, even during behaviorism's most dominant times. Physiological and comparative psychologists, such as Yerkes, Lashley (a student of Watson!), Hebb, Beach, and Lehrman managed to maintain a respectable opening for inborn propensities. Personality theorists have similarly for the most part continued to respect biological variables. Both, however, have been beyond the wellspring of experimental psychology since Watson's manifesto and have, until lately, influenced it but lightly. Beach, for example, complained in 1950 that experimental psychology appeared to be devoted almost exclusively to the study of rats. Whalen found that not much had changed by 1961, concluding that "the unfortunate possibility exists that animal psychology will remain the science of rat learning" (1961a). Much the same was true in the relationship between experimental psychology and personality theory; if there was any give, it was from the former to the latter.

This, as noted, is changing. The rise of ethology, the failure of behaviorism, and the long, sometimes lonely, resistance of physiological, comparative, and personality psychologists contributed strongly to the readiness of some of those formerly committed to the environment to consider a new paradigm. As important has been the spectacular success achieved through care, patience, and, finally, wisdom, by those in related fields.

Physical anthropology, as we shall see, contributed in major way to the shift of view currently at hand: Recent finds of hominid remains and artifacts, and in particular the developing consensus on the influence of hunting, has provided reason to eliminate some modes of thought and to institute quite different concepts. Add to all this the astonishing accomplishments of molecular biology in the past 20 years and the case becomes well closed. The discovery of the chemical nature of hereditary material (DNA) and the description of the genetic code lent a kind of substance to the genetic argument not otherwise to be obtained. The establishment of behavioral genetics indirectly drew upon this success as well as that of evolutionary and quantitative genetics. Behavioral genetics then contributed from within to psychology's new consciousness. Finally, the exponential growth of the neurosciences and their concurrent interest in behavioral problems has forced psychologists to recognize both the validity and the burgeoning reality of approaching the study of behavior from a more balanced position.

Figure 2.1 presents a chart of the relative emphasis placed upon biological and environmental determinants of behavior by various approaches in the recent past. One thus has some notion of the role that organismic variables have played in a number of systems and disciplines bearing directly upon behavior. Some names associated with particular positions are also given; they belong to but a few, though some of the better known, individuals concerned. As is clear (I trust), those on the left side of the chart tend(ed) to insist most strenuously upon the relevance of biological factors, whereas those on the right side stress(ed) the environment. Personality theorists, as described by Hall and Lindzey (1970), cover the entire range.

CONCLUSIONS

Psychological thought since its recorded inception has followed a circuitous path, but one whose study is necessary if psychology's present condition is to be comprehended. We have accepted, abandoned, and accepted again first one cause and then another, just in recent times touching the philosophical limits of environmental determinants, and running the methodological gamut from a complex, high-order, formal model of scientific structure to a simplistic fiducial technology. These have been quantum leaps but, unhappily, with little to show other than distance between points on the same dimension.

However, it is not the case that the experimental psychology of the past 50 years, despite its relative lack of scientific progress, has been without societal effect. Quite the contrary: Psychology's commitment to the environment has been proffered to and adopted by the public to the extent that it has almost achieved the status of a political right. We have come to believe in the capacity of the environment to control behavior with nearly the same matter-of-factness that we believe in the legitimacy of the guarantees of the Constitution.

This is understandable, given our history and social priorities. We prefer, and properly, a system wherein manipulations of the environment are made in order that no one will be denied basic opportunities. But, alas, an enormous chasm lies between the efficacy of our knowl-

BIOLOGICAL INFLUENCE

ENVIRONMENTAL INFLUENCE

EVOLUTIONARY APPROACH	PHYSIOLOGICAL AND COMPARATIVE PSYCHOLOGY			LEARNING APPROACH

ETHOLOGY

Lorenz
Tinbergen
Hinde
"NEW"

ANTHROPOLOGY

Campbell
Washburn
Leaky
Philbeam
Van Lawick-Goodall

PHYSIOLOGICAL AND COMPARATIVE PSYCHOLOGY

Lashley
Hebb
Beach
Lehrman

BEHAVIOR GENETICS

Fuller
Thompson
McClearn
Hirsch

LEARNING APPROACH

Thorndike
Waston
Hull
Spence
Guthrie
Tolman
Skinner

PERSONALITY THEORIES

PSYCHOANALYSIS

Freud
Jung
(Adler)

CONSTITUTIONAL

Sheldon

NEUROSCIENCES

Schildkraut
Kety
Sperry

FACTORY THEORY

Cattel
Eysenck

HOLISTIC

Goldstein
Maslow

PERSONOLOGICAL/
INDIVIDUAL

Murray
Allport

REINFORCEMENT

Miller
Dollard
Sears
Mowrer

EXISTENTIAL/
PHENOMENOLOGICAL

Rogers
May
Biswanger
Boss

SOCIAL

Horney
Fromm
Sullivan
(Adler)

FIELD

Lewin

Figure 2.1

edge and the urgency of our needs, and we have attempted to bridge that chasm by what amounts to intellectual self-seduction. We have transformed assumptions into axioms and made inalienable warrants of tentative hypotheses. The result was that we did not cross the chasm; we closed our eyes and fell in.

We are struggling now to get out. In the meantime, ironically, the very society that psychology seeks to serve has at its initiation determinedly built walls at the edge. The environmental doctrine has become cultural dogma, to the point where it sounds a little unethical to question it. Indeed, it sounds a little treasonous, as if to suggest that all problems may not be soluble by simple environmental alterations is to imply the necessity for a renegotiation of our political charter.

Nothing of the sort follows. The two domains are quite dissimilar. Political rights are declarations of intent, agreements to behave in certain ways. They do not inhere in nature. Scientific relationships, in contrast, are our best guess about what does. The rights to life, liberty, and the pursuit of happiness, therefore, do not have the same existential status as do atoms, genes, and the rotation of galaxies.

Whether or not the environment controls behavior is a scientific question and should on no account be related to our view of fundamental freedoms. Yet in ways both subtle and blatant, it is. It is as if we have come to believe that it is our right that the environment have the capacity to do what some psychologists have led us to believe it can. Just, then, as "Social Darwinism" dominated the political and intellectual scene in the early decades of this century, so has its presumed opposite, environmentalism, dominated the middle years. We are aware of the precepts and the residual of the former, because they derive from another age for most of us, and because historians have had time and motivation to examine them at length. However, the lack of distance between us and the events of psychology's recent past makes it far more difficult for us to see the zones that separate the reality of our experiences, scientific as well as other, from the expectations for which we have been trained. The impact of the theory of environmental determinism upon our lives is thus now not nearly so well understood as is the impact of the theory of evolution, as it was applied by Spencer and those who thought as he did. We are, on the contrary, hardly conscious, in the literal, articulate sense, of environmental determinism as *theory*, much less as *policy*.

This, we have noted, is normal, for whereas the story of "Social Darwinism" has been neatly packaged and set aside, radical environmentalism is a tale still in the telling. We do not know what the final outcome will be, although at this stage we can probably make an adequate guess. There are, however, a large number of dicta derived from environmentalism, the merits of which do not require prolonged deliberation. It has been taught at different times, for instance, that all infants should be fed only on a rigidly enforced schedule (the "Watson babies"), that physical punishment in any form is harmful to all children, that the technique of choice is to withhold affection, and that aggression occurs in children because it is learned from models. It has also been taught that aggression is due to frustration, that infants should be fed on demand, and so forth. The list is long, and contradictory.

Finally, although evolutionary theory was misapplied to human society by some, causing it still to be viewed with apprehension, it was based upon what proved to be the most profound biological thought of all time. The history of evolutionary theory from 1859 is thus one of success, a fact that in no way detracts from its appearance. Beyond that, there is a certain grace conferred upon those historical events by the character of their principals, and this too cannot help but provide sympathetic texture to all but the driest of accounts.

Chapter 3
Tools for Understanding: Principles of Transmission Genetics

Several points should have emerged from the first two chapters. First, our image of ourselves is in some (increasing) measure dependent upon scientific opinion on the origins of our behavior. Political and social thought derives from this image and in turn affects further scientific research by reflecting the concerns of society, to which many scientists are properly responsive. Second, two major classes of variables have come to be recognized as entering into the determination of behavior: biological (organismic) and environmental. Each class has at one time or another taken hegemony in the scientific and public mind as the chief source of behavioral differences. Third, it is clear now that an adequate account of behavior not only must include variables of both of these classes, but must examine the ways in which they *combine* to produce behavior. The differences among organisms are thus to be understood as due neither to biology nor to environment alone. A scientific strategy that takes advantage of the modern view must therefore be elaborated, and the implications of such a view should be examined.

Our concern here is to argue for a broad research strategy that is consonant with modern scientific knowledge; that is, one that allows for input from biological and environmental sources and is specifically intended to study the manner in which variables from both combine in the development of behavior. In order to comprehend the essentials of such a strategy, however, it is necessary to examine one of the chief sources of biological variation: differences in genes. This brings us to transmission genetics.

"MENDELIAN" GENES AND POLYGENES

All inheritance that has relevance to our concerns is Mendelian; that is, it follows the principles proposed by Gregor Mendel. There are differences, however, in the approaches used to study the transmission of genes, depending upon whether or not their individual effects can be followed from generation to generation. Genes that had effects so large that they could be individually identified were used by Mendel to demonstrate the law of segregation. These genes usually produce differences large enough to be considered qualitative, as with those for "tall" and "dwarf" in garden peas, wherein the dwarf variety is $\frac{1}{8}$ to $\frac{1}{4}$ the size of the tall. A "tall" × "dwarf" cross produces all tall progeny, but the progeny of matings made among this second generation segregate into two classes, three-fourths of which are tall and one-fourth of which are dwarf. The genes can thus be followed through their effects from one generation to the next.

In describing the transmission of such genes, it does not matter to us how many there are, so long as their individual effects can be identified. Although the system may become complex, it remains a case of "Mendelian" inheritance because we may sort the phenotypes into ratios of the kind described by Mendel, using adequate breeding tests. In the instance above, for example, at least one-fourth of the progeny of the second cross are known to have the "dwarf" gene. There is thus a phenotypic ratio of 3:1. (With further tests we could find a genotypic ratio.)

However, most genes cannot be individually identified through their specific effects. They are inherited in exactly the same way as "Mendelian" genes, but their effects, rather than being large enough to divide the population into qualitatively distinct categories, are small and cumulative. Such is the case, for example, in the usual distribution of a character like height. Height is not (except in rare cases of pathology) separated into discrete classes; instead, it varies from short to tall in nearly infinitesimal steps. The genes that affect this distribution are not few, each having a large influence, but are many, each having a small influence—so small, in fact, that individual "height" genes cannot be separately identified. Such genes are often described as "polygenes," and their inheritance is therefore called polygenic.

In dealing with characters that are distributed quantitatively we cannot expect to use Mendelian ratios because the population does not fall into discrete categories. Rather, we must use quantitative measures, particularly proportions of variance, to describe the genetic situation. Mendelian ratios are reserved for cases where the effects of individual genes can be followed.

It should be emphasized once more that there are not two kinds of inheritance, but two methods by which the mode of inheritance may be studied, and the methods differ depending upon whether the genes have large (qualitative) or small and cumulative (quantitative) effects. It should also be emphasized that there are not considered to be two kinds of genes, but only genes that differ in the extent of their effects. "Mendelian" genes having large effects are inherited in the same way as "polygenes" having small effects; the difference we see lies in their relative place in the metabolic network.

INDIVIDUALLY IDENTIFIABLE GENES

Mendel's innovation lay, as noted previously, in choosing to study traits having large effects, in studying one or only a few traits at a time, and in actually counting the number of progeny in the various categories rather than using a verbal summary. In this way he was able to demonstrate the particulate, segregating nature of inheritance. From his basic ratio, 3:1, he inferred that there was a single determiner of the trait, with two alternative forms, one dominant over the other.

That single determiner we now call a *locus*. It refers, quite literally, to a location, often unknown but inferred to exist, on a chromosome. Alternate gene forms, *alleles*, may occur at the same locus on a chromosome, one at a time. Thus, one might speak of the "dwarf" locus on a chromosome in garden peas, at which could occur the allele for normal height (say, D) or the allele for dwarf stature (d).

But a single allele does not appear in the organism, Mendel discovered; rather, alleles appear in pairs. The *phenotype* (what is measured) of the individual is thus a result of the actions of the *pair* of alleles which are present at the locus on each of the pairs of chromosomes. "Locus" then finally refers to some site on a homologous pair of chromosomes.

The "tall" × "dwarf" cross might thus be diagrammed as follows:

Let D stand for the dominant, 'tall' allele.
Let d stand for the recessive, 'dwarf' allele.

In the soma of the organism two of these alleles will appear; it is the purpose of the breeding tests to discover which ones are present. If the tall and dwarf varieties each breed true for the trait, the alleles of each of the parents of the cross may be represented as:

	Tall	Dwarf
parental generation (P)	DD	dd

Each parent will contribute half of its genetic complement to the reproductive cells that unite to form the next generation. These cells are called *gametes*. All of the gametes produced by the tall parent will contain the same allele affecting height (D), and similarly the gametes produced by the dwarf parent will contain only one kind of allele (d). The only possible combination is Dd. (It is conventional to represent the female contribution first. Obviously both males and females from each variety contribute gametes to the next generation, but they would be the same in true-breeding lines for both sexes.)

This initial product of the cross is called the *first filial generation*, or F_1. In the present circumstance, where D is dominant over d, all of the F_1 are tall:

	Tall
F_1	Dd

The *second filial generation*, F_2, is produced by mating the F_1 inter se. In this case each parent will produce two kinds of gametes, one containing D and one containing d:

	Tall		Tall
F_1	D d	×	D d

The combinations may be represented by a small matrix (Figure 3.1), and the expected frequency of the genotypes determined by a simple count. The genotype ratio here is 1 DD: 2 Dd: 1dd; one pure tall for every two heterozygous tall for every one pure dwarf. Since D is dominant over d, the phenotypic ratio (how the progeny actually appear) will be 3 tall : 1 dwarf.

It was on the basis of this sort of experiment that Mendel concluded that the expression of traits is governed by two elements, one contributed by the male and one by the female, and that the elements do not blend in the progeny but rather remain individually intact. They are thus capable of segregating into new combinations with each generation. Mendel also concluded that the elements (alleles) are either dominant or recessive. We know this not to be true of all allelic relationships now. Shortly after the discovery of Mendel's work, for example, crosses between snapdragons with red flowers and those with white flowers were found to produce progeny that were intermediate (pink). In this case the phenotypic ratio was 1:2:1, which accurately reflected the genotypic ratio. Dominance is not a necessary condition.

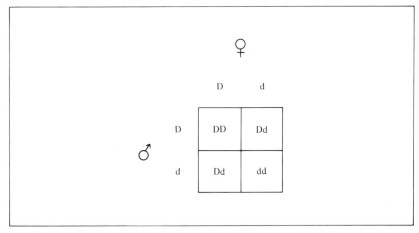

Figure 3.1   Representation of the outcome of a cross between true-breeding tall and dwarf garden peas, with dominance, demonstrating Mendel's law of segregation. (Source: E. J. Gardner, *Principles of Genetics*. New York: Wiley, 1964. By permission.)

The 3:1 (with dominance) or 1:2:1 (without dominance) ratios are used in the attempt to determine the mode of inheritance of a trait when the effects of individual genes can be identified. Thus, when the genotype of the parents is unknown, but the phenotypic ratio in the F_2 is 3:1, it would be concluded that there is one gene (locus) involved, with two alleles, one dominant over the others. If the ratio were 1:2:1, the interpretation would be the same, except without dominance. (Other breeding tests, such as crossing the F_1 to the suspected recessive parent, are also commonly used.)

When a number of traits are studied, the phenotypic ratio in the F_2 can be compared with some multiple of the 3:1 or 1:2:1 ratios. Suppose, for instance, a line having yellow, round seeds were crossed with a line having green, wrinkled seeds, and the ratio in the F_2 were 9 yellow, round: 3 yellow, wrinkled: 3 green, round: 1 green, wrinkled. There are two traits, one having to do with the color and another having to do with the texture of the seeds. Loci are usually named for recessive conditions (since it is through the discovery of recessives that we know of their existence). Hence, letting g stand for the recessive, green, G for the dominant, yellow, w for the recessive, wrinkled, and W for the dominant, round, the matings could be reconstructed as shown in Figure 3.2.

The $F_1 \times F_1$ cross is dihybrid, but may be resolved to two monohybrid crosses, that of yellow $\times$ green, and that of round $\times$ wrinkled. Results such as these are what led Mendel to postulate his second law, that of independent assortment. We know now that independence is due to the genes being on different chromosomes. If the genes influ-

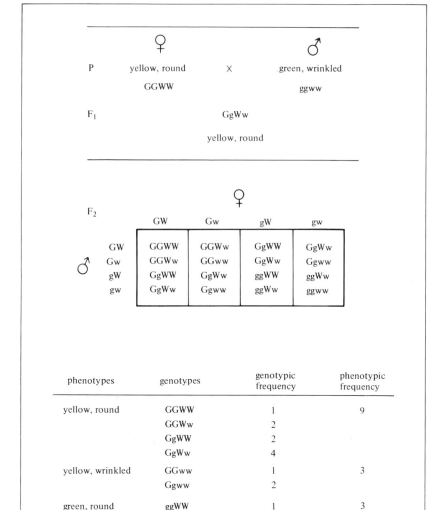

phenotypes	genotypes	genotypic frequency	phenotypic frequency
yellow, round	GGWW	1	9
	GGWw	2	
	GgWW	2	
	GgWw	4	
yellow, wrinkled	GGww	1	3
	Ggww	2	
green, round	ggWW	1	3
	ggWw	2	
green wrinkled	ggww	1	1

Figure 3.2   Representation of a cross between true-breeding garden peas having yellow, round seeds or green, wrinkled seeds, with dominance, demonstrating Mendel's law of independent assortment. (Source: E. J. Gardner, *Principles of Genetics*. New York: Wiley, 1964. By permission.)

encing the traits under study are on the same chromosome, the combinations will not of course be derived independently, and this will be reflected in a departure from the expected ratio.

The dihybrid ratio where there is dominance may be treated as an expansion of the monohybrid ratio using the binomial expression $(A + a)^n$, where $A = 3$, $a = 1$, and $n =$ the number of loci. There are then $(3 + 1)^2 = 4$ classes in the ratio 9:3:3:1 resulting from a dihybrid cross. (The 4×4 matrix from a dihybrid cross of course generates 16 cells, but many of the combinations therein are functionally equivalent, depending for differentiation only upon whether the alleles came from the maternal or paternal side. Thus, GgWw and gGwW may be classified, both genotypically and phenotypically, as the same.)

Keeping track of events beyond this becomes somewhat complicated. The number of classes in a trihybrid cross with dominance at each locus is found by expanding $(3 + 1)^3$, which gives a ratio of 27:9:9:9:3:3:3:1. A cross involving four loci produces a number of classes difficult to follow on paper (computers are good at this sort of thing). Any number beyond that should be left to numerical masochists.

The principle is, however, clearly the same, no matter how many loci are involved. Complications are introduced when the loci are not in fact independent of one another—that is, when one or more are on the same chromosome. When this occurs, the loci are said to be linked. The test for linkage is, as mentioned, a significant departure from what is expected under independent assortment.

When dominance does not occur the monohybrid cross produces three phenotypic classes rather than two. Some progeny in the F_2 are thus distinguishable from the two parents. This is found, as mentioned, in the case of flower color in snapdragons. If the red and white lines both breed true and are crossed, the results may be represented as shown in Figure 3.3.

The phenotypic ratio in the F_2 thus accurately reflects the genotypic ratio. The *heterozygous* RR' is intermediate between the *homozygous* (RR or R'R') stock. (A zygote is the product of the union of two gametes. The dihybrid ratio, involving two loci, in such a case is 1:2:1:2:4:2:1:2:1.) Obviously such an analysis may be extended to as many loci as one wants. The mode of inheritance in this system is said, appropriately, to be intermediate.

The above sounds very straightforward. But you must know by now that it could not remain so for long. Only a small proportion of traits whose inheritance is understood are in fact controlled by genes that behave (from the geneticist's viewpoint) so well. Linkage, the presence of two or more loci on one chromosome, has already been mentioned as one complicating factor. Another, which should be evi-

P	red RR	X	white R'R'
F₁		pink RR'	
F₂	1 RR :	2 RR' :	1 R'R'

Figure 3.3   Representation of a cross between true-breeding plants having red and white flowers, without dominance.

dent from the concept of dominance with which Mendel began, is that of *genic interaction.*

The most simple case is that of intralocular interaction. Here the effect of a locus varies according to what *combination* of alleles is present. Thus, in the tall × dwarf cross, the presence of one "tall" allele is all that is necessary to produce a tall plant. Only in the homozygous (dd) condition is the dwarf character expressed. With intermediate inheritance allelic interaction is evident, for different phenotypes result from all three possible allelic combinations.

It is also the case that the products of *different* loci interact with one another in a way that can be detected using genetic tests. Interaction of different genes rather than different forms of the same gene (alleles) is called *epistasis.* One of the most common kinds of epistasis occurs when the effect of one gene pair is to mask that of another, as often happens in the expression of color. In mice, for example, there is a single locus that, in the recessive condition, inhibits the expression of all pigment and produces the familiar albino animal. Crosses between albinos and pigmented mice homozygous for the color gene (CC) but lacking a colored hair banding pattern called "agouti" (therefore being homozygous recessive, aa, for agouti) produce progeny that are pigmented and show the aguoti pattern. This is clear from Figure 3.4.

It is evident that the albino mice carried the agouti gene, but its expression was masked by the effects of genes at another locus. Only in the presence of the dominant color (C) allele can the agouti pattern be expressed.

The agouti pattern is that normally found in wild mice, and is presumably favored by natural selection because of the concealment it offers. Usually the part of the hair nearest the skin is gray, and next comes a yellow band, and then either a black or a brown tip, depending on the presence of another pigmentation gene (which can only be expressed in the presence of C). Production of nonagouti but pigmented (usually black or brown) lines in the laboratory is possible by interbreeding mutants for the homozygous recessive (aa) condition at the "agouti" locus. Production of totally nonpigmented lines (albino) is made possible by the same procedure. When homozygous recessives at the color locus (CC) appear, they are maintained through interbreeding, so that all the individuals are albino. Crosses between pigmented, nonagouti animals and albinos wherein the agouti gene is present but not expressed give agouti progeny. Since the agouti condition is that generally found in the wild, it may be considered the natural state.

The presence of reversions or "throwbacks" to an ancestral condition was first discussed scientifically by Darwin. He of course had no concept of Mendelian genetics, but the phenomenon occurred often enough among the crosses of highly selected lines to seize his attention. It is understood now to be explained by epistasis. Ancestral traits appear in domestic animals or plants because a combination of genes occasionally comes together that allows expression of something hidden but carried for many generations, something that was masked by the interaction of genes at different loci.

The concept of genic interaction is only sparsely presented here. It can become extraordinarily complex, as in the case of genes whose only known effect is to modify the expression of other genes. The specific effects of only a few such *modifiers* have been identified. Usually their effects cannot be traced to given loci, and the distinction between their effects and environmental influences is difficult to establish. Their presence can, however, be demonstrated in certain instances.

Genic interaction can also be deadly. The effects of some genes

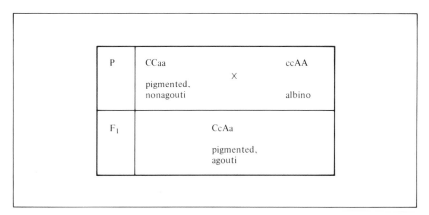

Figure 3.4   Representation of a cross demonstrating interlocular interaction (epistasis).

are so strongly negative as to kill the organism, and they are known, appropriately, as *lethals*. If the lethal allele is dominant the individual will die prematurely, although not necessarily immediately, for some lethals have a delayed effect. Lethals that are recessive may be carried in the heterozygous condition (and may affect viability), but have their full deleterious impact when in the homozygous state. One-fourth of the offspring of matings between carriers will thus be nonviable zygotes, but one-half will be carriers like their parents. Thus the deleterious recessive allele tends to remain in the population in spite of strong selection against the individuals expressing its effects.

The existence of genic interaction not only explains some otherwise incomprehensible outcomes, but, perhaps more important for our purposes, leads to an emphasis upon three crucial points. First, it is obvious that no single gene is responsible for the whole development of a trait. Many genes are required to operate the complex chemical relations involved in the development of even the simplest character. We know only a few of them, and then often only because a particular gene causes a disruption in those relations. Independence of gene transmission thus by no means implies independence of gene action. Second, a single gene does not necessarily have a single effect. Rather, one gene may affect many different characters and therefore be associated with a syndrome. Many such genes are in fact known and some examples will be given later. Genes with multiple effects are said to be *pleiotropic*. Third, the concept of genic interaction forces us to distinguish between genes, which are chemical units in the cells, and phenotypic traits, which are attributes that we choose to measure and result from the combined action of many genes (and from environmental influences). It is easy, and this mistake was made early, to begin thinking that there is a particular gene "for" a particular character, and that this gene somehow on its own is responsible for the form of that character. But such a view ignores both the effects of other genes and of the environment in the development of the character. It is an error that at all costs should be avoided.

INDIVIDUALLY IDENTIFIABLE GENES IN POPULATIONS

The study of the individually identifiable genes may be extended to units larger than the progeny of Medelian crosses to populations (or samples therefrom). There are many advantages to studying genes in populations; the chief one is that the evolutionary implications of gene changes may be elucidated. In addition, the breeding tests used in Mendelian analyses are often impractical for some organisms (as, for instance, for humans). It is of interest therefore to examine the events that tend to maintain or change the relative frequencies of genes in populations. Such is the subject matter of population genetics.

To describe the genetic constitution of a group it is necessary to specify the genotypes of the individuals and to calculate how many of each genotype there are. Suppose for simplicity that we chose a locus, A, on a homologous pair of chromosomes, and that a combination of two alleles, A_1 and A_2, were possible at the locus. There would be three genotypes, A_1A_1, A_1A_2, and A_2A_2, as we saw in the preceding section, obtainable for the locus. The genetic constitution of the group with regard to the locus would be completely described by the proportion of individuals having each of the genotypes, that is, by the relative frequency of the genotypes among the individuals. Thus, for example, if we found the relative frequencies of the three genotypes to be .25 A_1A_1, .50 A_1A_2, .25 A_2A_2, we would have described the genetic constitution of the population with reference to the A locus.

But a population is not interesting merely in itself; it is of interest also as the progenitor of the next population. We are thus concerned with the transmission of genes from one generation to the next, for differences in the relative frequencies of genes in populations over time underlie the phenotypic changes we see in evolution. But it is the *genes* of a parent that are transmitted to its offspring, not the genotype. As described in the preceding section, the alleles at a locus in a given parent go to different gametes, and either allele can combine in the zygote with an allele from the other parent. The genotypes of the parents are thus in effect broken down, and a new set of genotypes, some of which may be those of the parents and some of which may not be, is constituted in the progeny.

If this is the case, then our ultimate interest is in the relative frequencies of the alleles rather than in the relative frequencies of the genotypes. If, for example, A_1 is an allele at the A locus, then the relative frequency of A_1 is the proportion of A_1 alleles. The same is true for all other alleles at the A locus, so that the sum of the relative frequencies of all the alleles at the locus is 1. Taking two alleles, A_1 and A_2, suppose we were able to classify 100 individuals genotypically as shown in Figure 3.5.

Genotype A_1A_1 yields 60 A_1 alleles because there are 30 such genotypes having two A_1 alleles each. Genotype A_1A_2 yields 60 A_1 alleles, because there are 60 such genotypes having one A_1 allele each. Genotype A_1A_2 also yields 60 A_2 alleles for the same reason. Genotype A_2A_2 yields 20 A_2 alleles because there are 10 such genotypes having two A_2 alleles each. Summing, there are 200 alleles because each individual has two, and of those alleles 120 are A_1 and 80 are A_2. The relative frequencies are thus: $A_1 = 120/200 = .6$; $A_2 = 80/200 = .4$. A more general statement of the relationship is shown in Figure 3.6.

Since p is the relative frequency of allele A_1, q is the relative frequency of allele A_2, P is the relative frequency of genotype A_1A_1, and

genotypes				
	$A_1 A_1$	$A_1 A_2$	$A_2 A_2$	total
number of individuals	30	60	10	100
number of alleles A_1	60	60	0	120
A_2	0	60	20	80

Figure 3.5   Hypothetical example of genotypic and allelic frequencies. (Source: D. S. Falconer, *Introduction to Quantitative Genetics.* New York: Ronald, 1960. By permission.)

	alleles		genotypes		
	A_1	A_2	$A_1 A_1$	$A_1 A_2$	$A_2 A_2$
relative frequencies	p	q	P	H	Q

Figure 3.6   Representation of relationship between genotypic and allelic frequencies. (Source: D. S. Falconer, *Introduction to Quantitative Genetics.* New York: Ronald, 1960. By permission.)

so on, the relationship between the relative frequencies of alleles and those of genotypes is:

$$p = P + \tfrac{1}{2}H$$
$$q = Q + \tfrac{1}{2}H$$

If the breeding population is large, mates randomly, and selection, mutation, and migration are not operating, the relative frequencies of the alleles and of the genotypes will be constant from generation to generation. The relative frequencies of the genotypes will thus be determined entirely by the relative frequencies of the alleles, and the population is said to be in genetic equilibrium. This property was first described by Hardy and by Weinberg in 1908 and is suitably known as the Hardy-Weinberg law.

Deduction of the law is conveniently demonstrated (Falconer,

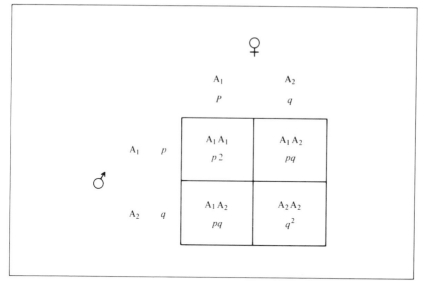

Figure 3.7   Representation of genotypic and gametic frequencies.

1961, pp. 9–10) by assuming that the parental generation has relative allelic and genotype frequencies of p, q, P, H, and Q, as described. Two kinds of gametes are produced, those bearing the A_1 allele and those bearing the A_2. The relative frequency p of gametes containing A_1, we have seen, is $P + \frac{1}{2}H$. Similarly, the relative frequency q of gametes containing A_2 is $Q + \frac{1}{2} H$.

We may think of random mating as a condition wherein there is a pool of gametes to which each individual contributes equally, and wherein zygotes are formed by the union of two gametes in a way in which each gamete has an equal opportunity to unite with any (sexually appropriate) other. This situation is represented in Figure 3.7, which shows male and female gametes, their frequencies, and the resultant combinations.

Summing, the genotypic frequencies are $A_1A_1 = p^2$, $A_1A_2 = 2pq$, and $A_2A_2 = q^2$. These genotypic relative frequencies may be used to determine the relative frequencies of the alleles in the next generation. Thus, the relative frequency of allele $A_1 = P + \frac{1}{2}H = p^2 + \frac{1}{2}(2pq)$ $= p^2 + pq$. Factoring, we have $p(p + q)$. Recall, however, that the relative frequencies of the alleles at a locus must sum to 1; therefore we may substitute $1 - p$ for q, and arrive at $p(p + 1 - p) = p(1) = p$. This is the same relative frequency with which we began. The same applies to the relative frequency of A_2. Therefore in a population in Hardy-Weinberg equilibrium there is no change in allelic frequencies over generations, and the genotypic frequencies are determined solely by the allelic frequencies.

Although the Hardy-Weinberg equilibrium is of interest in itself, it is actually of greater use as a idealized standard against which real situations may be compared. For a population to be in equilibrium a number of conditions must hold. A departure from equilibrium may then be used as a measure of the extent to which the conditions are not met.

A simple example of the Hardy-Weinberg equilibrium in human populations is found with the blood groups M and N. Three genotypes are formed: the homozygote MM, the heterozygote MN, and the homozygote NN. Let p represent the frequency of the M allele, and q represent the frequency of N. From the above it follows that the expectation is: $p^2 + 2pq + q^2$; that is, the proportions of the phenotypes MM, MN, and NN should reflect the distribution of alleles in a large population mating randomly without differential reproduction according to genotype.

Suppose now we sampled an actual population and obtained the following distribution of 6129 individuals (Gardner, 1964): MM = 1787; MN = 3039; NN = 1303. The number of M alleles will be all of those contributing to the MM phenotype plus half of those contributing to the MN phenotype; that is, $2 \times 1787 + 3039 = 6613$. Similarly, the number of N alleles will be all of those contributing to the NN phenotype plus half of those contributing to the MN phenotype; that is, $2 \times 1303 + 3039 = 5645$. The proportion of M alleles in the sample thus equals $6613/12,258 = .54$. The proportion of N alleles equals $5645/12,258 = .46$.

The probability is therefore .54 that a given chromosome will carry M and .46 that it will carry N. What should we expect if the population is in Hardy-Weinberg equilibrium? Where the relative frequency of M is represented by p and that of N by q, the distribution should be as $p^2 + 2pq + q^2$. Thus $p^2 = (.54)(.54) = .29$; $2\,pq = 2(.54)(.46) = .50$; $q^2 = (.46)(.46) = .21$. Turning to the sample figures, this is very much what we find. We conclude then that with regard to this character the population mates randomly, with no differential reproduction according to genotype.

Hardy-Weinberg thus provided a model for the genetic statics of a population. The dynamics—that is, events that lead to changes in the relative frequencies—of alleles, were described by Ronald Fisher, Sewall Wright, and J. B. S. Haldane. They are:

1. *Assortative mating:* a form of nonrandom breeding, defined as the occurrence of matings among like or unlike genotypes more often than would occur by chance. In the former case it is called "positive"; in the latter, "negative." The effect in the progeny population of positive assortative mating is to increase the relative frequencies of the homozygotes and reduce that of heterozygotes.

2. *Genetic drift:* a random fluctuation in the relative frequencies of

alleles due to the small size of the population. Not all possible progeny that a parental generation could produce are actually produced. Each generation is but a sample of the last. All else equal, it will be a representative sample, one that accurately reflects the distribution of genotypic frequencies of the parental generation, if the population is large, because the probability is very high that all alleles, even the rare ones, will be passed on. With large numbers of gametes involved, even alleles in small proportion to others will be maintained. However, when the population size is small, certain alleles may depend for their continued existence upon a few individuals and their gametes. If some of those individuals should not reproduce, the effect upon the next generation would be drastic. Even alleles at high relative frequencies will be affected for the same reasons. The total condition then is one in which the genetic architecture of the population changes over time, or drifts, as a function of chance fluctuations.

3. *Mutation:* the appearance of a new or different form of a gene. An example of a mutation is the "dwarf" allele in garden peas. The dwarf line does not exist in nature, but only in domesticated varieties; hence, it may be considered a mutation which, whenever it occurred, became fixed in domestic lines. Many striking mutations are known, and some will be discussed later. Mutations are important for two reasons: First, they provide new material upon which selection may operate; second, in themselves they alter gene frequencies. If A_1 mutates to A_2, for example, the relative frequency of A_1 is reduced, and if this continues over a long period of time and no other forces modify it, A_1 will disappear and be replaced by A_2.

However, mutation is also reversible, so that A_2 may mutate back to A_1, and an equilibrium may be established eventually such that a mutation that occurs often (a recurrent mutation) and the mutation back to the original condition balance each other.

4. *Migration:* the introduction of alleles from another population, or the loss of alleles from the given population because the individuals leave and are no longer part of the breeding structure.

5. *Selection:* differential production of viable offspring among parents having a particular allele or constellation of alleles. This will lead to an increase in the relative frequency of the favored allele or constellation, resulting in the nonrandom change in allelic frequencies that we recognize under natural conditions as evolution. Mutation supplies the new allelic material, and new and more viable allelic constellations may come about through interbreeding. Neither of these brings about evolution, however; selection must act before evolution can occur.

An interesting example of selection involves "industrial melanism" In 1850 it was found that most of the moths in nonindustrialized parts of England were light in color, less than 1 percent being dark. However, as industrial wastes darkened the trees and buildings, in these communities, there was a concomitant change in the color of the moths. Thereafter more than 90 percent of them were found to be dark. What occasioned the change?

The moths whose color was better adapted to the environment survived and reproduced at a greater rate. Thus, as the countryside darkened, the lighter moths were more visible to predators and the alleles leading to dark color were selected for. This hypothesis was tested later when dark and light moths were released in industrial and nonindustrial areas. In industrial Birmingham the darker moths survived in a ratio of more than 2:1 over the lighter ones, whereas in nonindustrialized Dorsetshire the lighter moths survived over the darker ones by 3:1.

It is important for us to understand the events leading to changes in gene frequency so that the results can be predicted when genotypes are measured accurately. The last condition is, however, most crucial. Phenotype must be a reasonably accurate guide to genotype or obviously any attempt to measure allelic frequencies, much less follow changes in them, will founder. And it is unhappily the case that behavioral phenotypes that accurately reflect individual alleles are quite rare.

Behavior, for one thing, ordinarily does fall into discrete categories. For another, it seems at least in higher organisms to have evolved to be environmentally adaptive. Differences among individuals even on a continuous distribution are by no means attributable only to differences in genes. Thus it is impossible in most cases to apply population genetics directly to behaviors, for behaviors are largely unsuited to the model upon which population genetics is based.

Nevertheless, study of population genetics provides insight for the behavioral scientist. Genes that influence behaviors are presumably no different from those influencing other traits; therefore, knowledge obtained from the examination of nonbehavioral characters is in principle applicable to behavior. In particular, the evolution of behavior is thought to the controlled by the same events that control the evolution of other characters.

NON-INDIVIDUALLY IDENTIFIABLE GENES

To this point we have considered genes with effects powerful enough to allow us to associate them with particular loci and with alternate forms that divide the population into qualitative classes. But most genes, we well know, do not exert such effects upon the phenotype, for most characters vary only by degree. Variation of this sort, without natural discontinuities, is called *continuous,* and traits that exhibit it are called *quantitative* because their study depends upon measurement instead of counting. The genetic principles underlying the inheritance of quantitative traits is, again, no different from those outlined above, but since the segregation of the individual genes cannot be followed, other methods of study must be applied. This is the business of quantitative genetics.

Quantitative characters form a continuous distribution for two reasons: First, many genes, each with a small, cumulative effect, segregate simultaneously, and second, the action of the environment tends to smooth over the discontinuities that might occur on a genetic basis alone. This was, as mentioned in Chapter 1, initially shown by Johannsen and by Nilsson-Ehle and East. Together their work demonstrated that continuous variation is due to the combined effects of genes and environmental factors and that many genes may jointly affect a single character.

The quantitative characters that may be examined in higher organisms are clearly quite numerous. Any trait that shows continuous variation may in principle be studied, provided it can be measured. Measuring a large number of individuals, however, requires that we use statistics to describe and analyze the data. Thus, we use means, standard deviations, and, most importantly, proportions of variance, to describe the genetic situation.

The *mean* is the familiar average. It is simply the sum of the scores involved divided by the number of such scores, or $\Sigma X/N$. This is one measure of the central tendency of a distribution; it offers a single number with which to describe a set of scores, in terms of a "center of gravity" or a "balance point" around which the scores fall. The mean is in fact that value from which the sum of the deviations is zero. That is, if one gave a positive sign to all scores having a value greater than the mean and a negative sign to all those having less than the mean, they would sum to zero.

A measure of central tendency alone, however, is inadequate to describe a distribution. We need to know also how far or how close the scores tend to be to the mean. This is given by another descriptive statistic, the *standard deviation,* defined as the square root of the summed, squared deviations about the mean, divided by the number of scores, that is,

$$\sigma = \sqrt{\frac{\Sigma(x - M)^2}{N}}$$

where σ (sigma) is the standard deviation, M is the mean, and N is the number of scores.

This gives us an index of the "spread" or "dispersion" of the distribution. Distributions with a very small standard deviation will, by definition, tend to have most of the scores close to the mean value, whereas distributions with a large standard deviation will tend to have many that are not. (We also obtain from the standard deviation a small idea of the cleverness of mathematicians. Why, one might ask, square something that one will immediately take the square root of? Because as just noted, the sum of the positive and negative deviations from the mean is zero, and hence of little use. But we need those deviations to measure dispersion. If they are squared they all become positive and do not cancel one another out. We can then remove the square.)

The *variance* is the most commonly used measure in quantitative genetics, for ultimately one wishes to describe what proportion of the differences among individuals is due to the contributions of genes, and then further what proportion of the genetic variance is due to the additive, dominance, and epistatic effects of genes. Portioning out these proportions may become complex and will be treated here in the briefest possible way. For the moment we shall merely indicate that the total phenotypic variance (the differences due to all causes, genetic and environmental) is given by the summed, squared deviations from the mean, divided by the number of scores involved; that is,

$$\sigma^2 = \frac{\Sigma \, (x - M)^2}{N}$$

The square root of the variance is, obviously, the standard deviation.

To discover the role played by genes upon quantitative characters we must first separate the genetic from the environmental influences. This is best done by thinking of the score one actually obtains on an individual (i.e., what is measured: height in inches, weight in pounds, and so on) as his or her *phenotypic value*. The phenotypic value can then be thought of as composed of two components: (1) the *genotypic value*, that score which the individual would obtain under genetic influences alone, and (2) the *environmental deviation*, which causes the phenotypic value to be different from the genotypic value. Thus, symbolically:

$$P = G + E$$

where

P is the phenotypic value
G is the genotypic value
E is the environmental deviation

In a population of individuals there will of course be many phenotypic values, many genotypic values, and many environmental deviations. Equally clearly, what is proposed in principle may be difficult to accomplish in reality: We cannot simply see in an individual or a population what is due to genotype and what is due to environment. We must therefore deal with proportions of variance among the individuals that we find to be attributable to certain influences. That is, our basic fact is that individuals differ; we must find a way to account for the difference in terms of proportions of the total variance so that the influence of each component may be estimated.

This is done by carrying the conceptual partitioning of the components farther, and by putting them in variance terms. We have already seen that

$$P = G + E$$

that is, the phenotypic value may be considered to be equal to the genotypic value plus an environmental deviation. In variance terms we would write

$$V_P = V_G + V_E$$

that is, the total phenotypic variance in the distribution is composed of two components: Differences among individuals due to their differences in genotype and differences among individuals due to differences in the environments to which they have been subjected.

However, this is clearly incomplete. We have seen in the case of genes with effects powerful enough to be individually identified that there is both intralocular and interlocular interaction, in other words, alleles at a locus do not act singly, but in pairs, allowing for the possibility of both intermediate inheritance and dominance. Moreover, more than one locus may affect a character, allowing for the possibility of interaction among loci, or epistasis. We may thus decompose the variance due to genotype into variance due to the additive, dominance, and epistatic effects of the genes.

The additive genetic variance is based upon the average effects of the alleles in the population. Average effect refers to the average deviation from the population mean value of individuals who receive a particular allele, say A_1, the other allele at the A locus coming at random. Suppose, that is, that one wished to determine the effect of an allele. There is no way to tell the effect it has except by comparing the individuals who carry it with those who do not. But the effect of allele A_1 will be different in individuals with otherwise different genotypes, and in individuals subjected to different environments. Therefore, A_1 will not have a single, set effect upon the scores in a distribution but will have an effect that varies with the individuals who carry it. To obtain a single value for A_1 we then take the average effect it has across those individuals. To determine the value of the effect we of course need a standard with which to compare it. The mean of the population is used for this purpose. The average effect of an allele is thus defined as the average deviation of the individuals who carry it from the mean of the population, the other allele taken at random.

"Average" then refers to the effect of the allele as measured in many individuals. "Additive" enters the vocabulary because in the polygenic systems that influence continuous traits the effects of individual alleles cannot be identified. The *additive effect* of alleles in a polygenic system is therefore defined as the *sum* of the average effects over the pair of alleles *at each relevant locus.* "Additive" therefore refers to the fact that the average effects of all the alleles are added. "Additive genetic variance" refers to the variance among individuals due to the differential additive effects of the alleles they carry.

If the effect of an allele in an individual depended solely upon that allele and were not influenced by the other allele at the locus or by other loci, then all genetic variance would be additive. One could in principle simply take the average effects of all the alleles and add them together to determine where on a genetic distribution an individual would fall. But we know this to be not so. Interaction between alleles at a locus (dominance) and among loci (epistasis) alters the effects of the alleles from what they would be under strictly additive conditions.

Dominance variance is based upon the difference among individuals due to genetic differences not attributable to additive variance and is derived from the interaction of alleles within loci. It is variance attributable to the fact that some alleles show dominance (partial or complete) over others. Thus there will be deviations from what would be expected if genetic variance were all additive—that is, if the average effects of the alleles could simply be summed without regard to the effect of one allele upon another within loci.

Epistatic variance is genetic variance which arises from interactions among loci. That is, if the presence or absence of a particular allele at, say, locus B, had no effect upon the alleles at locus A, there would be no opportunity for interlocular interaction. But we know this not to be the case. The alleles present at one locus do alter the effect upon the phenotype of alleles at other loci. This causes a further deviation from what would occur if all the genetic variance were additive. Summarizing, then, we see that total (phenotypic) variation is composed of two chief components, genetic and environmental variation. Genetic variance may be further decomposed into three main components: additive, dominance, and epistatic; that is,

$$V_P = V_G + V_E \quad \text{and} \quad V_G = V_A + V_D + V_I$$

where

V_A is additive variance
V_D is dominance variance
V_I is genetic interaction (epistatic) variance

Two points should be made now: (1) Additive effect in genes may not be taken to imply additive gene action; that is, additive effect is a statistical concept, and one cannot infer from its existence that the actual gene products combine additively. For conceptual purposes we consider that the gene products act *as if* they combine additively, but that does not mean they do. (2) The same set of genes may have additive, dominance, and epistatic effects. Thus genes that show dominance and epistasis may also contribute to the additive portion of variance, and genes which act additively may show dominance and epistasis as well. There are not then "additive genes," "dominance genes," or "epistasis genes" corresponding exclusively to the three components of variance, but rather all three kinds of variance may be derived from the same set of genes. It is possible, of course, for most or even all of the genetic variance to be associated with one or two of the components. Genes that show no dominance or epistasis will generate no dominance or epistatic variance, and so on. But this is relatively rare.

The above discussion, even in the very limited treatment presented, may seem unduly involved. Such distinctions are necessary, however, to understand the role played by the genes in changing or maintaining phenotypes. Additive variance, for instance, assumes the major importance when it is understood that additive effects are the chief cause of resemblance among relatives and therefore the main determinant of the observable genetic properties of the population. Moreover, the additive effects of genes are accountable for the responses of the population to selection and are therefore accountable for evolution. This makes a good deal of sense when considered carefully. The additive effects are those that do not depend upon interaction either within or among loci, do not depend upon certain alleles combining in the genotype, and thus are more responsive to the forces bringing about evolution, a process that acts over generations and necessarily requires the breakdown of the parental genotypes and the constitution of new ones in the progeny.

Further, an immediate use may be put to the concept of additive variance in the statement of a derivative concept: *heritability*. The heritability of a character is often of the greatest interest to quantitative geneticists and in particular to those concerned with behavior. It is the measure best known in psychology for its association with questions regarding the "inheritance" of intellectual and emotional factors and with questions of racial differences. It is, therefore, a concept with which one should, at some meaningful level, become familiar.

Heritability is defined as the ratio of the additive genetic variance to the phenotypic variance, i.e.:

$$h^2 = \frac{V_A}{V_P}$$

The usual symbol, h^2, represents heritability itself and not its square (for historical reasons, which need not trouble us). Heritability is thus the proportion of total variance that is attributable to the additive effects of the genes. As such it is the most important measure of resemblance between relatives and also the chief predictor of the response to attempts at selection. This occurs, again, because the additive effects of genes do not depend upon the combinations in which the alleles occur.

Another way to state the importance of heritability is to note that it expresses the reliability of the phenotype as a guide to *breeding value*—that is, the reliability of the phenotype of the parents as an indicator of what will be passed on and appear in succeeding generations. It should be emphasized that evolution works directly upon phenotype, not upon genotype or individual genes. For evolution to occur, therefore, the phenotype must fairly reliably reflect what the individual can produce by way of progeny, or there can be no natural selection, no differential production of viable offspring favoring the parents most adapted to their environment. (Neither can there be artificial selection. Imagine the task faced by, say, the cattle breeder if matings between animals with desirable characteristics continually produced offspring having a random assortment of traits.)

A final, and critical, point about heritability—one that is often not considered in discussions of the "inheritance" of behavior—is that heritability is not a property of a trait, but of a trait in a given population. Thus, one must always specify the population within which heritability is estimated, because the estimate holds only for that population. The reason for this is found in the definition of average effect. Recall that the average effect of an allele is the average deviation of individuals having the allele from the population mean, with the other allele at the locus taken *at random*. The "at random" is specified because we know that the other allele may affect the expression of the one in question, so we require that its effect be measured when in combination with a wide variety of others. Similarly, the average effects of a number of alleles may be influenced by alleles at different loci. What alleles are in the population, and their relative frequencies, in part determines the average effect of an allele or the average effects of a number of alleles in a polygenic system. Thus, the average effects of the same set of alleles may be different in one population from what they are in another simply because the alleles taken at random differ in frequency from one population to another. If average effect is dependent upon the genetic structure of the population, then of course additive effect, additive variance, and heritability must be. It is quite impossible to say, then, something of the order "visual acuity in humans has a heritability of X" unless one has sampled the entire world.

In the sense above, average effect, and hence heritability, do depend upon the combinations in which alleles occur, both intra- and interlocularly. By definition the average effect of an allele is dependent upon other alleles, and heritability is dependent upon the population as well as the trait. Decomposition of the genetic variance into additive, dominance, and epistatic components with reference to h^2 is an attempt to remove (statistically) as much of the nonadditivity as possible for a given population.

Environmental conditions also influence the estimate of heritability in a population. Recall that the different proportions of variance that compose the total variance must sum to 1. Therefore an increase in one component necessarily requires a decrease in one or more of the others. If environmental conditions are highly variable or unstable, the proportion of phenotypic variance due to environmental variance will increase, meaning that there must be a lowered estimate of the proportion of variance due to differences in genes and hence, very likely, in h^2. In contrast, if environmental conditions are uniform and stable, the estimate of the value of h^2 may be increased.

There is nothing biologically strange here. Again, it merely follows from the way in which total variance is decomposed. It does, however, offer another example of the variability of h^2. In fact, h^2 may be experimentally manipulated at the will of the investigator, provided he or she has control over certain conditions. The more genetically alike individuals are, for instance, the lower the value of h^2, because h^2 depends upon differences in phenotype due to differences in genes. An inbred strain of animals, within which the individuals are genetically identical, has an h^2 of zero. On the other hand, the estimate of h^2 may be raised in genetically unlike animals by reducing the environmental variance, as noted.

Heritability in addition depends upon population size. In a small population, where a number of alleles have been lost through sampling fluctuations, there will be fewer genes to contribute to additive genetic variance, and hence h^2 is likely to be lower than in a large population in the same environment. Finally, h^2 estimated in different ways often gives different values, partly due to sampling fluctuations and partly due to real differences among populations and/or the conditions under which they are studied.

Given all this, one clearly should interpret estimates of heritability with caution. Estimates of h^2 have their uses, of course, but particularly in psychology those uses are limited, both by the conditions upon which the estimates are based and by the ignorance that surrounds the understanding and measurement of behaviors. Heritability is thus neither a concept with the established value of, for example, gravity nor one of which an informed investigator need be ashamed. It is a concept that must be treated as any other scientific tool, to be judged by its productivity.

Heritability is expected to show some relationship with the kind of character measure. In general, characters most closely related to reproductive fitness should be lower in h^2 than those less fitness-related, because natural selection should have "used" the available additive variance of fitness traits during evolution. On an a priori basis at least, this seems to be true for morphological and biological traits. Falconer (1961, p. 167) presents data from several studies on a number of different species wherein the heritability of fitness characters is considerably less than that of traits presumably non-fitness-related.

One example is Friesian cattle. Heritability was .95 for amount of white spotting, .60 for percent butterfat in the milk, .30 for milk yield, and .01 for conception rate at first service. Conception rate seems an obvious fitness trait, as, in fact, does milk yield. Percent butterfat in the milk probably does not have a great deal to do with fitness, and evidently white spotting has almost nothing to do with it. On the basis of studies such as this, h^2 assumes a fair degree of empirical respectability. The same has not, however, been forthcoming in psychological data, one reason being that it is far more difficult, particularly in humans, to determine what behaviors do or do not relate to reproductive fitness.

Heritability may be estimated in a number of ways. The technique of choice obviously depends upon the experimental circumstances and/or the data available. Heritability estimates generally are derived from regression or intraclass correlation coefficients. This is, however, a subject that is not germane to our purposes. Treatments may be found in quantitative genetics texts; it is sufficient here to understand the principle of variance decomposition and the conceptual foundation of the most common measure of the "inheritance" of behaviors.

INBREEDING AND CROSSBREEDING

Selection is one means of altering the relative frequencies of alleles in a population. In the usual case, individuals with desirable characteristics are mated so that the genes that they possess will be passed on, and at the same time individuals with undesirable characteristics are not allowed to mate so that the genes that they possess will not be represented in the progeny generation. The obvious consequence of such a procedure is to increase the relative frequencies of the genes of the favored parents and to decrease those of the unfavored parents.

Another method of changing allelic frequency is to breed on the basis of relationship rather than upon the basis of phenotype. Related individuals will have more genes in common than will a group drawn randomly from the population; therefore, matings among such individuals will produce offspring wherein the probability is high that more genes are shared than among the parents. If such offspring are then mated inter se, they will produce progeny having even more genes in common. When this process is continued over a sufficient number of generations, a set of individuals is produced wherein the probability is very high that they share the same genes. Such a group is generally known as an inbred strain.

Inbreds have great commercial use, particularly in domestic plants, for often the best characteristics of two lines can be produced in the cross. However, our interest lies in inbred strains as a research tool, for they are uniquely valuable in this respect. The reason is not far to find: One of the difficulties in attempting to study how genotype and environment combine to produce behavior is that whereas initially we have a fair amount of control over environment and can manipulate it, in genetically random populations we have no control over genotype. Inbred strains allow us to control and to manipulate genotype.

"Control" here is used with reference to experimental design and means only that we can recognize certain genotypes and subject them to the environmental treatments we wish. The advantage that inbred strains offer is that within them the individuals are virtual genetic replicates of one another. Thus, genotype does not vary among individuals of a strain—in that sense it is controlled—and we may use the groups which experimental design and analysis normally call for to estimate the magnitude of effects and the probability that they hold for the populations from which the samples were drawn.

In the simplest design, one might subject one group of animals from strain A to an environmental treatment, and use another group of strain A, untreated, as a comparison. One would then learn what the effect of the treatment was for the genotype. In a slightly more complex design, one might use two strains, A and B, and thus have four groups, one treated and one comparison, in each strain. One could then determine whether the treatment affected the two genotypes in the same way. In this manner one would be manipulating genotype as an experimental variable.

Obviously, strains and treatments could be added almost ad infinitum, although such intricate designs usually generate data that prove unwieldy to interpret. The point, however, is that in no other way than by using genotypically replicate individuals can genotype be manipulated and controlled in designs intended to answer questions about the manner in which genotype and environment combine to produce behavior. And this, we have proposed, is the fundamental strategic issue in behavioral science.

Inbreeding (and the maintenance of inbred strains) thus becomes of interest to the modern behavioral scientist. The general topic of inbreeding is covered in detail in genetics texts and will not be elaborated here. If we are to comprehend the present and future attempts to integrate the biological and psychological approaches, however, we should understand at some level the derivation and uses of inbreds, and their advantages, as well as their disadvantages, in biobehavioral research. Table 3.1 gives the inbreeding coefficients (probabilities that identical genes occur within loci in the lines) over generations for different inbreeding systems. Note that, as you would expect, self-fertilization produces an inbred strain most quickly. Brother-sister

Table 3.1 PROBABILITY THAT ALL GENES ARE SHARED UNDER VARIOUS SYSTEMS OF INBREEDING

Generation (t)	A	B	C	D
0	0	0	0	0
1	.500	.250	.125	.250
2	.750	.375	.219	.375
3	.875	.500	.305	.438
4	.938	.594	.381	.469
5	.969	.672	.449	.484
6	.984	.734	.509	.492
7	.992	.785	.563	.496
8	.996	.826	.611	.498
9	.998	.859	.654	.499
10	.999	.886	.691	
11		.908	.725	
12		.926	.755	
13		.940	.782	
14		.951	.806	
15		.961	.827	
16		.968	.846	
17		.974	.863	
18		.979	.878	
19		.983	.891	
20		.986	.903	

SOURCE: Adapted from D. S. Falconer, *Introduction to Quantitative Genetics.* New York: Ronald, 1961. By permission.
Column A: Self-fertilization, or repeated backcrosses to highly inbred line.
Column B: Full brother × sister, or offspring × younger parent.
Column C: Half sib (females half-sisters).
Column D: Repeated backcrosses to random-bred individual.

mating, which is the usual system in mammals, gives an almost completely inbred set of individuals in 20 generations. (Most inbred strains of mammals used currently have been brother-sister mated for more than 100 generations.) Note also the effect of repeated backcrossing to random bred individuals (column D).

Inbreds, then, provide us with a means of manipulating and controlling genotype in ordinary research. Another desirable feature, which also derives from the replicate nature of inbreds, is that they offer us standard material. Any experimenter can send away to, say, the Jackson Memorial Laboratory in Bar Harbor, Maine, and obtain one or a number of strains that are genotypically identical to those used by other experimenters, including those at the Jackson Laboratory. Or one can establish one's own colony from the same stock. Whichever, the research community may be assured that with proper care it can follow the results obtained on certain genotypes and distinguish them from results obtained on others.

Specification of genotype is a necessary condition for modern animal research, but it is not one often satisfied in psychology. Typically, the departmental colony in a university consists of a small population, usually of rats of unknown genetic origin, whose breeding program is administered by an animal caretaker. Like any small population, it will have been subject to genetic drift and may not be representative of the species in general, even in the laboratory. It will also have become partially inbred, the animal caretakers over the years having maintained the colony through relatively few breeding pairs. Selection will have occurred for those things about which animal caretakers tend to be concerned: fecundity and docility.

The problem lies in the fact that there are many such colonies that have undergone the same process. In each, however, there will have been a different degree of inbreeding, and different alleles will have become homozygous because of random fluctuations. This results in genetically different animals in each colony, making it likely that various university departments have in effect sponsored different genotypes, many under the appellation "white rat," in their research. Little wonder then that the results of one psychological laboratory are often not reproducible in another. For one thing, they are not using the same animals.

Inbreds do, however, have several disadvantages, which are more or less important depending upon what is measured. First, they are unusual in that they are homozygous at every locus. (At least they are thought to be. There may be a few heterozygous loci in long-term inbreds, but if so they have not been detected except for recent mutations.) Behaviorally, in mammals at least, this appears not to make a great deal of difference. When healthy inbreds are compared with crossbreds there is nothing especially remarkable in the inbreds that is attributable to their homozygosity. With regard to other variables, however, the differences may be large and are well documented. Second, inbreds are subject to *inbreeding depression,* which is a reduction of the mean phenotypic value of characters connected to reproductive fitness. Inbreds tend to have fewer fertile matings, fewer and smaller offspring, and fewer offspring surviving to reproduce. The opposite of inbreeding depression occurs when inbreds of different genotypes are mated. Their progeny often show *heterosis* (the term is a contraction of heterozygosis), which is eividenced by a rebound in reproductive fitness.

Heterosis is in part accounted for by the fact that inbreeding tends to bring together in the homozygous condition deleterious recessive alleles that lessen the viability of the organism, whereas crossbreeding between strains restores the heterozygous state, and, most important, allows the dominant allele in the other strain, which is connected with greater viability, to act. Two inbred strains may thus be homozygous for dominants and deleterious recessives at different loci, as for example AAbb for one and aaBB for the other, and the cross will restore the more viable dominant at both loci—that is, AaBb.

Obviously these deleterious recessives cannot be too potent in their effects or the strain would never have survived. Indeed, most do not. The strains that do survive seem to have avoided homozygosity for too many or too powerful deleterious alleles.

The effects of inbreeding and of heterozygosity are still not completely understood. In physiological and anatomical traits, for instance, inbreds may be more variable than their crossbred progeny. Lerner (1954) has studied this problem extensively and concluded that such variability may be a simple function of the number of loci in the homozygous state. Waddington (1957) considers this not to be the case. Behavioral characters sometimes show the opposite results from physiological and anatomical ones, as in the "Tryon effect." Tryon (1940), in his familiar selection experiment, found that the variance of the F_1 cross between the "maze-bright" and "maze-dull" rats was as large as the F_2 (the product of interbreeding the F_1), which should have been larger due to genetic segregation. No convincing explanations have appeared to put the matter in order.

Inbreds are, nevertheless, living, reproducing organisms that in most cases show little in the way of behavioral aberrations, and they and their crosses can be used with great profit in behavioral research. Their advantages thus far outweigh their disadvantages. With just a normal amount of scientific caution, we may use them for what they are: a vast improvement over the unspecified material commonly found in the colonies of rats serving psychological laboratories.

GENES AND CHROMOSOMES

Geneticists often appear to treat genes as if they were the unit of inheritance when in fact they are not. This is merely a matter of terminological and analytical convenience. Genes, we well know, do not float around separately in the reproductive cells, to be passed on entirely independently of one another. Rather, they exist in definite structures, and it is these structures that are the units of inheritance.

The structures are of course *chromosomes*, and it was a major triumph of the union of cytology and genetics when the somewhat vague "elements" and "factors" of the early geneticists were located in physical forms that could be seen and described and that acted upon transmission in ways consonant with Medelian principles. In the process a new science was created, that of *cytogenetics*.

Mendel had established the particulate and duplicate (two alleles) nature of inheritance several years before chromosomes had been named or described in detail. Indeed, even the fact that a sperm fertilizes a single egg to produce (usually) a single individual had not been established beyond doubt when Mendel conducted his research. His "factors" were, as are so many things in science, inferred to exist because of their effects. But during the years while Mendel's work lay buried by authority, cytology advanced, and chromosomes were observed with great care and enthusiasm. The fact of their regular and precise duplication and distribution was appreciated, both with regard to the autosomal and reproductive cells. Further, their behavior during fertilization and development showed their unique importance to the cell.

It was not long after the rediscovery of Mendel's principles that the commonality of the behavior of genes and chromosomes in transmission from one generation to the next was recognized. The Mendelian pattern of inheritance requires that the genes be transmitted from cell to cell as the cells divide. How could transmission occur so that each daughter cell becomes genetically complete and not contain merely a part of the genetic structure of the parent cell? Wilhelm Roux first speculated upon this question. His solution involved the lining up of the transmitted structures in the cell nucleus, their duplication, and then the passage of exactly the material in the mother cell to the daughter.

Theodor Boveri later pursued these ideas. It seemed clear that in many respects the transmission of chromosomes paralleled that of the genes. It was not long before the two were formally put together. In 1902 an American graduate student, W. S. Sutton, recognized the parallelism. Boveri and Sutton became the cofounders of the chromosome theory of inheritance.

Mitosis

When a sperm fertilizes an egg a single cell begins a process that may result, as in the case of an adult human, in 10^{14} cells. Each of these cells contains the genetic complement of the original. Furthermore, new cells are being formed constantly, even in adults. Erthrocytes and epithelial cells (including those of the cornea of the eye) have a notably high rate of replacement. How does this occur?

Mitosis is the name given to the process whereby the normal genetic compliment of a cell is duplicated, and half of the material, containing exactly what was in the original cell, is passed on to another during cell division. The whole procedure is visible when appropriate staining techniques are used. Several stages are recognized.

· *Interphase.* The cell nucleus shows little definable chromosomal structure. The chromosomes are not found in the coiled appearance of the other phases. During interphase the chromosomes have duplicated.

· *Prophase.* As the nucleus prepares to divide, the reticular network of chromosomes resolves into distinct chromonemata, which are visible as double threads. As the threads shorten, the chromosomes become distinct. Each chromosome has duplicated, but the sister *chromatids* (duplicate chromosomes) remain connected by a common *kinetochore*. By late prophase the nuclear membrane has disappeared.

· *Metaphase.* The chromosomes move to the center of the cell and line up more or less linearly. Menwhile, two *centrioles* have formed at opposite sides of the cell, and *spindle fibers* appear which can be seen to be attached to the kinetochores of the chromosomes. The kinetochores duplicate.

· *Anaphase.* The sister chromatids are pulled apart by the spindle fibers to opposite sides of the cell. Each side then contains the same genetic material, either duplicated or "original."

· *Telophase.* The diffused character of the chromosomes reappears, two nuclear membranes are formed, the cytoplasm is split in half, and two outer membranes appear. Two cells now exist.

A simple outline of mitosis is shown in Figure 3.8. Mitosis (the name comes from the Greek *mitos*, for thread, after the appearance of the chromosomes at the early stage of cell division) is thus a simple and elegant way of ensuring that each daughter cell receives the exact genetic complement of the mother cell. All of the somatic cells, by this process, contain (with certain exceptions) exactly the same genetic

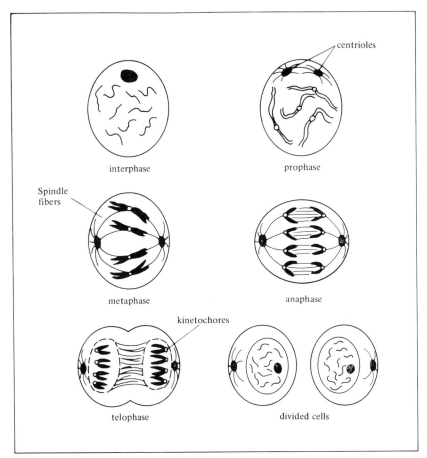

interphase

prophase

centrioles

Spindle
fibers

metaphase

anaphase

kinetochores

telophase

divided cells

Figure 3.8   Representation of mitosis.

material as the fertilized ovum. (That different cells form organs and function differently is another consideration.)

Meiosis

If the fertilized ovum is to contain a chromosome complement normal for the species, then the sperm and the unfertilized ovum must each contain one-half of that complement. To this end, the reproductive cells undergo a special kind of cell division known as meiosis (after the Greek for diminution) to form gametes.

Meiosis proceeds along much the same lines as mitosis until just before the chromosomes are pulled to opposite sides of the cell. That is, the pairs of homologous chromosomes, one of paternal and one of maternal origin, have duplicated, and lie side by side along the metaphase plate. The cell then has twice as many chromosomes as is usual.

At cell division, however, the replicated and "original" chromosomes do not separate from one another as in mitosis, allowing for duplicate genetic material in each of the product cells; rather, homologous chromosomes, each in the replicated condition, are pulled to opposite sides of the cell. Since each homologue came from either the father or the mother, the effect is to separate the chromosomes of paternal and maternal origin, *but,* and this is a very important consideration, in a fashion that is essentially *random.* The process is outlined in simplified form in Figure 3.9, in which two sets of replicated homologous chromosomes are represented, those stippled having come from the father and those open having come from the mother. At the first meiotic division the paternal and maternal homologues are thus separated.

The results of that separation, however, depend upon how the homologues have lined up at metaphase. This is, as noted, a random affair, with no systematic events governing the side upon which homologues from the mother or the father will be found. This random separation of maternal and paternal chromosomes leads of course to the consideration of the *probability* of a gamete containing a given number of chromosomes of paternal or maternal origin. We shall return to that later.

Having once divided such that replicated homologues are separated, there is in meiosis a second division wherein the sister chromatids separate, as in Figure 3.10. In this illustration we have shown two kinds of gametes produced, one containing chromosome A of maternal origin combined with chromosome B of paternal origin, and one containing chromosome A of paternal origin combined with chromosome B of maternal origin. Obviously this is not the only combination, even in a case where only two pairs of chromosomes are involved. Both paternal chromosomes could have lined up on the same side of the

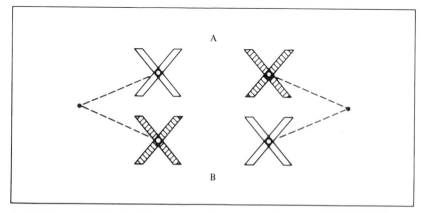

Figure 3.9   Representation of homologous chromosomes of paternal and maternal origin in the process of being drawn randomly to different gametes in the first meiotic division.

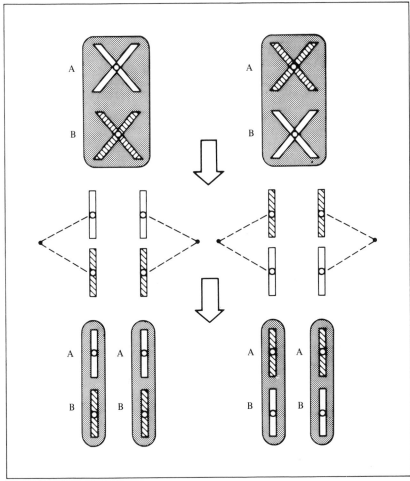

Figure 3.10   Representation of second meiotic division.

metaphase plate and have been drawn to the same side of the daughter cell, producing then gametes containing only the paternal complement. (The same would then necessarily be true for the maternal chromosomes under this condition.)

As the number of chromosomes increases, the number of potential combinations clearly increases as well, and the probability of any particular combination thereby decreases. In general, if there are n pairs of chromosomes, there will be 2^n possible chromosome combinations in the gametes, the combinations of course referring to whether the chromosomes are of paternal or maternal origin. In the case outlined, we would expect $2^2 = 4$ different kinds of gametes, which is what we obtained. In humans there are 23 chromosomes; therefore there are 2^{23}

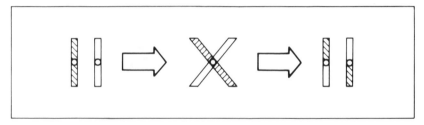

Figure 3.11   Representation of crossing over.

or 8,388,608 gametes having different combinations of chromosomes of maternal and paternal origin that can be produced by an individual.

The process of meiosis assumes its major significance in originating genetic variance. With more than eight million gametes that differ from each other on a chromosomal basis alone being potentially generated by each individual, people of unique combinations of genes will be conceived in each generation. Consider, for example, the probability that one set of human parents will produce two offspring having the same genotype (other than identical twins). The probability of any chromosome of a pair appearing in a gamete is $\frac{1}{2}$. Given 23 chromosome pairs, the probability of a gamete having a particular chromosome complement is $(\frac{1}{2})^{23}$. The probability of the union of two gametes having given chromosome complements is thus $(\frac{1}{2})^{(2)(23)}$, or something less than one in 70 trillion. This makes the union of a second pair of gametes identical to the first pair highly unlikely.

Of course, not all gametes unite with others to form zygotes. A woman produces only about 400 mature egg cells during her reproductive years; obviously, few of these eggs are fertilized, form zygotes, and are brought to term. A man is somewhat more prolific, generating approximately 1000 billion gametes during his lifetime, but again only a few find their way to the ova. The huge potential numbers are thus reduced in actuality by the biological necessity to sample from the gametes of each generation in order to form the next. But, unless the sampling is blatantly nonrandom, the genetic variance will remain large.

Another source of genetic variance derives from the fact that chromosomes do not entirely retain their integrity. It is possible, even likely, that homologous pairs of chromosomes will sometimes exchange portions, so that some of the genes which were on the paternal chromosome are now on the maternal chromosome, and vice versa. This process, called *crossing over*, is represented in Figure 3.11.

The importance of crossing over for us is not its cytogenetic description, but again as a means whereby genetic variance comes into being through the production of new combinations. Crossing over breaks up combinations of genes originally found on the paternal and maternal chromosomes, allowing then for new possibilities in the gametes. Obviously, the extent of the genetic differences so produced depends upon the degree of heterozygosity. If the maternal and paternal homologues carry exactly the same genes, then crossing over produces no new combinations. But if allelic pairs are carried in a heterozygous state by the parental homologues, then crossing over will be functionally meaningful. These situations are outlined in Figure 3.12. In the first case, crossing over has no functional significance with reference to new combinations, because both homologues contain the same genes (represented by portions A and B). In the second case new combinations result, and these may be passed on to the offspring.

Just how many loci exist on human chromosomes in unknown. Estimates range between 10,000 and 100,000. Similarly, just how many loci carry alleles in the heterozygous condition is unknown. However, according to Stern (1973), at least 10 percent of the loci may be heterozygous, and, using the lower figure for total number of loci, 1000 loci may be heterozygous in humans. On the average, then, each of the 23 chromosome pairs would be heterozygous for almost 40 loci.

In different cells of the same individual, crossing over could occur at any one of the 39 regions of a chromosome delimited by 40 heterozygous loci. Given only that single crossovers occurred for each pair, the gametes produced could contain 80 different combinations for each chromosome. Because crossing over in one chromosome pair is mainly independent of crossing over in others, the total number of combinations is the product of the possible combinations for each pair. The resultant figure is staggering: 80^{23}. The capacity of our species, and that of others, to put together different combinations of genetic material borders upon the infinite.

This is, again, not to say that all possible combinations are realized.

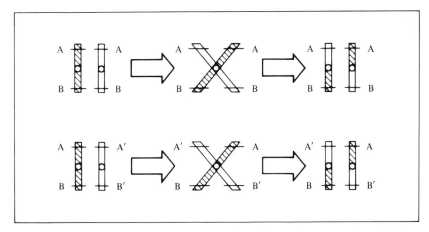

Figure 3.12   Representation of the results of crossing over when alleles are carried in the heterozygous condition.

Fortunately there are biological limits upon our profligacy. Nor is it to say that all the different combinations that do occur are biologically or behaviorally meaningful. Merely to *be* a species we must have most genes in common. Rather, it is important to note first that there is an evolutionary program that in the normal course of things guarantees genetic variance. The system of meiosis is the initial variance generator, and crossing over (as well as other chromosomal events) enhances its accomplishment in this regard. Genetic variance is in turn necessary for evolution. Without genetic differences that are expressed in the phenotype natural selection cannot operate, for the genes of the more adapted must differ from those of the less adapted for there to be selection, which alters allelic frequencies over generations. Meiosis, crossing over, and so on are then, loosely, evolution's way to further evolution—or, more aptly, the evolved means at the level of gene transmission to evolution at other levels.

Second, even though not all possible genetic differences are expressed, those that are expressed are usually enough. By "enough" I refer not only to the genetic variance necessary for evolution but to the fact that the differences among us attributable to differences in genes are sufficient to warrant not only our scientific but also our social and political attention. We have about all the differences with which we can cope, if indeed we can cope with them at all.

It is thus irrelevant to remind us, as some incessantly do, that the differences among men are small compared with their similarities, if one were to sample from a large variety of forms and systems. The differences that do exist, however small in the grand taxonomic scheme, are supremely important, so important now that the very existence of the species, not to mention its relative well-being, is on their account sometimes in doubt. On any given set of physical, biological, and behavioral variables, for instance, Adolf Hitler more closely resembled Mahatma Gandhi than he did a toad. Yet the distance beween the two men is overwhelming. It is therefore of little issue how large is some imagined ratio of differences to similarities. It is a matter of understanding the differences that are there, for they are entirely adequate not only to determine the ultimate destiny of humanity but possibly to determine the destiny of life as we know it.

Humans behave in a manner most lovely and most vile, most true and most foul, most sublime and most horrible. Within the species we do the very best and the very worst. However, it is rare that we do both as individuals. It seems that this has always been so, and the question is why. Although it is not suggested here—indeed it is contrary to my intent to suggest—that "goodness" or "badness" is controlled by genes, it is nevertheless the case that the existence of genetic variance and the relation of genes to behavioral propensities argue for the inevitability of behavioral differences. These differences will of necessity form a distribution, and individuals in certain portions of that distribution will have tendencies to act in ways that we recognize, and have generally recognized, as exemplary or as deplorable. We should not be surprised then at our own fascination with eternal recurrence, nor with our participation in it. The ancient metaphor of the struggle between virtue and abomination is reified in the struggle among the individuals of each generation. Eternal vigilance is the price of justice. This we know well, and eternally forget.

The Karyotype

Thus far we have seen that genes are on chromosomes, and that chromosomes occur in homologous pairs, with one allele of a pair influencing the same function located on each homologue. One-half of the homologous set derives from the male and one-half from the female parent. The chromosomal habitat of genes provides a physical basis for the principles espoused by Mendel with regard to the transmission of genes from one generation to another. It also explains many exceptions. The law of independent assortment, for instance, holds only when the genes are on different chromosomes. Otherwise the characters will be expressed together.

If chromosomes are the structures that hold the material responsible for the continuation of species, then we should expect those structures to be relatively stable, and, within a species, to be generally invariant with reference to their number, size, shape, and so forth. This is indeed the case. Each species may be characterized by the particular number and constitution of its chromosomes. The chromosomes of a species, arranged according to differences in their size and shape, is called a karyotype.

Most multicellular organisms are, as we have discussed, *diploid;* that is, they contain a chromosome complement consisting of two of each kind of chromosome (homologues), with the exception of the sex chromosomes in the male. *Haploidy*—half of the normal complement —is found in males of certain insects, such as bees, wasps, and other hymenopteran forms. But this is not a mammalian feaure and need not concern us. (Mammalian gametes are of course haploid but unite to form a diploid zygote.) Neither should we be concerned with another possible condition, polypolidy, which consists of some multiple of the normal diploid number of chromosomes. This is rare in animal nature, occurring chiefly in plants.

The normal diploid chromosome number in humans is 46. There are 22 pairs of *autosomal* chromosomes (those that do not determine, although they may influence the expression of, gender), and two sex chromosomes, the constitution being XX in the female and the XY in the male. Seven classes of autosomes are recognized, according to their length and the position of the kinetochores. A representation of the karyotype of a normal male is shown in Figure 3.13. The seven groups of chromosomes are designated as follows: A = 1–3, B = 4–5, C = 6–12, D = 13–15, E = 16–18 F = 19–20, G = 21–22.

Placing the chromosomes in a karyotype is an uncertain procedure, especially in groups C and G, for it can be questioned whether chro-

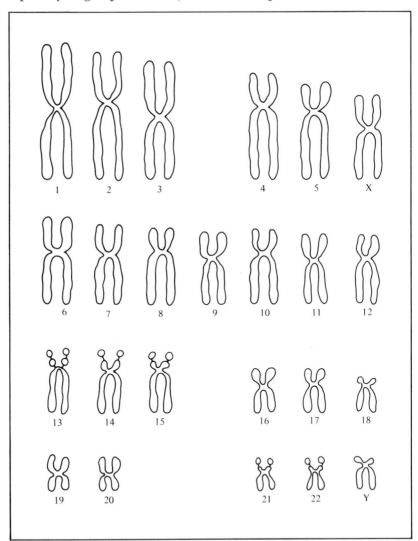

Figure 3.13   Representation of the human male karyotype. (Source: Adapted from C. Stern, *Principles of Human Genetics*, 3rd ed. San Francisco: Freeman, 1973. Copyright © 1973, W. H. Freeman and Company.)

mosomes that appear identical in overall length and location of kinetochore are in fact homologous. With a good preparation, however, the
chromosomes may be assigned to the seven groups reasonably well.

The two X chromosomes of the female are identical in size. But
interestingly, the Y chromosome of the male tends to vary, with close
similarity in the males of a family. Further, the variations may reflect
not merely differences in degree of contraction, but amount of chromosomal substance. The pedigree of males of a family having a short Y
is presented in Stern (1973, p. 861). Also, there are, according to Stern,
racial differences in the average length of the Y chromosome. In a
study of five populations, the length of the Y among the Japanese was
greater than that of Asian Indians, American blacks, Jews, and non-
Jewish whites. The Y of the non-Jewish whites was significantly shorter
than those of the others. The functional meaning of these differences,
either among individuals or among populations, is not known.

GENERAL INFORMATION AND EXAMPLES

More than 1800 genes have been identified in humans, and new ones
are regularly being discovered. Some merely distribute us into categories that have little evident functional meaning, as with the petal
color of Mendel's peas. Many others, however, are known only because
of their harmful effects, which may range from lethal, to the extent
that the embryo aborts, to mildly deleterious. There are few known
genes that affect behavior exclusively (although there a number
that have strong behavioral consequences); hence, a proper treatment
of human genetics is restricted to texts on that subject. However, we
may get an idea of the range of traits affected and the mode of inheritance of the genes influencing them by briefly examining a few cases,
which we shall do in what follows. Then we shall consider chromosomal aberrations.

Dominants

An example of simple autosomal dominance is seen in a trait of descriptive interest, though, happily, hardly of much importance. A
pedigree in a Norwegian kindred (Figure 3.14) presented by Stern
(1973) shows the inheritance of "woolly hair," a curled or frizzy type
of growth that breaks off at the end and never gets long. What appears to be the same condition has been observed in families of other
countries.

Nothing is known about the first incidence of the trait. However, the pedigree extends for five generations and shows the direct
transmissions from affected parent to half the children, which is expected for autosomal dominants (provided, of course, that the other

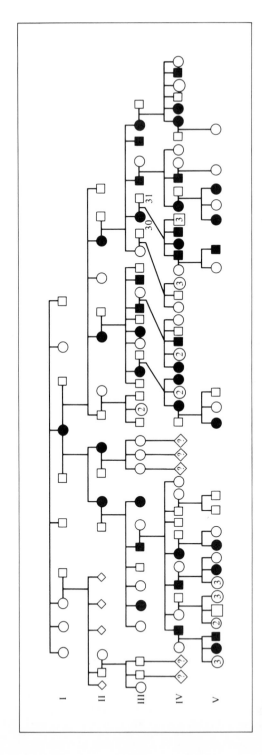

Figure 3.14 Pedigree of woolly hair. (Source: C. Stern, *Principles of Human Genetics*, 3rd ed. San Francisco: Freeman, 1973. Copyright © 1973, W. H. Freeman and Company.) After Mohr.

parent does not show the trait). Thus, when we examine the 20 marriages wherein one parent had woolly hair, we find 38 individuals with the trait and 43 without it (the condition of three long-deceased persons could not be ascertained). This is close enough to fall within the 1:1 ratio expected for dominant inheritance. Studies of other pedigrees support this conclusion.

Another case involving the hair is "white forelock." This is a streak of white, usually down the middle of the head. It shows the same mode of inheritance as woolliness. Monilethrix (beaded hair), and pili torti (short, twisted, fragile hair) do the same.

More serious instances are those of brachydactyly (short-fingeredness); dwarfism (achondroplastic type) in which the size of the trunk and head is similar to that of unaffected people, but the limbs are very short; porphyria, epiloia (which involves mental deficiency), congenital amputation of the arms and/or legs; ostogenesis imperfecta; polydactyly; and others.

Of particular interest to psychologists (and a genetic condition that bears directly upon behavior) is dyslexia. People who suffer from this defect appear clearly to perceive the individual letters of a word but cannot meaningfully combine them. Every child may experience such a difficulty early in school, but dyslectic children usually do not overcome it, in spite of an intelligence that is normal otherwise.

A more familiar dominant condition to those of you who are helpless immediately after entering a dark theater from the sunlight, is stationary night blindness. The retinal rods take an inordinately long time to become dark-adapted. This becomes more than merely annoying when the rods fail to adapt for several hours, as they do in some cases. Night blindness is not uncommon, as a number of noncombatants discovered during the blackouts of World War II.

Perhaps the most famous case of a family resemblance, the "Hapsburg lip," is also described to an autosomal dominant. The condition is actually due to a slight malformation of the lower jaw, which results in a protruding underlip and a mouth that gives the appearance of being permanently open. The trait can be traced to the fourteenth century in the Hapsburgs, being present in Emperor Charles V (1500–1558), Maria Thresea of Austria (1717–1780), and Alfonso XIII of Spain (1886–1941), among others.

Recessives

A recessive trait will be expressed only when the relevant alleles are in the homozygous state, in contrast with a dominant, which is expressed when the alleles are either homozygous or heterozygous. Pedigrees of recessive traits therefore show few afflicted individuals unless there have been a number of generations of matings among persons of close relationship. While brother-sister and parent-offspring matings are infrequent in humans, cousin marriages for some time were not, and these led to instances among the offspring of homozygosity for the deleterious recessive alleles carried by the parents in the heterozygous state. As mentioned in the section on inbreeding and crossbreeding, inbreeding increases the probability that recessives will become homozygous and considering the *genetic load* (deleterious mutations accumulated over a series of generations) in humans—is a dangerous practice.

Albinism is inherited as a recessive autosomal trait. In nonalbinos, numerous small granules containing the pigment melanin are deposited in the skin, hair, and iris, and give these color. Albinos are very nearly unable to transform the amino acid tyrosine into melanin, and hence have very light skins and hair. Their eyes sometimes appear pink because reflected light passes through the blood vessels. They may also experience eye abnormalities of different kinds.

Albinism is rare in Europeans and their relatives, occurring in one in 20,000. In other populations, however, it is relatively more common: one in 3000 was reported in Nigeria, and one in 132 among the San Blas Indians of Panama, for example. Others noted for a high incidence are the Hopi and Zuni Indians of the United States.

Sexual ateliosis (midgetism with normal sexual development) is also a recessive condition. Charles S. Stratton (better known by the name bestowed upon him by P. T. Barnum: General Tom Thumb) was such a midget, as was his bride, Lavinia Bump. Midgets, in contrast with dwarfs, are normally formed, but, of course, are small. Stratton was 3 feet 2 inches and Lavinia was 2 feet 8 inches tall. Stratton's parents were of normal stature, as were Lavinia's. Significantly, his parents were first cousins and Lavinia's parents were third cousins.

Midgetism without normal sexual development (asexual ateliosis) may also be due to a recessive gene. Several cases have been found among Hutterite communities of the plains of Canada and the United States. Among the Hutterites consanguinous marriages are common. The small stature of both kinds of ateliosis has been traced to a deficiency in growth hormone.

Total color blindness (achromatopsia) is also a recessive trait. Persons affected with this rare form see the world in shades of gray only. The cones of the retina that normally control color vision may be missing or defective. Therefore, only the rods, which are not color-sensitive, are functional. Since rod vision is superior to cone vision only when the light is of low intensity, affected individuals see better at night than during the day. Their condition is thus sometimes known as "day blindness."

Many other autosomal recessive traits are of course known, among them Tay-Sachs and Spielmeyer-Vogt diseases (amaurotic idiocy), diabetes mellitus, galactosemia (involving mental impairment), hypoglycemia (sometimes involving mental impairment), Niemann-Pick disease, Wilson's disease, and others. Perhaps the most widely known and best studied, however, is phenylketonuria. Figure 3.15 presents the pedigree of a group of related families in which phenylketonuria appeared. The individuals marked with a cross died young and were probably affected. The families lived on an isolated group of small islands in Norway.

We shall discuss phenylketonuria at length later. It has generated much research interest not only because it is a human problem but because the primary biochemical origin of the disorder was discovered fairly early. Phenylketonurics typically are severely mentally retarded, lightly pigmented, and have a musky odor. The latter, strange as it seems, was responsible for the syndrome first being described. Phenylketonuria accounts for approximately 1 percent of all institutionalized mental detectives. It is estimated that one in 50 of us carries the allele responsible.

Intermediate Inheritance

Probably the best known example of intermediate inheritance in humans is sickle-cell anemia. This will also be discussed in detail later. For the moment we need only note that the sickle-cell gene occurs in a number of populations, particularly those dwelling in warm lowlands, and most particularly in those dwelling in the part of sub-Sahara Africa stretching from the east coast to Gambia on the west coast.

Sickle-cell anemia is a disease of the hemoglobins, and is so named

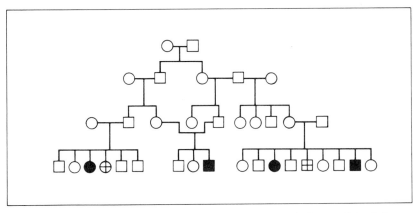

Figure 3.15   Pedigree of phenylketonuria. (Source: C. Stern, *Principles of Human Genetics*, 3rd ed. San Francicso: Freeman, 1973. Copyright © 1973, W. H. Freeman and Company.) After Folling, Mohr, and Rund.

because affected blood cells take on a sicklelike shape when exposed to low oxygen tension outside the body. In individuals homozygous for the sickle-cell allele, more than one-third of the red blood cells are abnormal. The anemia is often fatal before the individuals reach reproductive age. In persons heterozygous for the sickle-cell allele, less than 1 percent of the blood cells are abnormal. Persons so affected are said to have the sickle-cell trait. Identification of the heterozygote is of course requisite for intermediate inheritance. The sickle-cell trait offers some protection against malaria and may have developed in conjunction with certain agricultural practices.

Lethals and Sublethals

It is estimated that 25 to 40 percent or more of all human conceptions fail to develop normally and are aborted spontaneously during the early months of pregnancy. Many of the very early abortions are due to embryonic malformations caused by lethal genes. Some alleles act to reduce viability in single dosage but are lethal when homozygous. Others vary in their effect upon the phenotype, some individuals surviving and some not.

A most insidious condition is one in which the effect is not expressed until the individual has had time to reproduce and so pass on the allele. Most persons having a dominant lethal allele die early, but some who have the dominant for epiloia survive and do pass on the allele. Epiloia is associated with severe mental retardation, epilepsy, and tumors of the heart, kidneys, and other organs. Abnormal growth of the skin also occurs, producing the "butterfly rash" on the face by which the condition is sometimes recognized.

Another example of a dominant, but this time with a relatively late average age of onset, is Huntington's chorea, which is a progressive deterioration of the nervous system leading to mental impairment and, finally, death. The age of the individuals showing the first sign of the condition is usually between 40 and 45, although it may be first expressed as early as the teens or as late as the 60s. In some cases, obviously, the affected individuals have had ample time to reproduce before they learn of their affliction.

A number of recessive alleles, when homozygous, affect infants who are brought to term but whose condition is evident immediately or shortly thereafter. Ichthyosis congenita (leathery skin with deep, bleeding fissures), a special type of dwarfism (thanotophoric), and Werdnig-Hoffman's and Tay-Sach's diseases are some examples. Tay-Sach's disease results in mental deterioration and blindness, among other defects. It is of medical interest for one reason because the heterozygous carriers of the allele, who appear normal otherwise, may be detected by appropriate biochemical tests. It is therefore possible to determine the risk that a particular mating will produce affected progeny. The Tay-Sach's allele is much more frequent among Ashkenazi Jews than other populations.

Some alleles that are expressed as dominants in the heterozygous state act as recessive lethals when homozygous. An example is achondroplastic dwarfism. Two families are known wherein both parents were affected. Each family had one child with skeletal deformities similar to, but more severe than, those of the parents. Both children died shortly after birth. It is likely that they were homozygous for the allele that in the heterozygous state in their parents produced the dwarfism.

Multiallelism

In the cases discussed above we have seen examples involving only two alleles, a "normal" allele and one influencing the expression of some unusual condition. As suggested earlier, however, this circumstance by no means exhausts the genetic possibilities for even one locus. Many instances are known wherein more than two alleles (although only two at a time) can occur as a locus.

The most thoroughly examined case in humans involves the blood groups. If the red blood cells are removed from a sample of a person's blood and then reintroduced into the fluid part from which the clotting agent has been removed, the cells will distribute evenly throughout the fluid serum. If the red blood cells of one person are introduced into the serum of another, one of two things may happen: (1) The cells may distribute throughout the serum or (2) the cells may clump together. Whether or not the cells clump together depends upon the antigens and antibodies contained in the cells and the serum.

The red blood cells of an individual may possess one, the other, both, or neither of the two substances called antigens, A and B. An individual's blood serum may possess one, the other, both, or neither of two substances called antibodies, anti-A and anti-B. Red cells containing antigen A are agglutinated by serum antibody anti-A and cells containing antigen B are agglutinated by serum antibody anti-B. Persons having type-A cells do not of course have anti-A in the serum, nor do persons having B cells have anti-B in the serum. It is possible, however, to have cells with both A and B antigens. Such individuals have neither anti-A nor anti-B in their sera. It is also possible to have neither A nor B antigens. Such individuals have both anti-A and anti-B in their sera.

Four groups of persons are described by the above, named on the basis of the antigens of their red blood cells: A, B, AB, and O. AB is the case wherein both antigens are present, and O is the case wherein neither is present. Combinations of three alleles, A, B, and O, at a single locus account for the various types. Genotpye AA produces an individual with only the A antigen. Genotype BB produces only the B antigen. Genotype AB produces both antigens. Genotype OO produces neither antigen. Alleles A and B are codominant when present together, and both A and B are dominant over O.

Further immunological tests have uncovered much finer distinctions, but the simple version of the ABO system is adequate to convey the concept of multiple alleles. Many other blood groups, incidently, are now known. Many are controlled by multiple alleles—more than 40 in the Rh system alone. Some of the more recently discovered blood groups are P, Kell, Luthern, Duffy, Kidd, Lewis, Diego, Yt, Dombrock, Auberger, and Stoltzfus.

All this makes for a great deal of *polymorphism* in a population. With numerous blood groups controlled by numerous alleles, there will be many different forms resulting from various combinations. One example of diversity of form is reported by Stern (1973). Using the sera of 132 persons and examining only nine different blood groups, 129 of the 132 had distinct combinations. With further work on other systems it may be possible soon to identify an individual unambiguously from the unique combination of groups in his or her blood alone.

Polygenic Traits

As discussed previously, when the effects of genes cannot be followed separately and yet there appears to be a genetic involvement, a polygenic system of many genes, each having a small effect, is often implicated. Each "polygene" is postulated for purposes of analysis to carry a positive or negative "weight," that is, to move the phenotype a certain distance on a scale. In a distribution of height, for example, tall people are assumed to have a large number of polygenes which are "+" for height. Average individuals would have about an equal number of "+" and "−" genes, etc. This accounts for the generally normal distribution of polygenic characters; the further the distance from the mean the more rare the combination. Individuals with more "+" than "−" genes and those with more "−" than "+" genes would occur less frequently than individuals with more nearly equal combinations. A few (of many) traits that appear to meet these assumptions are height, weight, fingerprint patterns, length of life, and IQ.

It is of course possible for dominance and epistasis to occur in polygenic systems, just as it does in systems wherein the effects of individual genes may be followed. It is the purpose, as indicated earlier, of quantitative analysis resulting in variance decomposition to ascertain the proportions of variance due to the various effects of polygenes.

Perhaps one of the most interesting applications of polygenic theory, particularly with regard to the combined effects of genotype and environment, has to do with traits that do not appear to be continuous in the population, but rather distribute individuals into discrete categories. Usually a discrete distribution of phenotypes would suggest that a few individually identifiable genes were acting. However, it may be the case that a polygenic system underlies the character, and that the categories are created because individuals with a certain number of "+" or "−" genes cross a particular *threshold* for expression in the phenotype.

There are stages in the development of many traits where growth and differentiation may take one of several alternative paths. Either a limb bud forms the normal number of digits, for example, or too many or too few. The lateral part of the embryonic face may grow together and form a normal upper lip and palate, or the fusion may be incomplete and a cleft palate and a harelip will be formed. Behaviorally, we are of course very given to distinguish between abnormal and normal, and, although such classifications may sometimes be arbitrary, most would agree that there is a point on almost any behavioral distribution beyond which one set of actions may be said to be qualitatively different from another. This would be true, for instance, for severe cases of mental retardation and schizophrenia.

The concept of polygenic systems underlying discrete distributions of phenotypes was first developed by Sewell Wright in his study of polydactyly in guinea pigs. Using two lines, one having the normal number of toes on the hind feet (three), and another having an abnormal number (four), he made various crosses and concluded that four additive loci accounted for the distribution. Figure 3.16 illustrates his hypothesis. Four additive loci produce eight alleles, each of which may be considered to be "+" for polydactyly. In A, line I has zero of these alleles on the average, whereas line II averages the maximum number, eight. The cross between them produces a distribution with a mean number of "+" alleles of four, but some individuals in this distribution will have enough to cross the developmental threshold for polydactyly. This in fact occurred. In the F_1, nearly all individuals were three-toed, but a few had four toes. In the F_2 (B), we see again a continuous polygenic distribution with the threshold value at between five and six "polydactyly" alleles. This produces a discrete distribution at the phenotypic level.

Such *threshold characters* with underlying polygenic systems are of great importance in the study of genotype-environment interaction. In behavioral traits, which are often quite labile anyway, the environ-

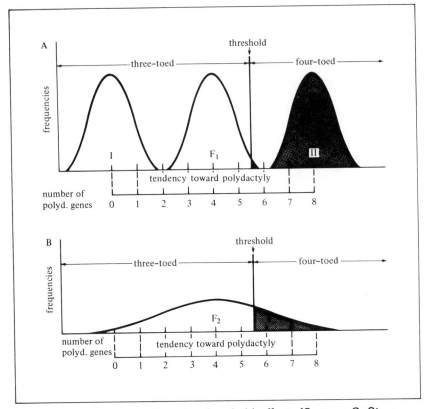

Figure 3.16   Representation of a threshold effect. (Source: C. Stern, *Principles of Human Genetics*, 3rd ed. San Francisco: Freeman, 1973. Copyright © 1973, W. H. Freeman and Company.) After Wright.

ment may act to determine the threshold that separates one category of behavior from another on the same dimension. This is in fact, as we shall see, the basis of what appears to be the most reasonable theory of the etiology of schizophrenia.

Chromosomal Abnormalities

It has been noted that between 25 and 40 percent of human zygotes are spontaneously aborted. Some of these of course result from poor intrauterine conditions. Others result from the presence of lethal genes. Between 20 and 50 percent of the abortions, however, are due to chromosomal abnormalities, often an abnormal number of chromosomes.

Approximately 15 percent of spontaneously aborted fetuses have 69 chromosomes, thus being *triploid*, and some have 92, being *tetraploid*. The origin of such *polyploid* embryos is not known with certainty. They may result from multiple fertilization, from abnormal fusion of cells, or from abnormal duplication of chromosomes within cells. Whatever the cause, polyploidy in mammals seems very rarely to be compatible with fetal survival. No polyploid human embryos have developed beyond a few days after birth, although several children who were mosaics of diploid and triploid cells have survived, with serious defects.

Another unfortunate situation occurs when an abnormal number of a single chromosome appears in the complement of an individual. This is usually due to *chromosomal nondisjunction*—that is, the failure of homologous or replicated chromosomes to go to different cells, either during gametogenesis in meiosis or during development of the zygote in mitosis. Obviously, the more developed the organism when the nondisjunction occurs the fewer aberrant cells it will have, since the aberrant chromosomal condition will be passed on to daughter cells.

Nondisjunction occurs earliest in the gametes that go to make up the zygote. If, during the meiotic divisions by which the haploid gamete is produced, homologous or replicated chromosomes should fail to separate and be pulled to one side, a condition will have occurred wherein one daughter cell has too many chromosomes and the other too few. The situation where replicated chromosomes (sister chromatids) fail to separate is illustrated in Figure 3.17.

Thus, as a result of nondisjunction of one of the pair of sister chromatids at the second meiotic division, two of the gametes, A and B, have aberrant chromosomic sets. (Nondisjunction could also occur at the first meiotic division, in which case the homologues would have failed to separate.) If either one of these gametes should fertilize, or be fertilized by, a normal gamete, a zygote with an abnormal chromosome number would result. This situation is shown in Figure 3.18.

In the first case, the resulting zygote is triploid for the relevant chromosome. In the second case it is monoploid. Autosomal monoploids are aborted spontaneously, as the triploids for most chromosomes. There are, however, three instances of autosomoal trisomy that are brought to term. Individuals having trisomy-13 (Patau's syndrome) or trisomy-18 (Edward's syndrome) die early. Those having trisomy-21 may not. They are severely mentally retarded, as well as having some physical stigmata, such as spotting of the iris, the "simian crease" of the palm, cardiac defects, and the fold in the eyelid that Langdon Down in 1866 mistook for evidence of a relationship to the "Mongol" race, leading him to describe the condition as mongoloid idiocy, or as it was known until recently, "mongolism." Since the term "mongolism" is neither accurate medically nor acceptable socially, the condition is now called Down's syndrome or trisomy-21.

Down's syndrome accounts for approximately 15 percent of the cases of mental deficiency requiring institutionalization in the United States. The overall incidence of births in Caucasoid populations is 0.15 percent, but many affected individuals do not survive. The life expectancy of those who do is about one decade, although survival to an

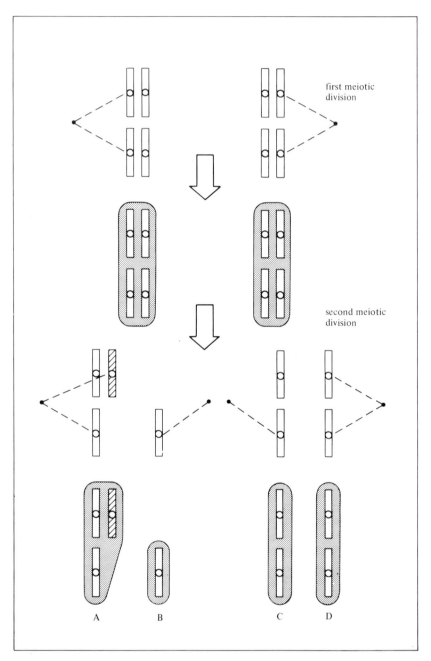

first meiotic
division

second meiotic
division

A   B   C   D

Figure 3.17   Representation of chromosomal nondisjunction.

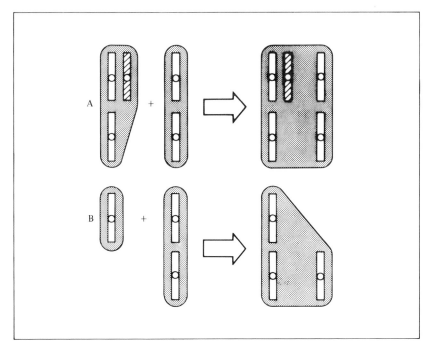

Figure 3.18   Representation of abnormal chromosome numbers in zygotes, resulting from nondisjunction.

advanced age has been noted. Cardiac defects account for much of the mortality.

The affected children have an average IQ of between 25 and 50, with an appreciable number in a higher category. They are cheerful and friendly, but although they may learn to feed and dress themselves, they remain for the most part dependent upon others. There are no specific neurological findings, although brain weight is below normal, the brain is smaller than usual, and the patterning of the sulci and gyri of the cerebral cortex is poorly developed. The entire picture is one of developmental retardation.

For many years after Down described the syndrome a relationship between the age of the mother and an increased risk of the condition was noted. This, it was assumed, was due to an unfavorable relationship between mother and fetus, with older mothers more likely to interact unfavorably with their fetuses. That the condition was not ascribable to the mother alone was seen from the data on twins. Monozygotic twins (two individuals developed from the fertilization of a single egg) were always *concordant* for the condition, both having it, where as dizygotic twins (two individuals developed from the fertilization of separate eggs) were *discordant*, there being one Down's and one normal. Since monozygotic twins have identical genotypes, it seemed clear that expression of the trait depended at least in part upon something in the zygote.

It was not until 1959 that the French researchers Lejeune, Turpin, and Gautier described the trisomy. (It was not until 1956 that Tijo and Levan in Sweden and Ford and Hamerton in England showed that humans have 46 chromosomes.) Thereafter, the mother-fetus interaction hypothesis was abandoned. However, the age of the mother at the time of the child's conception continues to be a most relevant variable. Among mothers aged 18, the relative frequency of affected children is approximately 1 in 2500, whereas among mothers aged 45 the relative frequency is more than 1 in 50. The risk in mothers of 45 is then 50 times what it is in mothers of 18.

In contrast, there appears to be very little relationship between the age of the father and the incidence of trisomy-21. It might be thought, since older women have older husbands, that the increase in affected children could be influenced by either the father or the mother. The British geneticist L. S. Penrose has demonstrated, however, that the incidence of affected children rises at the same rate with age of the mother even when the age of the father is held (mathematically) constant. On the other hand, there is no increase in number of affected children with increasing age of the father, when age of the mother is made constant. Figure 3.19 shows the relationship between mother's age and the relative frequency of trisomy-21.

Many cases of Down's syndrome are thus due to nondisjunction of the 21st chromosome in the gametes of an older mother. Why the relationship with age? Little is known. However, it is generally assumed that aging in the mother is correlated with nondisjunction in general (i.e., for all chromosomes), but that only certain trisomics survive. It is not then a matter of an increase in nondisjunction of a particular chromosome; rather, it is a matter of other trisomics spontaneously aborting.

Age, in turn, is assumed to affect oogenesis because the ovarian cells that become gametes exist in a woman in a "waiting" condition from before birth until just prior to ovulation. That is, a woman is born having all her primary oocytes, which through meiosis will become gametes, already developed. Each of these primary oocytes exists for decades in a kind of arrested stage of prophase, until, one by one with each menstrual cycle, they divide into an egg and the three nonviable polar bodies. The longer the oocyte has to wait before dividing, according to the hypothesis, the higher the probability of nondisjunction. In males, spermatogenesis is continuous from puberty on. There are no long periods when the germ sells are in an arrested state of development, and hence there is less opportunity for this sort of nondisjunction to occur.

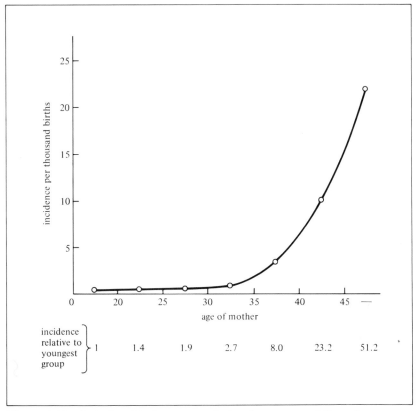

Figure 3.19   Relationship between relative frequency of Down's syndrome and age of mother. (Source: C. Sern, *Principles of Human Genetics,* 3rd ed. San Francisco: Freeman, 1973. Copyright © 1973, W. H. Freeman and Company.) After Collmann and Stoller.

The same trisomy could result from nondisjunction at the first cell division of the zygote rather than during gametogenesis. If a fertilized egg of the normal, disomic type suffered nondisjunction at the first cleavage division, two types of cells would be produced. One would be trisomic and the other monosomic. The monosomic cell (or its descendants) would not survive, leaving only the trisomic cell, which could form an embryo that would then develop into a Down's child.

It is not known what proportion of trisomy-21 children result from postfertilization, mitotic nondisjunction and what proportion develop from nondisjunction in oogenesis. Both are compatible with the age effect in the mother.

It is, of course, not possible to distinguish nondisjunction at oogenesis from that at the first cleavage in an affected individual. Evidence that postfertilization nondisjunction exists comes from *mosaics,* that is, individuals having cells with different numbers of chromosomes in them. If, for example, nondisjunction should occur in the blastomere of a zygote, which consists of more than two cells, then three types of cells would occur: (1) normal, disomic cells not involved in the nondisjunction, (2) trisomic cells, and (3) monosomic cells. The monosomic cells would be nonviable, leaving an individual with both disomic and trisomic cells. Mosaics of this nature have indeed been discovered. Proportions of the two types of cells vary, depending upon when during development the nondisjunction occurred. Significantly, the phenotypes of mosaics vary as well with regard to severity of symptoms.

Nondisjunction is, finally, not the only way in which Down's syndrome can occur. Chromosome parts occasionally undergo a *translocation* to a nonhomologue, and may be transmitted while attached to another chromosome. If a significant portion (or all) of the 21st chromosome becomes attached to a nonhomologue, then the translocation chromosome will carry the genes of both, and the zygote will, as in nondisjunction, have a triple set of 21st-chromosome genes. In this case, also, Down's syndrome becomes heritable. The translocation chromosome, carrying the extra genes, will segregate as it normally does, and whenever the other 21st chromosome appears with it in the same gamete and is fertilized by a normal gamete, trisomy-21 will result. Only very few men, again, are carriers of the translocated 21st chromosome. It is not understood why.

Down's syndrome results from too many autosomal genes. An instance, also uncovered by Lejeune and his associates, of a syndrome resulting from too few autosomal genes is that of the *cat's cry* (cri du chat). Affected individuals are physically and mentally quite abnormal, but the property of the syndrome for which it was named is a plaintive, continuous crying, particularly by the younger children, which resembles the cry of a cat. The difficulty results from the loss of about one-half of the short arm of chromosome 5. It is thus an instance of a *deletion*.

Deletions, translocations, and nondisjunctions are all chromosomal aberrations. Most result in spontaneous abortions, for which, it must be said, we can only be thankful. It is thus evident that the diploid autosomal complement is necessary for the development of a viable embryo, except in rare cases.

Aberrations of the sex chromosomes also occur. Two of the best known of these, Turner's and Klinefelter's syndromes, are due to nondisjunction. Normal females receive an X chromosome from both parents, and thus carry the sex complement XX. Normal males receive an X from their mother and a Y from their father, and are thus XY. Individuals with Turner's syndrome have only one sex chromosome, an X; their complement of sex chromosomes is described as XO. Individuals with Klinefelter's syndrome have an extra X chromosome, and are

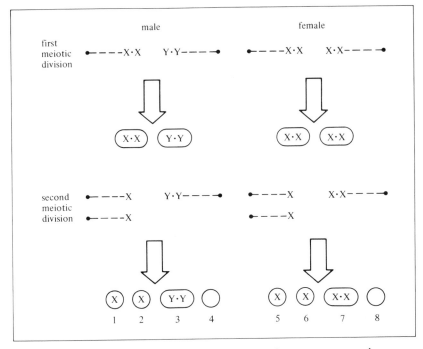

Figure 3.20   Representation of abnormal sex chromosome numbers resulting from nondisjunction.

therefore XXY. Some ways in which this could come about in gametogenesis are described in Figure 3.20. The autosomal complement is not represented.

Gametes 1, 2, 5, and 6 are normal. Should gamete 4 fertilize an X-bearing ovum, or should gamete 8 be fertilized by an X-bearing sperm, a case of Turner's syndrome would result. Should gamete 7 be fertilized by a Y-bearing sperm, a case of Klinefelter's syndrome would result. Should gamete 3 fertilize and X-bearing ovum, another syndrome, which has only recently been described and about which little is known with certainty, the XYY syndrome, would result.

What of the other possibilities? The OO combination is nonviable, as is the YO and the YYO. It seems that at least one X chromosome is necessary for survival. The XXX combination does survive, as indeed do some other, even more unlikely combinations.

At this point it should be recognized that in the example given the nondisjunction occurred at the second meiotic division. It may, however, occur in the first division, resulting in some chromosomal combinations within gametes not shown here. Thus, nondisjunction of the homologous, duplicated X chromosomes at the first meiotic division (as well as the second) could result in an egg bearing a sex complement of XXXX. Fertilization by an X-bearing sperm would produce an

XXXXX, whereas fertilization by a Y-bearing sperm would produce an XXXXY. Such individuals, as well as others such as XXXX, XXXY, XXYY, XXXYY, and XYYY exist, although they are rare.

As mentioned with regard to Down's syndrome, the effects of an increase or a decrease from the normal complement of chromosomes is obviously a matter of gene dosage. Too many or too few genes will very likely disrupt development to the extent that the organism will not live, particularly in the case of autosomal genes. We seem to be more tolerant of alterations in the number of sex chromosomes, perhaps because males and females are chromosomally dimorphic anyway, the total system having evolved to accommodate either XX or XY in the normal case. For this reason many more individuals with abnormal numbers of sex chromosomes survive, although they are not unaffected.

Individuals having one Y chromosome, no matter how many X chromosomes they have, are phenotypically male, although they are sterile if they have more than one X. Individuals having one or more X chromosomes (and no Y) are phenotypically female. XO (Turner) females are sterile. XXX females, however, are sexually normal, and usually fertile, although they tend to have menstrual irregularities, and may be subject to early menopause. In individuals with more than three X chromosomes the most striking clinical feature is mental retardation. It would seem that the developing human nervous system cannot accommodate too much imbalance, even of sex chromosomes.

Mosaicism is of course possible with sex chromosomes, as it is with the 21st autosome. Unusual types of fertilizations, fusions of initially separate embryos and loss of chromosomes, as well as nondisjunction, may produce individuals with two or more types of cells. There may thus be cells of a different choromosomal sex—that is, XX/XY, in the same person. These individuals are described as gynandromorphs, and as they develop, different parts of their bodies may show different sexual phenotypes. The occurrence of a breast on one side of the body of a young boy, has, for instance, been recorded (suspected genotype: XO/XY). The phenotype of the sexual mosaic in mammals is, however, complicated by the role of the sex hormones. This, also, will be discussed at length later.

One interesting proposal that has appeared fairly recently is that adult human females with the normal (XX) sexual complement may be functional mosaics since one of their X chromosomes may be inactive. Further, which one is inactive is determined randomly, giving a woman two different populations of cells. This hypothesis, proposed by Lyon and others, derives from several observations. First, XX females have a stainable body in the interphase somatic nuclei of their cells that normal males do not. Such "Barr bodies" (named after Barr, who in the late 1940s discovered them) have proved to be one of the X chromosomes. The "Barr chromosome" replicates later than its homologue, and instead of uncoiling during the interphase stage between cell divisions, it remains condensed. Since gene action appears to depend upon the chromosome being in an extended state, the condensed condition of the "Barr chromosome" would correspond to its being inactive.

Second, mice wherein the expression of coat color depends upon a gene on the X chromosome show a random pattern of coloration if they are heterozygous for the gene. This would suggest that in some cells the genes of the paternal X chromosomes are active, whereas in others it is the genes of the maternal X that are active.

Third, individuals having more X chromosomes than normal have a number of Barr bodies one less than the number of chromosomes. Thus, XXX females have two Barr bodies, XXXX females have three, and so on. This again suggests that only one X is active in the adult.

Fourth, women heterozygous (normal/deficient) for the gene on the X chromosome that controls the expression of an enzyme, glucose-6-phosphate dehydrogenase, have two populations of red blood cells, one with normal enzyme activity and one wherein activity is deficient.

Just when an X choromosome becomes inactive is not known. It is clear, however, that both X chromosomes in a normal female complement must be active at some time during development, or else the phenotypic anomalies associated with an abnormal number of X chromosomes would not occur. The XO genotype, for example, produces only rudimentary gonads and a primitive streak for a uterus. The other X must be required for normal development of female reproductive organs.

Similarly, the testes of the XXY male are tiny and spermatogenesis does not occur. The presence of the extra X must have an effect upon development of the male reproductive organs. The same sort of argument obtains for the mental retardation noted concomitant with the presence of an abnormal number of X chromosomes. Their effect upon the developing nervous system must occur before they become inactive.

Finally, it is appropriate here to discuss *sex linkage*. The term "sex-linked genes" usually refers to genes that are on the X chromosome. This nomenclature derives from the fact that historically the first identifiable genes governing nonsexual phenotypes and found to be on the sex chromosomes were on the X. The X is considerably larger than the Y, and carries genes related to traits other than those involving sexual development, which, with a few possible exceptions, appears not to be true of the Y. According to Stern, about the only candidate for Y linkage in man is the genetic determinant for long, stiff hairs on the rims of the ears.

In contrast, many X-linked genes are known—far more, in fact, than genes on autosomes. For one thing, such genes are more likely to be found, since only one X chromosome appears in the male, and the

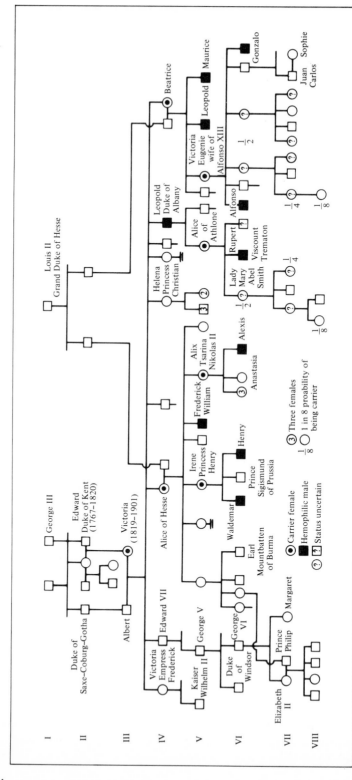

Figure 3.21 Pedigree of hemophilia among relatives of Queen Victoria. (Source: V. A. McKusick, *Human Genetics*. Englewood Cliffs, N.J.: Prentice-Hall, 1964. By permission.)

opportunity for masking is not there. The effects of genes on the X chromosome, thus, are more likely to be expressed in males.

More than 100 X-linked traits have in fact been discovered. Two of the best known are red-green color blindness and hemophilia. If a color-blind woman marries a normal man, the inheritance of the trait appears to follow a "crisscross" pattern such that all sons are color blind, like the mother, and all daughters are normal, like the father. If the daughters marry normal men, however, half of their sons will be color blind, and none of their daughters will be. These facts may be explained by X-linkage of a recessive genetic determinant for color-blindness.

A color-blind woman has two X chromosomes, each of which carries the defective allele; hence, the phenotype is evident in her. All of her sons would be color-blind because they received their X chromosomes from her, and the Y they received from the father did not carry the dominant, normal allele. All her daughters would be phenotypically normal, because they received one X chromosome from their father, and it carried the normal, dominant allele. One-half of her grandsons would be affected because her daughters carried one defective and one normal X, and gametogenesis is essentially random.

Figure 3.21 presents a pedigree of hemophilia among related members of royal families of Europe. A mutation evidently occurred upon one of the X chromosomes of Queen Victoria of England early in her embryonic life. She in turn passed on the defective chromosome. One of her sons, Leopold, died of hemophilia at the age of 31. At least two of her daughters, Alice and Beatrice, were carriers, as were her granddaughters Irene, Alix, Alice, and Victoria Eugenie. Three of Victoria's grandsons and six of her great-grandsons were affected.

One interesting genetic defect, of which little mention is usually made, is tone deafness. People who are tone-deaf often sing everything in a monotone or in the same off-key pattern. The genetic analysis of tone deafness is complicated by the fact that there is a strong environmental influence: In societies where musical competence is regarded as valuable, children are trained to overcome the handicap to some extent and can achieve reasonable proficiency on the piano. But, according to Kalmus, probably no such person has become involved successfully in a musical endeavor in which one has to find one's own notes, such as singing or playing the violin.

Charles Darwin was apparently tone-deaf. While at Cambridge he

. . . got into a musical set. . . . From associating with these men and hearing them play I acquired a strong taste for music, and used very often to time my walks so as to hear on weekdays the anthem in King's College Chapel. This gave me intense pleasure, so that my backbone would sometimes shiver. Nevertheless I am so utterly destitute of an ear, that I cannot perceive a discord, or keep time or hum a tune correctly; and it is a mystery how I could possibly have derived pleasure from music. (Darwin, 1882, p. 20)

Further, his musical friends

. . . soon perceived my state, and sometimes amused themselves by making me pass an examination, which consisted in ascertaining how many tunes I could recognize, when they were played rather more quickly or slowly than usual. "God Save the King," when thus played, was a sore puzzle. There was another man with almost as bad an ear as I had, and strange to say he played a little on the flute. Once I had the triumph of beating him in one of our musical examinations. (1882, p. 21)

Happily, this triumph was not his last.

CONCLUSIONS

It seems clear that genetic architecture has evolved in such a way as to be both conservative and open to change. It is conservative in that certain features of the system, developed early in the history of life, have been extended throughout all phyla. The gene and the chromosome are two of these. The diploid chromosomal condition is also common phylogenetically, and, as we have seen, is intolerant of much aberration. Meiosis is a precise and likewise a common means of generating the cells that will form new generations. When meiosis malfunctions, however, the results are usually incompatible with survival.

On the other hand, the system is one that is open to change at another level. Genes mutate, chromosomes exchange parts, and meiosis produces a huge number of different genic combinations, all of which results in genetic variance. At this level the system is in fact "designed" for change, for without it there would be no evolution.

Evolution is itself then hardly a haphazard affair, in spite of the occurrence of the word "chance" so often in conjunction with it. Evolution uses the conservative elements of the genetic system as the stable blocks with which to generate the new combinations that result in change. It requires stability at one level, in that there *be* genes, and variability at another, in that there be *differences* among individuals in the genes they carry. This is sometimes difficult to comprehend, but only when the dual tendencies underlying evolution are understood can the operation of evolution itself be understood. It is evolution, ultimately, that has put us where we are. It therefore continues to deserve our attention.

Chapter 4
Tools for Understanding:
The Biochemical Gene
and Its Products

THE STRUCTURE OF GENETIC MATERIAL

The "differentiating characters" of Mendel, the "factors" of the early Mendelians, the "genes" of Johannsen, and the "classical" gene of Muller were given chemical reality by Watson and Crick in 1953. Their accomplishment not only solved an ancient puzzle but opened research activities to new questions, the answers to many of which came quickly. More, it seems certain, are on the way.

The hereditary substance in the great majority of organisms (the exceptions being plant viruses and some animal viruses) is deoxyribonucleic acid (DNA). Watson and Crick did not discover DNA. Actually it was discovered in 1869 by Friedrich Meischer in experiments on the white blood cells in pus. Because it occurred in the cell nucleus, he called it "nuclein." "Nucleic acid" was a term applied later by Richard Altmann, and it is the term used presently in reference to the class of substance involving the hereditary mechanism.

Nucleic acid was not recognized as the biochemical entity of the hypothetical gene until the middle of the twentieth century. This may seem strange, because by that time fertilization had been described at the cellular level, mitosis and meiosis had been photographically confirmed, and even genes had been located on chromosomes in the nucleus. One reason is that nucleic acid is not the only type of molecule found in the nucleus, nor indeed in the chromosome. Chromosomes consist of nucleic acid together with a class of proteins, the histones. A second reason is that nucleic acid seemed too simple to direct the complex biochemical machinery of life. That is, although its exact structure was not known, DNA had been shown to consist of only a few simple constituent molecules that evidently occurred again and again, hardly the sort of things one would look for given the enormously complicated task undertaken by genes. Some workers in fact believed that the proteins associated with chromosomes were the only substances of sufficient complexity themselves to direct such an intricate series of events.

Watson and Crick's discovery of the exact structure of deoxyribonucleic acid strongly implicated DNA as the hereditary material. The molecule as they imagined it would have the chemical potential necessary both for continuation of a species and for development of the individual—namely, the capacity for self-replication and the capacity to direct the synthesis of other material. How this transpires, some examples of misdirection, and some conclusions to be drawn therefrom are the subjects of this chapter.

DNA consists of a double helix, each strand having a backbone of alternating units of a phosphate group and the sugar deoxyribose. To each molecule of sugar is attached one of four bases: adenine, guanine, cytosine, or thymine. The bases of the two helical strands project inward, so that the bases of one strand come in close contact with those of the other. The arrangement of the bases along each of the strands is such, however, that an adenine on one strand is always opposite a thymine on the other, and cytosine on one is likewise always opposite a guanine. Weak bonds are formed between the paired bases. Figure 4.1 shows the linear arrangement of the phosphate groups, the sugars, and the bases of one strand of DNA.

Figure 4.2 shows the double helical structure of a portion of DNA molecule. The double helix, with the bases turned inward and bonded to one another, offers the characteristic stability obviously necessary in genetic material. On the other hand, the bonds that bind the bases together are not strong. They can be broken when it is necessary for the molecule to replicate. This is in fact what occurs; the molecule "unzips," and a new strand, having bases complementary to those on the old, is formed on each of the original strands. The importance of the fact that a base of one kind will bind only with a base of another kind (adenine-thymine, cytosine-guanine) is now evident. It ensures the replication of the molecule.

The linearity of the bases on the strands aids in the self-replication

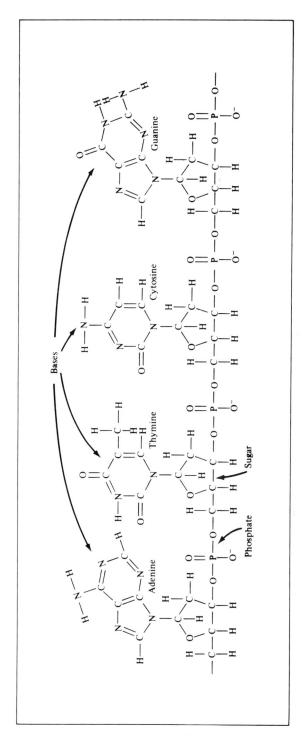

Figure 4.1 Arrangement of chemical groups on a portion of one DNA strand.

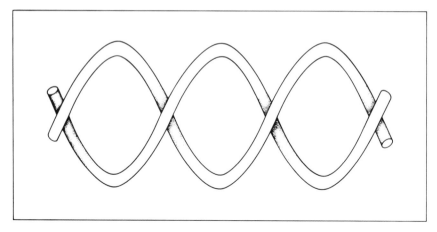

Figure 4.2   Representation of the double helical structure of DNA.

of DNA, and is essential for its other synthesizing activities. The molecule must contain, in some way, the instructions according to which the organism will be constructed and maintained. The bases in fact form a code, one which is read linearly, as are the words of an English sentence.

THE REGULATION OF GENE ACTIVITY

What, then, is a gene? It is some portion of a DNA molecule, clearly, but what portion? This question remains unanswerable, as such. A gene is still defined in terms of its effects, and its effects depend, among other things, upon its product. Since all gene products are not exactly the same, all genes cannot be expected to be exactly the same. One thus cannot say that a gene consists of a portion of a DNA molecule of a given length. The length of DNA involved and therefore the "size" of a gene depends upon the size (and hence complexity) of its product.

Furthermore, the function of all genes is not the same. In bacteria (used because of their relative simplicity), at least two classes have been discovered: *structural* genes and *regulator* genes. Structural genes are those that direct the synthesis of the substances necessary for growth and differentiation of the organism. Regulator genes direct the activity of the structural genes.

There are in addition a number of other types of genes that must be present to govern the timing and sequence of events within the hereditary mechanism itself. These are sometimes called architectural and temporal genes. Less is known about them, however, than about the structural and regulatory genes.

A structural gene is not under the immediate control of a regulator gene, but rather responds to an *operator*, which is located near it. The operator activates the structural gene, but its activity is in turn inhibited by the product of the regular gene. The operator and its structural gene has been called an *operon* by Jacob and Monod (1961), who described the system.

An example of this system is seen when the medium in which bacteria are growing is changed. *Escherichia coli* (the common colon bacillus) requires a carbohydrate source. When a colony is grown in a medium containing glucose, it does not produce the enzyme beta-galactosidase. However, when the medium is changed to one containing the sugar lactose rather than glucose, the bacteria, within minutes, begin to produce beta-galactosidase, which allows them to digest galactose.

How is this possible? How is the genetic structure of the organism induced by the change in medium to produce something it had hitherto failed to produce? According to the Jacob and Monod model, the regulator gene directs the synthesis of a repressor which inhibits the operator controlling the structural gene that produces the enzyme. The effect of the inducer (the substrate, lactose) is to inactivate the repressor, the effect in turn being to disinhibit the operator, which then activates the structural gene. Figure 4.3 diagrams these events. When the bacteria are transferred back to a glucose medium, the production of beta-galactosidase ceases. Without the substrate there is no need for it.

This may at first perusal seem logically to be a somewhat awkward arrangement. One is, however, often presumptuous when applying human logic too quickly to nature. Actually the basic relationship serves the organism well in a changing environment. It would not be optimal for all structural genes to be producing continuously if their products are not needed. It might in fact generate biochemical con-

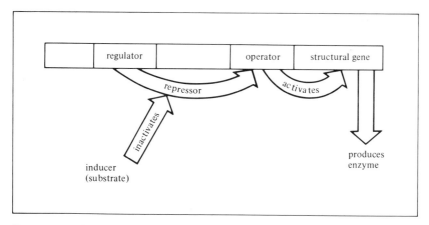

Figure 4.3   Representation of the possible relationship between substrates and regulator, operator, and structural genes.

flicts that could be damaging. Having a portion of the genome inhibited unless the proper inducers are available prevents this. It also goes some way toward explaining how the same genome produces different substances in different organs and different cells.

A most important point about such a relationship is that it shows the means whereby the genome and the environment may, almost literally, be in touch. It has been thought for some time that there must be a way for changes in the environment to be communicated to an organism's genetic apparatus so that it might react appropriately. Although it may not be exactly the same in mammalian cells as in bacteria, the work of Jacob and Monod nevertheless demonstrates the possibility. The genome should not be conceptualized then as a programmed device working in complete isolation from its surroundings, but rather it should be seen as a system that has evolved to receive input from the environment and to respond to that input. This view has great use, particularly when one considers some of the potential effects of environmental manipulations made early in life.

TRANSCRIPTION AND TRANSLATION

The primary gene products are proteins, used in development and life-sustaining processes. However, proteins are not produced directly by genes; rather, there are several intermediate steps, involving three kinds of another nucleic acid. Ribonucleic acid (RNA) very much resembles DNA, the differences being as follows: (1) RNA usually occurs as a single strand; (2) in the backbone of the strand another sugar, ribose, occurs in the place of deoxyribose; and (3) the base uracil occurs in the place of the base thymine attached to the backbone.

There are three types of RNA, and each plays a unique role in assembling the proteins specified by DNA. *Messenger RNA* is formed on one strand of the DNA as a chain of bases complementary to the sequence of the DNA bases. When the DNA "message" is thus *transcribed* onto messenger RNA, the messenger RNA moves into the cytoplasm and attaches to bodies located in the endoplasmic reticulum made up of *ribosomal RNA*. Within ribosomal RNA the message is *translated*.

Ribosomal RNA acts as the factory for the assembly of the units of which proteins are constituted, amino acids. The ribosomes move along the messenger RNA, exposing attachment sites for a complex of another type of RNA, *transfer RNA*, and an amino acid. There are approximately 20 amino acids, and for each amino acid there is at least one transfer RNA capable of "recognizing" it.

The amino acids are released by their transfer RNA at the ribosomal site and are bonded linearly to one another to form a chain according to the message specified by DNA. When the chain is complete, the protein is released. This process is outlined in Figure 4.4. Only one side of the DNA molecule is read.

A particular strand of messenger RNA, as one would suspect, does not have a very long life in the cell (in bacteria it decays in approximately 5 minutes). This is necessary because older messages cannot stay and possibly become confused with newer ones. Ribosomal and transfer RNAs, being constantly reused, last longer.

The amino acids which enter into the composition of proteins possess at least one amino and one carboxyl group, both attached to the same carbon atom. Their general structure is

where R represents various other groups. From the point of view of protein structure, however, the most important feature of amino acids is that they react with one another, the carboxyl group of one uniting with the amino group of another in a series of peptide bonds to form long, more or less coiled or spiraled bands, called polypeptides. A short polypeptide, with R again standing for other parts of amino acid molecules, appears schematically in Figure 4.5. A long polypeptide

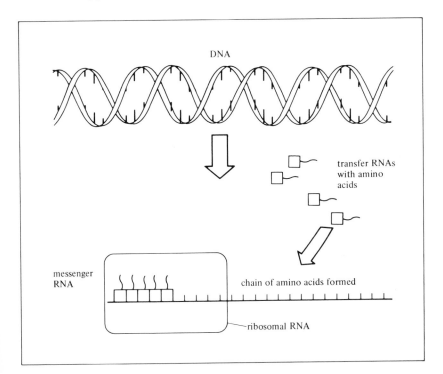

Figure 4.4   Representation of genetic translation.

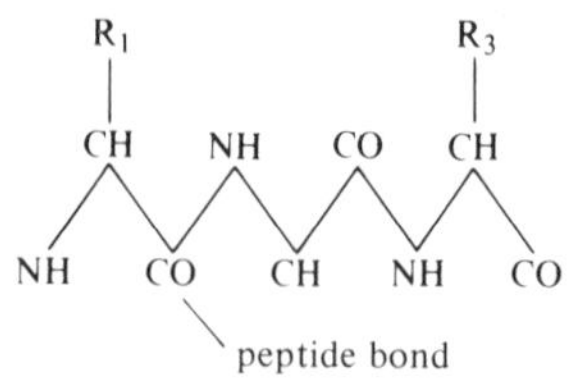

Figure 4.5   General representation of a polypeptide.

chain is a protein. (Just how long is an arbitrary matter. Many proteins contain hundreds of amino acids.)

Proteins are of course necessary for the formation of cell membranes, function in nervous, muscular, connective tissue, and so on, and are part of the chromosome itself. Unlike fat and carbohydrates, however, proteins cannot be stored in the body. A constant supply of protein (which when ingested may be metabolized to its constituent amino acids and then reconstituted in the desired form) is thus required. Should it not be available, the animal will metabolize the protein in its own tissues.

Of the 20 or so amino acids needed for the manufacture of animal protein, only about half can be synthesized by animals from other (e.g., carbohydrate) sources or are synthesized from such sources too slowly for normal development. Animals must rely ultimately upon the metabolic machinery of plants to obtain the protein they need. Plants can synthesize all of the amino acids from inorganic compounds.

Proteins are not merely long, loose threads of amino acids, but are often elaborately folded, with bonds between various parts of the chain. The nature of the folded structure that a protein assumes is, however, determined in large measure by the amino acid sequence. As a result of the folding, irregular, sometimes globular structures are formed, in which not all parts of the chain are presented equally to the surround. The result is that certain portions of the chain are more likely to react chemically to the substances in the surround than are others.

This property has great significance for the nature of the class or proteins with which we shall mainly be concerned: *enzymes*. Enzymes contain certain "active sites," which are in part determined by their folded structure and are the locations where the enzymes, in effect, do their jobs. It is important to understand this, for a mutation that affects an enzyme at an active site may render it ineffective, whereas a mutation that occurs in the gene specifying the same enzyme but affects it at another location may leave its activity unimpaired. There is a certain tolerance, therefore, depending upon which part of the enzyme is affected.

Life as we know it is dependent upon enzymes. Enzymes are biological catalysts; that is, they are substances that increase the rate at which a chemical system operates without themselves undergoing an ultimate chemical change. Catalysts are necessary in living systems because the reactions upon which the life processes depend would not otherwise proceed at a viable rate, given the temperatures and pressures that ordinarily prevail.

The function of a biological catalyst is to bring the reactant molecules together. (This of course can be accomplished in nonliving matter by increasing temperature and/or pressure.) The catalyst usually does this by forming a complex with the substrate, adding atoms to or removing them from a molecule, or rearranging it in some way.

"Enzyme" means literally "in yeast" (en-zyme) and was so named by two German chemists, the Buchner brothers, who in the late 1800s ground up live yeast with sand and prepared a juice, intending it for medicinal purposes. The juice, however, quickly went bad, and in attempting to deal with the problem of preserving it the Buchners tried the kitchen chemistry method of adding large amounts of cane sugar. But the yeast juice, although it contained no intact cells, vigorously set about fermenting the sugar into alcohol. Thus, for the first time, an active substance had been artificially separated from the cells that produced it. The Buchners and their yeast extract therefore laid the foundations for the whole of modern biochemistry. Although they thought ' of a single substance, en-zyme, we now know that there are more than 14 separate enzymes in such extracts, each responsible for one step in the conversion of sugar to alcohol.

It is characteristic of enzymes that they are highly specific; that is, their activity is confined to one reaction or a small group of similar chemical reactions. Enzymes therefore typically act at only one step of a metabolic chain. For each reaction of the thousands required for the growth and maintenance of an organism, a different enzyme is needed. The information that enables the organism to produce this enormous number of complex molecules is of course contained in DNA. The three-dimensional structure and the means by which its task is performed of at least one enzyme, lysozyme, which breaks open bacterial cells, is now known. Many others are, naturally, being studied.

THE GENETIC CODE

The linear arrangement of the bases along the DNA molecule provides the message to messenger RNA, which specifies the sequence of amino acids and hence determines protein synthesis. With only four different bases with which to work, however, it might at first seem a mysterious assignment. There are 20 amino acids, each requiring a specific designation. How can four bases spell out 20 amino acids? The answer, as with any alphabet, lies in combinations. Four bases taken one at a time could specify only four amino acids; four bases taken two at a time could specify only 16, not enough. But taken in combination of three, the four bases have a potential vocabulary of 64 words. It appears that in fact the genetic code consists of words three bases long. Table 4.1. shows the DNA triplets and the amino acids for which they code. Several triplets, as indicated, appear to code for the initiation or termination of a sequence.

Four things are immediately evident from inspection of the table.

Table 4.1 THE GENETIC CODE, SHOWING DNA BASES
AND AMINO ACIDS THEY SPECIFY

AAA phenylalanine	AGA serine	ATA tyrosine
AAG	AGG	ATG
	AGT	
AAT leucine	AGC	ATT "stop"
AAC	TCA	ATC
GAA	TCG	
GAG		
GAT	ACA cysteine	GCA arginine
GAC	ACG	GCG
		GCT
GTT glutamine	ACT "stop"	GCC
GTC		
	ACC tryptophan	TGA threonine
GGA proline		TGG
GGG	GTA histidine	TGT
GGT	GTG	TGC
GGC		
	TAC methionine	CAA valine
TAA isoleucine		CAG
TAG	TTT lysine	CAT
TAT	TTC	CAC
TTA asparagine	CGA alanine	
TTG	CGG	
	CGT	
CTA aspartic acid	CGC	
CTG		
	CCA glycine	
CTT glutamic acid	CCG	
CTC	CCT	
	CCC	

SOURCE: From *The Genetics of Human Population* by L. L. Cavalli-Sforza and W. F. Bodmer. W. H. Freeman and Company. Copyright © 1971.

A = adenine, G = guanine, T = thymine, C = cystosine.

First, in most cases more than one triplet codes for a given amino acid. In this respect, the code is said to be degenerate. Second, the triplets coding for the same amino acid are usually similar, particularly in the first two letters. Third, other than termination indicators, the code is unpunctuated. Fourth, there can be no room for overlap in the code. It must be read, triplet after triplet, along the DNA molecule. Any overlap would change the meaning.

The code is, so far as is known, universal; that is, the same triplets specify the same amino acids in all organisms. If the messenger RNA of one species is incubated with a preparation of ribosomes, transfer RNAs, and amino acids from another, the messenger RNA directs the synthesis of the protein it ordinarily would in its home cell. The messenger, ribosomal, and transfer RNAs and the amino acids all react in the same way to one another, regardless of species. What differs among species is of course the sequence of bases in DNA.

MUTATION

Mutational events may now be understood more clearly. If DNA specifies proteins by the sequence of bases in a triplet code, then any change in the sequence is a change in the code, potentially a change in protein structure, and potentially a change in protein efficacy. The effect of a change in DNA base sequence is potential because not all such alterations need have a meaningful effect. The DNA sequence specifying ACA in messenger RNA, for instance, could be altered to that specifying ACG, ACC, or ACU, and the amino acid threonine would still be called for. Similarly, CUA, CUG, CUC, and CUU all produce leucine. Other cases are obvious from Table 4.1. Such instances, of course, occur because of the degeneracy of the code, and the result is that not all base sequence changes necessarily lead to alterations in the sequence of amino acids.

Also, as mentioned previously with reference to enzymes, not every change in an amino acid sequence need have a significant functional effect. A large protein molecule such as an enzyme may be relatively unaffected by an amino acid substitution that does not alter the configuration of an active site. Thus it is possible to have a number of the same enzyme forms that occur through mutation for or against which there is no selective advantage.

On the other hand, a mutation affecting an amino acid at a more "sensitive" place may alter the functional capacity of a protein considerably. The folded structure of a protein, we have seen, depends largely upon the amino acid sequence, and that structure is in turn essential to its activity. Amino acid substitutions affecting such structure could produce an inefficacious molecule or one with less capability.

Finally, changes in base sequence could produce "nonsense words" that do not code for any amino acid. In such cases the polypeptide would likely be nonfunctional, and functionally the protein would not be present. Changes in base sequence have been found by Benzer (1962), who mapped completely a certain region of the T_4 virus. Mutational sites were found at the level of a single base pair.

One example of the substitution of a single amino acid in a protein molecule that has strong clinical effects is found in hemoglobins. Four different genes code for four different polypeptide chains: alpha, beta, gamma, and delta. Each of these forms a dimer (pair) with its own kind, and the dimers, in turn, form tetramers with pairs of another kind. The hemoglobin molecule consists of such tetramers and a heme group, which contains the iron that binds oxygen to be delivered to the cells.

The sickle-cell character, as noted in Chapter 3, is controlled by a pair of alleles. Normal individuals are homozygous for the allele Hb^A. Heterozygotes possess one sickle-cell allele and one normal ($Hb^A Hb^S$). In sickle-cell anemics both alleles are sickle cell, Hb^S. Approximately one-third of the red blood cells of sickle-cell anemics take on the sickle condition under low oxygen tension. Many homozygotes fail to reach reproductive age. Heterozygotes are at a slight disadvantage relative to normals under usual environmental circumstances.

Ingram (1958), using eletrophoretic and chromatographic techniques by which amino acids may be separated on the basis of the differential electric charges they carry, demonstrated that the sickle-cell trait is due to the substitution of a single amino acid, valine, for another, glutamic acid, at a particular site in the beta polypeptide. This change from one amino acid to another (out of 146 in the beta chain) is sufficient to alter the configuration of the entire hemoglobin molecule enough to produce the sickle-cell trait. In homozygotes both beta chains are affected, and sickle-cell anemia is the result.

At present more than 60 other abnormalities in hemoglobin beta chains have been discovered, as well as many abnormalities in alpha chains. Each is dependent upon a particular allele, and each is, in homozygotes for the same allele or heterozygotes for different abnormal alleles, responsible for anemias of varying degrees of severity.

PROTEIN DIVERSITY

It should be obvious at this point that the opportunity for differences among individuals in the proteins specified by genes is almost limitless. Omenn and Motulsky (1972), for instance, using 20 proteins (blood groups and enzymes) found in the blood, calculate that the probability that two people randomly selected from a population will have the same phenotype is 4.8×10^{-7}. Although this is a small figure, it still does not approach the figure that would be obtained if a larger number of variables were compared.

Omenn and Motulsky (1972) also provide an estimate of the percent of undiscovered protein polymorphisms in humans, based upon the following assumptions: (1) that the total number of nucleotide pairs in the haploid human chromosome set is three billion; (2) the maximum number of genes (at one gene per 1000 nucleotide pairs) is three million; (3) the probable number of structural genes (at 27 percent of the DNA) is 60,000; and (4) that the number of polymorphic genes (30 percent of the structural genes) is 20,000. Given that the number of protein polymorphisms known is approximately 50, the ratio of polymorphisms discovered to their probable number is 50/20,000, and therefore the percentage remaining to be found is 99.75. If this figure even approaches accuracy, we are thus only beginning to comprehend the extent of our diversity.

Racial differences in blood groups have of course been known for some time, and some are large. Among the Rh blood groups, for example, the cDe combination has a relative frequency of .00 in Japanese and .89 among African Bushmen. On the other hand, the CDe combination is .60 among Japanese and .09 among Bushmen (Dobzhansky, 1962). Stern (1973) presents some data on racial differences using other blood groups.

The functional meaning of such differences, however, remains unclear; they may or may not reflect environmental adaptations. The only thing of which we may be certain at the moment is that individuals with given blood groups have in the past tended to breed among themselves more frequently than they bred with people having different blood groups. Indeed, blood typing has become a standard anthropological technique for calculating the probable relationships among different peoples.

A more recent event, which has captured the attention of biochemists and geneticists alike, is the discovery of multiple forms among enzymes in the same individual. (For a detailed discussion see Shugar, 1970.) The different forms may be separated on a uniform medium in an electric field (that is, by electrophoresis), because they carry slightly different charges in accordance with the different amino acids they possess. Such "isoenzymes" were, as the name suggests, originally thought to have essentially the same activity (Shaw, 1965). Later evidence suggests, however, that this may generally not be the case. Multiple enzymatic forms may in fact react with the same substrate at different rates, and may react differently with different substrates.

Shih and Eiduson (1971), for example, found a difference in form (detected by band patterns in the electrophoretic medium) of monoamine oxidase in brain, heart, and liver of adult rats. Radioactive assay of the various bands indicated a differential activity in the conversion of substrate (in this case 5-hydroxytryptamine) to product. The same authors (1969) had shown previously that different forms of monoamine oxidase exist in embryonic and adult chicken brain. They found this also to be true in neonatal and adult rats (1971). Further, they found a differential affinity among the forms when different substrates (5-hydroxytryptamine, tryptamine, and benzylamine) were used. Collins, Sandler, Williams, and Youdim (1970) have reported multiple forms of monoamine oxidase in human brain. It is thus possible that at various times during development and in various tissues of an organism multiple forms of enzymes exist which have differential activities and differential preferences for certain substrates. This of course suggests that there may be meaningful differences in the functions of the forms.

Experimental manipulation during development may alter adult forms of an enzyme. In preliminary work with Eiduson and Shih (Eiduson, 1972), I found that injections of the male sex hormone testosterone into neonatal female mice of the inbred BALB/c strain resulted in a change in the pattern of monoamine oxidase bands following electrophoresis of a preparation of whole brains in adults, as compared with oil-injected controls. The effect was, in addition, substrate-specific; that is, it was present when benzylamine and 5-hydroxytryptamine were used as substrates, but not when tryptamine was used. The same type of effect appeared in a hypothalamic preparation of testosterone-injected and control female Swiss-Webster mice.

All of the above refers to structural genes only. When the possible differences in regulator, architectural, and temporal genes is considered, the potential for diversity becomes, quite simply, awesome. We are not as yet, except in a few instances, in a position to state the precise relationship between such diversity and the subtle or even the obvious behavioral differences among us. The realization of its chemical reality, however, offers hope both for a respect for the uniqueness of individuals and for an approach to the study of behavior within which the importance of that uniqueness is recognized.

GENES AND ENZYMES: SOME EXAMPLES AND CONCLUSIONS

The portrait we have thus far of the role of enzymes is that of biological catalysts that act upon specific substrates to produce specific products in a vast and complex chemical network. We have seen that at each step in such metabolic chains a specific enzyme is required.

We have also seen that, as in the case of variant hemoglobins, alterations in the DNA code may produce variant proteins that function less efficiently, or, in some instances, not at all. When this occurs in an enzyme, a particular reaction in a metabolic series is arrested. The consequences for the organism depend upon a number of considerations. If the end product of the metabolic series is essential and there is no other means of obtaining it except through the pathway affected, then the prognosis obviously is poor. The organism may simply be starved of a necessity. Equally unfortunate circumstances may occur even when another substrate may be used to generate an essential product, because the usual substrate can accrue at the point of the block, and at high levels may be toxic. An enzyme deficiency may thus lead to substrate poisoning.

All manner of other combinations may also occur. A complete alternative metabolic pathway may exist, for example, but be less efficient than the one that is blocked. As a result there is a slight deficiency of an essential product, while at the same time enough unmetabolized substrate remains in the system to be mildly toxic. On the other hand, a way around the block may be found through the induction of a very high level or production of another enzyme that catalyzes a similar reaction to the one blocked but is less efficacious. The iterations are clearly quite numerous.

The point, however, is that the effect of an alteration in enzyme activity depends upon the role that enzyme plays in the metabolic system. All enzymes, that is, have the same function, but some catalyze reactions that are more crucial to the organism than are others because of their place in the metabolic network. Mutations that alter the activity of enzymes may thus affect organisms to various degrees from devastating to hardly noticeable.

Much of this may appear to be recent scientific thought, and with regard to the specific chemical nature and function of genes, enzymes, and so on, of course it is. Some of the essential concepts, however, had their origin near the turn of the century. An English physician, Archibald Garrod, actually described in 1902 the first case of a hereditary metabolic disorder, alcaptonuria. The symptoms of this condition include the somewhat startling fact that the urine of affected individuals turns black upon exposure to air. (A concerned mother with a child in diapers brought it to Garrod's attention.) A normal substance in mammalian urine is urea, a product of amino acid metabolism. Instead of urea, Garrod's investigation revealed, the urine of alcaptonurics contained homogentistic acid. It was the reaction of homogentistic acid with air that led to the color change.

Garrod noted a familial relationship in alcaptonuria, and he and Bateson, whom he later consulted, concluded that the condition was indeed inherited. Garrod then gathered data on a number of other possible inherited metabolic disorders, and in 1909, in *Inborn Errors of Metabolism,* described four: alcaptonuria, albinism, cystinuria, and porphyria. Each, he concluded, was due to a block at some point in a metabolic sequence, a block due to the failure of one substance to be converted to another.

Alas, here the matter lay for many years. Geneticists were in the throes of resolving gene theory, and neither genetical nor biochemical techniques had advanced to the stage where the implications of Garrod's work could meaningfully be tested. In 1941, however, Beadle and Tatum found evidence in a simpler organism that Garrod's hypothesis was correct.

The mold *Neurospora* can synthesize all of the substances it requires for growth from a minimal medium including only certain inorganic salts, a suitable carbohydrate (such as the sugar sucrose) and a vitamin, biotin. Beadle and Tatum irradiated normal neurospora with the intention of producing mutations and found that a large number of progeny could not grow on the minimal medium. However, they could grow on a medium containing just one other ingredient. Backcrosses to the parental stock indicated that mutations had indeed been produced.

But the culmination of the research was the finding that each of the different mutants required the addition of a *different* substance in order to grow on the minimal medium. This suggested that the effect of each mutation was to block a different step in the metabolic processes that normally lead to growth.

The mutations subsequently were found to be expressed in the form of alterations in enzymes. When the missing substance was provided in the medium, the mutant could grow; otherwise the mutation was lethal. The concept that emerged then was one envisaging each step in the metabolic pathway of neurospora to be under the control of a particular enzyme, which in turn was specified by a particular gene. This led to the "one gene–one enzyme" hypothesis, which for a period both stimulated and dominated biochemical genetics. We know now that a number of genes may participate in the synthesis of a protein, as in the case of the hemoglobin chains. We might thus state the modern version as the "one gene–one polypeptide" hypothesis.

It is of interest to note that the assumptions which provided the basis of Beadle and Tatum's demonstrations of the gene-enzyme relationship derived from research on humans. Garrod had uncovered four instances of metabolic blocks. A number of others had been discovered in the interim, among them a related set involving the metabolism of the amino acids phenylalanine and tyrosine.

Figure 4.6 shows a part of the pathway for the degradation of these two amino acids. Blocks occurring at the places indicated lead to disorders of different types and severity.

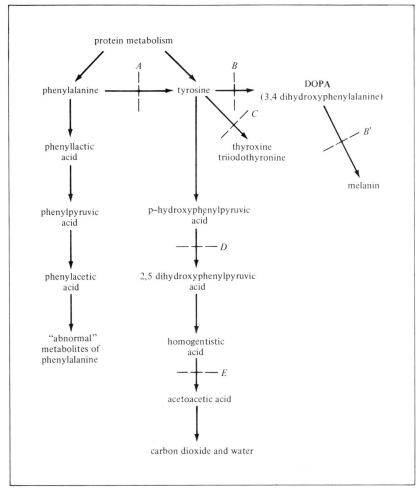

Figure 4.6   Representation of metabolism of amino acids phenylalanine and tyrosine. Blocks indicated result in disorders described in text.

A block at *A* in the conversion of phenylalanine to tyrosine occurs with a deficiency in the enzyme that catalyzes this reaction (phenylalanine hydroxylase). Phenylketonuria, a disorder whose main symptom is severe mental retardation, is the result. Phenylalanine is, as noted, an amino acid, and hence a natural product of the degradation of protein.

When the metabolism of phenylalanine is blocked, it accumulates in the system. Some of it may be degraded through another pathway, producing phenyllactic, phenylpyruvic, and phenylacetic acid, and these "abnormal" metabolites can be detected (as can the accumulation of phenylalanine). The secondary pathway is, however, not efficient enough to metabolize all the accumulated phenylalanine, which may be up to 60 times normal. The exact chemical cause of the disorder —whether it is due to the elevated level of phenylalanine or to the abnormal metabolites—is not known. Phenylketonuria is inherited as an autosomal recessive.

A block at B is known to produce albinism, resulting from a deficiency in tyrosine hydroxylase. Tyrosine hydroxylase converts tyrosine to DOPA, which is converted eventually to the pigment-forming melanin. A block at B' gives the same results. Albinism is inherited as an autosomal recessive.

A block at C leads to goiterous cretinism, with which small stature and mental retardation are associated, due to low levels of the hormones thyroxine and triiodothyronine. Several enzymatic defects as a number of stages have been decribed (Hsia, 1972).

A block at D produces tyrosinosis. Little is known of the mode of inheritance of this disorder. A deficiency of p-hydroxyphenylpyruvate oxidase is implicated.

A block at E produces alcaptonuria (the disorder with which Garrod began), inherited as an autosomal recessive and ascribable to a defect in the enzyme homogentistic acid oxidase. Besides blackening of the urine, darkening of the cartilage and sometimes arthritis is associated with this defect.

Study of this very small portion of a metabolic sequence provides a crude model for thought about the role of genes and enzymes and their possible involvement with behavior. Organisms are large and complex chemical networks. Alterations in some parts may be of great consequence, whereas alterations in other parts may produce only mild symptoms, if any at all. The differential effects of genic mutations that we see at the behavioral level are then related to where in the metabolic network the chemical effect occurs. Genes that can be individually identified obviously have large effects because of the relative importance of the role of the proteins they produce.

More than 1600 human diseases are associated with defects in content or expression of DNA. Indeed, it is estimated that 25 percent of the hospitalizations of children are for illnesses having a major genetic component (Friedmann, 1971). Of these, more than 40 instances involving mental retardation are known to be associated with a specific enzymatic deficiency (Friedmann, 1971; Hsia, 1972), some of which may be detected by aminocentesis (removal of embryonic cells from the amniotic fluid by means of a puncture) followed by a tissue culture wherein the genetic defect continues to be expressed.

A particularly forbidding instance is the Lesch-Nyhan syndrome, which involves retardation, involuntary movements, and self-destructive, impulsive behavior. Affected individuals sometimes bite off the tips of their fingers and their lips. The defective enzyme is hypoxanthine-guanine phosphoribosyl transferase (HGPRT), which has its highest activity in the basal gangalia of the brain (Hsia, 1972).

The Lesch-Nyhan syndrome is inherited as an X-linked condition, presenting the researcher an opportunity (albeit an unhappy one) once again to test the hypothesis of X inactivation in women. Women who appear normal but are mothers of Lesch-Nyhan males (hence, who bear one normal and one abnormal X chromosome, having passed the abnormal one to their sons) in fact show only 50 percent of the normal activity for HGPRT in skin fibroblasts. This of course suggests that half of the X chromosomes inactivated were normal and the other half carried the gene producing the deficient enzyme. There is as yet no information on a reduced HGPRT activity in cells from the nervous system (Omenn & Motulsky, 1972).

However, detection of the lowered activity of HGPRT in carrier women does more than support the X-inactivation hypothesis. This and similar cases wherein carriers of genetic defects appear to fall within the normal range at one level of measurement but are found to differ markedly at another (biochemical) level offer a refined model for a genetic involvement with some of the differences among us for which a biochemical origin had long been theoretically implicated but neither convincingly demonstrated nor understood.

Autosomal chromosomes, we know, come in pairs. Two alleles at a locus on homologous chromosomes thus specify a protein. If the organism is homozygous at the locus, the alleles are the same, and their products are likewise the same. If one allele should be defective and fail to produce the proper protein, however, a number of possibilities are open. The product of the other allele could be sufficient for normal functioning and the defect be completely undetectable. On the other hand, the product of the nondefective allele could be sufficient in quantity to lead to normal function at a gross level or in a certain environment but be found to produce detectably less product as a level closer to actual gene activity.

A question then arises whether carrying in the heterozygous state one or a number of defective alleles may in fact alter behavior or performance in ways more subtle than we have been prepared to discern. A carrier might, for instance, fall within the normal behavioral range but could have been at a higher point in the distribution were it not for the defective allele. How well one does in general—one's health, one's performance—could indeed be related to the number of defective alleles he or she carries.

This is clearly quite speculative, but considering that 20 percent of the individuals in a population carry one or more *newly arisen* mutations (Dobzhansky, 1962) in addition to those acquired from previous generations, the effect of the genetic load in humans is not to be dismissed. Many defective genes are eliminated in homozygotes because they are lethal or because the individuals fail to reproduce. Nevertheless, the number of carriers is not small. If a metabolic disorder is recessive and occurs homozygously in only one in 40,000 births, the associated heterozygous carrier frequency in the population is 1 percent. One out of every hundred individuals will carry a gene potentiating *that* defect. When one contemplates the number of genetic defects known, and, without undue foreboding, the number as yet undiscovered, the problem of genetic load does (as H. J. Muller vigorously noted) seem to deserve our scientific and social respect.

Several conditions are, as mentioned, already biochemically detectable in the heterozygote. The allele for phenylketonuria, for Duchenne-type muscular dystrophy, galactosemia, and Tay-Sach's disease may be found by means of substrate tolerance tests or actual enzyme assay. Carriers of the phenylketonuria allele metabolize high levels of phenylalanine more slowly than do homozygote normals. Heterozygosity for Duchenne-type muscular dystrophy is characterized by elevated levels of creatine phosphokinase, heterozygosity for galactosemia is characterized by a lowered activity of galactose-1-phosphate-uridyl transferase, and that for Tay-Sach's disease by a lowered activity of hexosamidase A. Genetic counseling is obviously greatly aided by the recognition of carriers.

We should not, however, think only of genes with large effects, or of deficits, but rather we should study the overall configuration that the gene-metabolism-behavior picture presents us. Some enzymes are "rate limiting"; that is, the whole chain of reactions can proceed no faster than they convert substrate to product. Other enzymes do not have this property. They seem capable of accommodating as much substrate as occurs. Alterations in a single, rate-limiting enzyme already operating at or near maximum capacity could have significant and immediate effect upon the organism. On the other hand, alterations in a number of non-rate-limiting enzymes affecting several metabolic routes over a period of time, such as might occur in a selection process, could gradually produce powerful results without any single protein change identified as the "cause."

Again, we may at this point conceive of organisms as enormously intricate, highly ordered, and specific sets of ongoing chemical reactions whose activities are under the control of genes. We differ from one another in a large number of those genes. The reactions they control proceed at different rates; perhaps even different metabolic routes are preferred. When this is understood, the potential for a relationship between behavioral differences and genetic differences becomes clear.

The chemical origin not only of large differences ascribable to single genes, w'ere evidence abounds, but also of subtle and complex behavioral distinctions influenced by many genes having a cumulative effect may thus be comprehended. The meaning of individual differences and the role of genes in participating in their determination thereby takes on some form of reality. We now know enough, at one level at least, to understand what it means to say: I am not you.

Chapter 5
Tools for Understanding:
Evolution and Behavior

Scientific evidence engendering a rapid change in the human self-image must to be effective be both convincingly derived and compatibly interpreted. To be convincing it must meet the conditions put by specialists upon its status as truth. To be compatible it must be consonant with (although not necessarily emphasized by) more comprehensive expositions of nature. Such evidence has recently been forthcoming, primarily from two sources: the study of primate behavior in natural environments, and the discovery and reclassification of fossils and implements associated with the lineage believed to have evolved into human beings.

The study of primate behavior in natural environments has produced data that both supplement and contradict those drawn from the laboratory. Even more important, it has provided insight into relationships between behavior, ecology, and social organization in animals living in groups that are thought in some ways to resemble those of hominids during a part of their evolutionary sojourn. Recent discoveries of the fossils and equipment of the first hominids indicate that they did in fact live in organized groups and that they lived in them under certain conditions for a period far longer than hitherto suspected.

Such primatological and paleological studies have quite simply revised the science of physical anthropology. Most relevant to us, the revision centers upon the reinterpretation of the role played by behavioral change in the process of evolution. It had been concluded previously that behavioral changes were made possible by, and therefore followed upon, changes in anatomy: Humans were seen as first having evolved into animals who could behave in a given manner and then did so behave. However, behavior is understood now always to have been in a selective feedback relationship with anatomy. Selection for *behavior* is indeed presently held to be the focus within which the neuronal, physiological, and skeletal alterations characterizing human physical evolution coalesce.

The implications of such a concept for psychology, particularly as contrasted to those attributable to radical environmentalism, are enormous. Humans cannot be understood merely as large-brained, bipedal, tool-making creatures who can learn. Rather, they are animals whose past behavior has evolutionarily formed their present potential. That is, if selection for behavioral traits were a central part of the human evolutionary process, if it were in fact the defining feature of that process, then humans must be treated as the behavioral as well as the physical descendants of their ancestors, with all the limitations, alterations, and additions they have intermediately accrued. Humans are not—in fact, cannot be—a tabula rasa nor a piece of blank paper, indiscriminately open to environmental impress. They are distinguished not by their unalloyed ability to learn (although that is considerable), but, like any other organism, by the kinds of things they *must* learn.

This reinterpretation, and the evidence that prompts it, is obviously convincing to the specialists from whose hands it disseminates. It is also obviously compatible with a larger, highly validated exposition of nature: evolution through natural selection. That is, one need adduce no new principles nor alter any basic precepts to accommodate it. One need only reorder priorities, as we comprehend them, within the existing evolutionary framework—reorder them, in fact, in accord with Darwin's original formulations, for he dealt with behavior as readily and as well as with any other character. The emphasis upon anatomical change, and the subsequent notion that behavior was in some way added on, came later. The reinterpretation, though based upon more sophisticated data, is thus a theoretical return. As Campbell (1972) notes: "The history of human evolution is meaningless without its behavioral dimension, and it has taken us nearly 100 years to see this as clearly as Darwin did. In the ninety years since Darwin died, we have unlearned a lot that he can teach us" (p. vii).

In the present chapter we shall examine, very briefly, the modern view of human evolution and its implication for the study of some recurrent human problems.

HUMAN EVOLUTION AND THE PRIMATE MODEL

Humans are primates and evolved from primates. Therefore, the study of living primate groups, although it cannot directly inform us of the behavior of our progenitors, is of value both in the analysis of the meaning of fossil remains and in the attempt to comprehend the social behavior and organization of early hominids. That is, although behavior per se has not been fossilized, if anatomical structures can be related to function and function to behavior and ecology in living species, then we have some basis for associating structure (known through fossils) with behavior and ecology in extinct forms. Combining this knowledge with that derived from study of certain modern peoples, we have reasonable hope for reconstructing the behavioral and other changes that have occurred and for understanding why they occurred.

Primates are an order of the mammalian class, consisting in turn of two suborders: prosimians and anthropoids. Prosimians are relatively little changed descendants of animals that originated between 70 and 40 million years ago. The early prosimians were apparently frugivorous, unlike their insectivorous descendants, and had comparatively large brains. They were arboreal, had evolved binocular vision, and climbed in trees by grasping.

The latter was a more important step than it may at first seem, for the development of grasping endowed the prosimians and thus the later primates with dextrous hands equipped with a mobile thumb and feet having a big toe. Both have major functions in human evolution, relating particularly to tool use and bipedalism.

The early prosimians moved in a unique way. The body was maintained at rest in a vertical position, with the feet grasping the tree trunk and supporting most of the body's weight while the hands held the trunk and prevented the body from falling over backward. When the animal moved from one place to the next, it did so by leaping with its powerful hindlimbs. Contemporary clingers and leapers show many skeletal adaptations to this form of locomotion.

The number of prosimian species has decreased since prosimians first evolved, the remaining species now being confined either to isolated geographical areas (especially Madagascar) or to a nocturnal life. The decline of the prosimians is attributed to an ironic but not uncommon fact of evolution: They were outcompeted for food and space by the creatures that evolved from them, the higher primates. Though of less intelligence (as measured by standard tests) than monkeys or great apes, at least some modern prosimians live in permanent social groups. It is thought now that the primate type of social organization evolved first in certain prosimians and was preadaptive to the evolution of the level of intelligence associated with higher primates.

Two infraorders of higher primates are recognized, one consisting of New World monkeys and the other of Old World monkeys, apes, and humans. Of particular interest are the Old World monkeys (cercopithecoids), lesser apes (hylobatids), great apes (pongids), and, of course, the hominids (humans). Very little is known of the earliest monkeys as compared to the hominoids (apes and humans). By the Miocene epoch (25 million years ago), however, numerous forms had evolved, and they became highly diversified and successful during the Pliocene (10 million years ago). The cercopithecenes are mainly frugivorous, although they eat a wide variety of foods, sometimes including other animals when they can get them. They live in diverse environments, including forest, woodland, and savanna, and have different social structures, each seemingly appropriate.

The hominoid and cercopithecoid lineages split about 35 million years ago. Biochemical as well as anatomical studies demonstrate the relatedness of hominoids and their distinctness from all other primates, apes, and humans sharing blood proteins as well as skeletal features not shared with monkeys (Sarich, 1971). In addition, hominoids have brains that are relatively larger than those of monkeys, the expansion mainly involving the association areas of the cerebral cortex.

Hominoids have similar dental and postcranial characteristics, probably because they evolved originally from small, arboreal, frugivorous forms that exploited the small-branch environment of the forest canopy. Early hominoids seem to have been quadrupedal arm swingers, which is reflected in their relatively long arms, reduced lumbar regions, shallow, broad thoraxes, scapulae on the back (as opposed to the side as in monkeys), and a series of muscular adaptations associated with arm raising. Skeletal differences between humans and the other hominoids are particularly marked in the shoulder and pelvic girdle, although the changes may have appeared only in the last few million years.

The hylobatids (gibbons and siamangs) are the smallest of the hominoids, weighing no more than 25 pounds. They live in the tropical forest of Southeast Asia and have been distinct from other hominoids for perhaps 30 million years. The living species are arboreal frugivores characterized by arm swinging and leaping. They rarely come to the ground.

The pongids are considerably larger than the hylobatids, and even the most arboreal of them (orangutans) are not nearly so acrobatic. Chimpanzees arm-swing but do not leap as gibbons do, and gorillas arm-swing only when they are young.

Orangutans are confined now chiefly to Sumatra and Borneo, living in trees in areas of swampy rain forest. They are frugivorous and, being large (males may weigh 200 pounds or more), must distribute their weight carefully to reach the food on terminal branches. This they do by using all four limbs in climbing. To this end, their feet are similar to their hands, having considerable grasping ability, with long, curved phalanges.

The other apes are confined mainly to regions of tropical rain forest in the Congo Basin of Africa. Gorillas live in the lowland areas of the west and elevated areas in the east. They are ground-living foragers, eating for the most part herbivorous foods such as shoots and stems, rather than fruit. They are very large, with the males weighing 400 pounds or more. They move quadrupedally, with a specialized form of locomotion known as knuckle walking, in which the forelimbs contact the ground on the backs of the middle phalanges of the flexed fingers. This adaptation is, however, fairly recent, because a number of features of the trunk, forelimbs, and thorax suggest a period when arm swinging and suspension formed a greater part of locomotion.

Chimpanzees are more widely distributed than gorillas, being found in forest fringe and open woodland areas as well as in the forest. They are basically frugivorous, with most of the feeding occurring in trees. The diet is, however, supplemented with insects and small game. Males may weigh 100 pounds. Chimps, like gorillas, are knuckle walkers.

Similarities in Primate Behavior

Nonhuman primates differ dramatically in some aspects of their behavior. There are, however, a number of features common to all of them, and these have great relevance to human behavior, being, in essence, the background from which it evolved (Dolhinow, 1971). First, primates are social. No primate lives alone during development, and few are forced to be solitary thereafter. Each animal spends most of its life in a social environment, one toward which, even when forced outside of a group, it orients its activities. Each individual must thus accommodate itself to the patterns of behavior that are necessary for group existence and hence to survival.

Second, the majority of behavioral items and the variations on each behavioral pattern must be acquired and practiced by members of the group. Some behaviors are specific to sex, some to age, and so on. A few social roles require behaviors that are unique to those roles. The behaviors are acquired and practiced during maturation, until each assumes adult form and meaning in the context of social relationships.

Early learning may depend largely upon the normally very close bonds between mother and infant, which are, in fact, the most important and strongest bonds in the life of a nonhuman primate. Rarely, if ever, is the father of a particular animal known. The personal identification between father and offspring thus does not exist. (This is not to say that adult males do not behave in ways that might be considered fatherly—they often do. It is to say that the "my" in the relationship is lacking. To infants and juveniles there is a class of adult males, some individuals of which may be friendlier than others, but none of which is behaviorally recognizable as fulfilling the special role of male progenitor. Similarly, to adult males there is a class of infants and juveniles, none among which is recognizable as "own offspring," and for that reason there is no special relationship with any member of that group.)

It follows then that the mother's status in the group, her social preferences, and her temperament are important in determining the young primate's experiences. An infant with a very confident high-ranking mother will have experiences different from those of an infant whose mother constantly behaves in a subordinate and yielding manner. The quality of mothering itself is also an obvious factor. Some mothers are conscientious and protective (regardless of their status); others are not.

Genealogical relationships may be significant in determining the course and nature of social interactions for long periods. Bonds between mother and offspring may persist into adult life, to become a nucleus from which other bonds ramify. This seems certainly to be true of chimpanzees (Van Lawick-Goodall, 1968). Such a nucleus includes several generations of offspring from the same female, who then form alliances that may aid all members of the family. A high-ranking female with several adult and subadult progeny (particularly males) can be a formidable force in a relatively small group. It should be noted that there is no indication that genetic relationships are recognized; rather, the social bonds are determined by the close association among siblings having the same mother (Washburn, Jay, & Lancaster, 1971).

Of course, not all patterns of behavior are learned from the mother. Males, for instance, must learn some activities that differ from those of females. These are things that are learned easily, however, granted normal predispositions and exposure to male behavior. Both male and female young spend many hours each day in play, which serves as a means of acquiring and practicing both motor and social skills.

It should be remarked that anthropologists do not conceive of "learning" in quite the same way as would behaviorists in this context. Among the former, the concept revolves around the fact that an organism is genetically programmed to be relatively sensitive to certain kinds of stimuli, and relatively insensitive to others. Behaviorists although they must use this fact, tend to act as if they do not; instead, they promulgate the view that learning is a process for which genetic differences have little relevance. All students of psychology doubtless are at this point familiar with this supposition. The contrasting anthropological position is well stated by Washburn, Jay, and Lancaster (1971):

> Animals are able to learn some things with great ease and other things with greatest difficulty. Learning is part of the adaptive pattern of a species and can be understood only when it is seen as the process of acquiring skills and attitudes that are of evolutionary significance. (p. 299)

Third, every primate group has a location, a home range, in which all events of life occur. Everything the group requires, including food, water, and sleeping places, is located in the home range. The extent of the range, in addition to being determined by physical features, is under constant adjustment as a result of contact with other groups of primates. In some instances groups may attempt to maintain exclusive use of their range (it is then often called a territory), but frequently a group within its range has little contact with others of the same species.

Fourth, each group has a social organization that in part structures the behavior of every member. Individuals do not behave autonomously, and few actions occur that do not have social implications. An elaborate set of relationships exists among individuals, such that whatever occurs in one segment of the group has repercussions in the group at large. Many groups are stable over time. By far the majority of animals that do move from one group to another are males. Normally, a set of relationships is derived and each group has a shared history of experiences.

An important part of the social organization is the dominance hierarchy, the structure of leadership patterns that determines which individuals have precedence over others in certain situations. Adult males tend to rank themselves serially from first to last. Adult females, however, are probably best considered as members of the upper, middle, or lower strata (Dolhinow, 1971). Dominant males may control the movement of the group, and they generally have first or principal access to food and estrous females. Dominance is not, however, a concept simply expressed among other primates, even as it is not among humans.

Mention has been made that sons of very dominant females often have a great advantage over other males of the same age. This occurs first because their mothers' power is available to them in social encounters. In addition, they are often raised among other dominant animals, where they are likely to acquire a fair measure of confidence. A male must nevertheless maintain or advance his position by his own abilities; an easily intimidated male is unlikely to acquire high status, no matter how strong his ties with dominant kin.

It should be clear that communication is essential in species having such complex social structures. Communications, as would be expected, can become quite complicated and rarely involve only one sense modality. In the normal context the group has an ongoing set of relationships. The members know one another well, and their past experiences are important in determining reactions to the expressions of others.

Many of the signals are rather subtle. An animal's posture, its movement, and its tone of voice are monitored by others and responded to appropriately. A dominant male, for example, is studied constantly and closely; he does not need to send out strong signals to be heard. On the other hand, a very low-ranking individual may expend great energy and produce very strong gestures and still be effectively ignored.

A large amount of information on the internal state of individuals is shared by recognizable (to us) gestures or glances. A dominant chimpanzee may reassure a subordinate by a kiss on the hand or a pat on the head. Chimps also use touches and kisses as submissive gestures and gestures of greeting (Van Lawick-Goodall, 1968). Among some monkeys, an individual may place his or her hand momentarily on the back or the side of another to indicate that a particular behavioral sequence is concluded (Dolhinow, 1971). Many more such instances will doubtless be discovered as research continues.

Ecological-Behavioral Differences Among Primates

Although primates share a number of behavioral adaptations essential to their existence, there are many differences, both across and within species. There is thus no general model of primate behavior that can be usefully applied to all situations. The picture we do have, while extremely informative, is too broad to be adequately predictive. All primates do not behave alike; indeed, all prosimians, all monkeys, and all apes do not behave alike.

For one thing, prosimians, as we have noted, are not as intelligent as monkeys, and monkeys in turn are not as intelligent as apes. These differences certainly will be reflected in the complexity of behavioral repertoires and social relationships. Equally significant, if not more so, the habitat in which the animals live has lately come to be recognized as an important determiner of the type of social organization found.

At the simplest level, one may distinguish between primarily arboreal and primarily ground-living forms. Animals that spend most of their time in the trees do not move over long distances and do not require behavioral defense mechanisms against ground predators. Therefore, they are not either anatomically or behaviorally adapted for ground life. Adult males do not ordinarily defend the females and young of the group.

Some ground-living forms, on the other hand, especially those that live in the open, have evolved elaborate defensive systems, the most notable of which is found in savanna baboons. In transit, a savanna troup somewhat resembles a military column. Females, infants, juveniles, and the most dominant males remain at the center, whereas the younger males stay at the periphery and are thus the first to encounter potential predators. If a predator should appear, the dominant males rush to the scene and position themselves between it and the troop, making all the while loud barks and exposing their huge canines. It is a formidable commotion—enough, evidently, to deter all but lions.

It should not be thought, however, and this is a most important point, that even all ground or arboreal forms are behaviorally adapted in the same way. Spider monkeys of South America, gibbons, and orangutans are all arboreal, for example, but only spider monkeys live in large groups. The basic gibbon group consists of one male and one female. Gibbons defend their small territories, even excluding their own offspring when they are grown. Less is known of the social behavior of orangutans, but they appear to live in dispersed units consisting of mothers with infants, to which adult males become attached only infrequently.

Similarly, both savanna baboons and patas monkeys live on the ground. The patas male spends a large portion of his day acting as a sentry; if a predator approaches, he begins a noisy display, apparently attempting to draw attention to himself, and then runs away, leaving the rest of the group very inconspicuously and quietly at rest in the tall grass. This is very different from the maneuvers of the savanna baboons. It helps the cause of the patas male that the troop does not wander too far from the safety of trees and that he is usually a swift runner.

Gorillas are also ground dwellers, traveling in groups of from 12 to 20, under the leadership of a single dominant male. The troop consists basically of the "silverback" male (so known because the hair of his back turns gray after a certain age), females, and young. Other adult males may join the troop, but few are as permanent as the silverback, and all are subordinate to him. Individual males are often found unattached to any group.

The size of gorillas alone is something of a deterrent to most predators. The male in addition displays ferociously when his group is threatened, charging at a high speed, waving and shaking branches, and of course standing erect and resoundingly thumping his chest. The appearance of his size is enhanced by very hairy arms and legs and the large sagittal crest on his head.

Other adaptations to territorial life may be found in other environments. Whereas the gorilla lives in dense tropical forest, the savanna baboon lives in the open (but where trees are often available for sleeping), and hamadras and gelada baboons inhabit the far more arid regions of Ethiopia. Both latter species have developed a basic social system that is different from those of other ground dwellers: the one-male, multifemale group.

This group is a reproductive unit, which has obvious genetic implications. The females follow their male closely and seldom, if ever, leave his side. At night many such units come together and sleep upon steep cliffs for protection, but there is no mixing, and aggression among males is remarkably low. Evidently each male is relatively content in his own domain.

One-male groups may be viewed chiefly as an adaptation to the location and scarcity of food. The food items are scattered evenly over the ground, in contrast to that in other environments, requiring that the animals disperse widely, yet maintain social order. The one-male grouping accomplishes this. At the same time, food is not plentiful, and the one-male group ensures that females and young do not suffer nutritionally at the hands of males. The male at the head of a group usually does not tolerate other males (except in the sleeping areas). As he ages, however, a male may take on a younger "junior partner," who actually serves the reproductive function and eventually acquires the harem.

It should be obvious from the above that different species of primates, adapted to different environments, do not behave alike. It is also true that the same species may behave differently under different environmental circumstances. This is exemplified by rhesus monkeys of India.

Where the monkeys are subject to trapping, harassment, and a high incidence of infectious disease, the group members appear to be tense, and aggression is high. This occurs at places where the animals are in contact with humans, and there is intense competition both within the group and between groups for food and water. In forest environments, where normal spacing is possible and food is more plentiful, aggressive behavior serves more to maintain the stability of the group (i.e., to maintain the dominance hierarchy) and to keep groups apart. The forest rhesus, in contrast to those living in the city, show substantially fewer wounds, and disease appears to be rare (Washburn, Jay, & Lancaster, 1971).

Similarly, the common baboon found on the African savanna (*Papio cynocephalus*) behaves somewhat differently in a woodland habitat. In the woodland, almost all males and many of the subadult females change troops during their lifetime, and the integrity of the troop is maintained by adult females with their progeny. Aggressive encounters among troop members are comparatively infrequent, and the dominance hierarchy is not rigidly enforced. On the savanna there is little interchange of members among troops, the dominance hierarchy is stoutly affirmed, and aggressive encounters are frequent (Philbeam, 1972).

The foregoing information is particularly relevant because the organization and behavior of *Papio cynocephalus* upon the savanna has been taken as a model for that of early hominids, assumed to have evolved in a similar habitat (Tiger & Fox, 1971). However, some anthropologists now believe that the treeless savannas, at least in Africa, are of recent origin, due to the agricultural activities of modern humans who have converted woodland savanna, woodland, and even forest into open grassland (Philbeam, 1972). The behavior of baboons upon the open plain may thus be a new adaptation of basically woodland creatures.

More important, most anthropologists now recognize that the total biological and behavioral matrix from which an adaptation evolves must be considered in each form before extrapolation from one to the other is justified.

Although the same ends may have to be attained by different species in similar environments, it is by no means necessary that differing species evolve responses that attain those ends in the same manner. Evolutionary adaptations are not generated de novo: Evolution, though sometimes marvelous to behold, is not magic, and it cannot produce something from nothing. Rather, the new adaptations come from the old. What is has been shaped from what was. If the biological differences between two species are large, the chance is remote that they will solve problems, especially complex ones, in the same way.

The evidence indicates that even the earliest hominids did not resemble present-day baboons, either anatomically or behaviorally. (We shall review the data briefly later.) It is thus unlikely that baboons—whether the African savanna is of recent or ancient origin—offer the best insight into the origins of hominid behavior and social organization. The animal favored for that purpose today is the chimpanzee.

The Gombe Stream Chimpanzees

The chimpanzees studied in most detail in the wild are those observed by van Lawick-Goodall (1968, 1971) in the Gombe Stream Reserve of Tanzania, near Lake Tanganyika. It was estimated that there were from 100 to 150 animals in the 30 square miles of the reserve, although not all of them were identified or followed individually. The animals that were identified were seen to exist in small, temporary groups, consisting of various combinations of age and sex classes. The one group that is possibly stable over several years is one containing a female and her infant and juvenile offspring.

The reserve area contains both dense forest and deciduous woodland, with some grassland. The chimpanzees, chiefly frugivorous, traverse their home ranges daily in search of food. They follow no regular circuit, the distance and direction of travel corresponding with the availability of food. Each night they construct sleeping platforms in trees. They sleep alone, except in the case of mothers with infants.

Groups, when formed, are (with the exception of females with offspring) associations of individuals lasting no more than a few hours or days. Adolescent and mature males frequently move about alone, as do females, though apparently less frequently. Mutual attraction among individuals appears to be the basis for the composition of small groups, including some consisting of siblings.

It should not be thought, however, that the temporary nature of chimpanzee groups indicates a loose status structure. On the contrary, the social status of each individual is well defined; the individuals all recognize one another's rank, and this comes much into force during social encounters.

There are a number of ways in which the dominance of one individual over another may be seen, although the expression of dominance depends on the other animals involved and on the situation. A high-ranking animal, for example, may walk toward a subordinate, who will immediately move away from a desirable resting place. If two animals meet on a narrow branch, the subordinate normally gives way. When the animals are eating, or about to eat, a dominant may threaten, chase, or attack a subordinate. Low-ranking individuals may avoid close contact with dominants or, if contact is necessary, will make gestures of submission or appeasement.

Some males appear to form alliances, with the result that the dominance situation changes, depending upon which group members are present. The highest-ranking males known at the Gombe Stream were Mike, Goliath, J.B., Hugo, David, Leakey, and Huxley. Each of these males dominated other males, females, adolescents, and juveniles. Within this male group, however, status fluctuated. Mike was dominant to all at all times. J.B., who often traveled with Mike, was subordinate to Goliath when Mike was absent, but when Mike was present he was superior to him (as well as to others). Similarly, David, when he was with Goliath, frequently dominated Leakey; but when Goliath was not there, the two were equal. One reason underlying such alterations in status lies in the fact that the higher-ranking male will sometimes go to the aid of his less-dominant partner. Once, for instance, J.B. was attacked by Goliath. He screamed, Mike came over, and both threatened Goliath.

Females also form alliances, but they tend to be less permanent. Moreover, females frequently attempt to enlist the aid of a high-ranking male when they are threatened or attacked. One female, Melissa, was notable for this type of behavior, often hurrying toward a dominant male and barking or screaming in the direction of the offender, while holding her hand toward the high-ranking male or laying it upon his back. However, it was rare that the males responded to Melissa's entreaties by attacking the offender.

Chimpanzees have a large number of vocal and gestural means of communication, which are thought to express their internal states. These have been and are being studied at the Gombe Stream and elsewhere. Only a few will be mentioned here. One of the most interesting is the display behavior of males, which consists of charging rapidly, shaking and dragging saplings or branches, throwing objects, stamping, and drumming on trees. These displays are seen frequently when two groups containing males meet.

They are often directed at other males in what appear to be status quarrels (although they are sometimes directed at no individual in particular). Females and adolescents may bear the brunt of such displays, and infants have been gathered up and used as objects in them. Displays also occur when the individuals appear to be frustrated. When one adolescent was unable to obtain bananas because of the presence of adult males, he "frequently whimpered, hurried some 100 yd. away from the group, broke into a run, drummed on a tree, swayed branches, and stamped and slapped on the ground, meanwhile uttering 'pant hoots' and screams; he then returned, apparently relaxed, to his group (1968, pp. 273–274).

The function of the displays is not always clear. They may be the product of sheer emotional excitement or—in some instances, as above —of redirected aggression (see pp. 339–340). Another interesting pattern, not restricted to chimpanzees but most highly developed in them, is reassurance behavior. In many circumstances one animal appears to be calmed by the touch of another, particularly following a period of emotional involvement. Often a dominant individual, having attacked a subordinate, reassures it by touching, patting, embracing, mounting, or kissing it. Further, the subordinate appears to require such gestures on the part of the dominant before it can be content. An example is given by van Lawick-Goodall (1971):

> And Flo, after Mike's vicious attack, and even while her hand dripped blood where she had scraped it against a rock, had hurried after Mike screaming in her hoarse voice until he had turned. As she approached

him and crouched low in apprehension, he had patted her several times on her head and then, as she calmed, had given her a final reassurance by leaning forward to press his lips to her brow. (p. 127)

On other occasions, a subordinate will seek reassurance from a dominant in situations where it is possible that a threat or attack might occur. At one point Pepe, an adolescent male, approached Goliath, who was feeding on a pile of bananas. He did not take one, but, crouched, whimpered, and screamed, looking at Goliath and reaching his hand toward him. Finally Goliath patted his face and head. After that Pepe reached slowly toward a banana, but then suddenly pulled back his hand and began to scream. Goliath patted him again for several seconds. At last Pepe stopped screaming and, watching Goliath carefully, gathered up a few fruits and moved away with them.

Dominant animals, when aroused, are often calmed by others. On several occasions a high-ranking male, having attacked some individual or performed a violent display, was touched, patted, or groomed by other males. Two aroused individuals may even calm one another by contact, as was the case when Mike finally wrested the highest rank from Goliath, following a long display on both their parts. Goliath had "lost his nerve" and rushed toward Mike with submissive gestures, and began to groom him vigorously. Shortly thereafter Mike began to groom Goliath. They remained thus for an hour.

One animal may initiate physical contact with another (often a higher-ranking one) in a number of situations when it is apparently afraid, agitated, or intensely stimulated by social activity or the presence of a favored food. The gestures used most often are touching, holding the hand toward the other, embracing, mounting, and grooming. This occurred particularly when the animals were startled by a noise or movement, when males were displaying or fighting, or when a large box of bananas was being opened (by researchers). During the latter event the chimpanzees touched, patted, kissed, and embraced one another continuously.

Food, which chiefly consists of wild fruits but is augmented by insects, eggs, birds, and other mammals, is frequently shared, particularly by relatives. The animals were never seen to have been food-deprived, so it is difficult to know what the outcome would have been under that condition. Under the circumstances that prevailed, however, chimps were often seen begging for a desirable item, and were usually rewarded, at least by a tidbit.

Young chimpanzees spend a considerable period of time with their mothers, not being weaned until the fourth or fifth year. After that they may continue to travel with their mothers (and possibly her more recent offspring) during much of their adolescence (from the seventh to the twelfth year). Males thereafter leave their mother's company, although a close relationship appears to persist. Adolescent and mature males often aid their mothers in social situations but apparently do not copulate with them.

This long association between female and offspring provides time for the young to learn social roles while still being protected. Adult males are generally quite tolerant of the actions of infants, up to the age of about 4. The long association also provides time for sibling ties to develop, which are important in many ways. For one, in cases where the mothers of infants are known to have died, it was their older siblings rather than other adult females who cared for them. In this regard, it should be noted that adolescent females often "kidnap" infants from their mothers for a short time, playing with them, pushing them to their breasts, and carrying them in typical motherly fashion.

Chimpanzees use naturally occurring objects as tools. Leaves, for example, are used as one would a sponge to obtain water from hollows in trees and to wipe the body, and branches and rocks are thrown in the direction of offenders. More importantly, in at least one instance natural objects are *modified* for use as tools. This latter is an important difference, because it represents a substantial amount of what would be described as forethought. (A number of animals use tools, including a species of finch, members of which hold thorns in their beaks in order better to dig after insects.) The modification of an object for a particular use implies a comprehension of relationships between nonmodified object, modified object, and the use to which it will be put. It is a case of the capacity to "see" the tool in the object and make the suitable alteration.

What chimpanzees actually do in this regard is, on the surface, hardly dramatic. They select twigs and grasses, break and strip them to the proper length and diameter, and gather termites by pushing the twigs or grass down holes in the termite nests and pulling them out again. However, perceiving the relation between breaking and stripping and the idea of proper length and diameter is a major step. It was heretofore seen only in humans.

The preparation of grass or twigs for use as tools is also a behavior that is learned by one generation from the last and is, in that sense, cultural. (However, the capacity is obviously biologically based. Baboons habitually watch the chimpanzees catch termites but are themselves unable to learn how to obtain the evidently delicious creatures.) During the termite season, chimpanzee infants not only observe the adults at work, but also attempt to make the necessary tools. Their first efforts are unproductive, the twigs being too short, too wide, or the like, but by the age of 4 they have usually mastered the technique—which includes a bit of patience in that the termites must be allowed to crawl on the twig and be withdrawn while still attached to it.

The picture that emerges of chimpanzee life in the wild is, then, one of great behavioral complexity and subtlety. This is hardly to say that a chimp is in any sense to be considered some sort of reduced human. It is not. The hominid and pongid lines are estimated to have diverged 15 million years ago, and chimpanzees have changed through evolution since. It is to say rather that chimps are our closest living relatives and that there are certain features of chimpanzee life that lead us to believe that it is worth studying for what it can tell us of the possible behavior of the first hominids.

The flexible social organization (characterized by both aggregation and dispersal), the existence of all-male groups that may wander widely in search of new food sources, the strong emotional ties (both between a female and her offspring and between adults), the existence of occasional food sharing, and the fact that males collectively hunt other animals are all consonant with the current view of the activities of early hominids. Food sharing and organized hunting are of special interest since they represent the development of rudimentary cooperative behaviors of a sort that do not appear in other primates (although baboons hunt also) and that probably depended for their emergence upon the evolution of a sufficiently large brain (Philbeam, 1972).

HUMAN EVOLUTION

Mammalian Adaptations

Human evolution clearly is, as it must be, an extension of the theme of the evolution of mammals. According to Campbell (1974), changes in four large complexes characterize mammalian evolution, and, although we shall not dwell upon them, it is well that they be reviewed in order that the continuity between human and general mammalian evolution be recognized. The four are homoiothermy, mastication, reproductive economy, and determinants of behavior.

1. *Homoiothermy.* We have already noted (Chapter 4) that the chemical reactions that underlie life would not occur at the rates necessary were it not for biological catalysts. Enzymes allow reactions to proceed at a temperature and pressure compatible with living matter. However, this obviously does not imply that temperature is an irrelevant factor. On the contrary, the enzymes themselves require a very specific temperature in order to be optimally active. All this has been evolutionarily incorporated into the business of the organism, including that of behavior, such that the organism strives to maintain a temperature at which life may be maintained.

The oldest animals in the fossil record lived approximately 600 million years ago. The first amphibians appeared about 400 million years ago, and the reptiles about 350 million years ago. It was not until 200 million years later (that is, about 150 million years ago) that the capacity for homoiothermy, a constant body temperature despite fluctuations in the environment, evolved. It occurred in mammals, and was a distinct advantage for it enabled them to maintain a high level of activity during colder times when their competitors, the reptiles, became sluggish. It also enabled mammals to move north from the tropics and exploit the huge forests of the temperate zones. Homoiothermy is achieved by many mechanisms, among them fur or hair, sweat glands, dilation and constriction of capillaries, and special behaviors.

2. *Mastication*. The dentition of amphibians and reptiles is homodont; that is, all the teeth are similar in shape and use. The reptile-amphibian jaw and its dentition are of course related to their function, which is mainly to trap prey, or, in the case of herbivores, to tear foliage. There is little in the way of chewing. Thus some animals, such as snakes and alligators, usually swallow their food whole.

However, swallowing food without prior chewing is inefficient. For one thing, more nutriment is released, and more quickly, when food is broken into small particles and a greater surface area exposed. For another, a wider variety of foods may be exploited if they are prepared with the teeth. Seeds, roots, and particularly the plant proteins in tough vegetation are all more readily processed.

But chewing requires a different jaw structure and different teeth from those of amphibians and reptiles. Mammals thus evolved a heavier, much-strengthened jaw, and heterodont dentition suitable for cutting and grinding. (A reversal to homodont dentition has occurred in toothed whales.) The development of a masticatory apparatus then played an extremely important role in the mammalian expansion, having to do with available food resources and food processing.

3. *Reproductive Economy*. The early reptiles became fully adapted to the land through the development of fertilization by copulation, thus removing the requirement of water as a medium in which sperm and egg could combine and increasing the probability of such combination. However, reptiles still reproduce like aquatic organisms in one respect: They produce large numbers of fertilized eggs and leave them to develop and survive on their own. Both eggs and young are often easy prey to small carnivores, especially mammals.

The mammalian innovations lay in the evolution of a system that improved reproductive economy by reducing the losses of eggs and young. The first of these innovations was viviparity, the retention and nourishment of the fertilized egg within the mother until it has achieved a fairly advanced stage of growth. The second was the evolution of parental care after birth (resulting in a period of dependence of the young upon the parents and, eventually, in a period of postnatal learning by the offspring).

4. *Determinants of Behavior.* As postnatal dependence and learning increased, a new element appeared in mammals, one not evident in reptiles and amphibians—namely, behavior that does not satisfy the immediate needs of hunger, sex, or self-preservation. The element is sometimes called curiousness, but the terminology need neither confuse nor detain us. It is a fact that young mammals, especially those most phylogenetically advanced, engage in exploratory and playful activity. Such activity leads them to discover how the environment may be manipulated and what the consequences are from such manipulation. Obviously, particular species would evolve a sensitivity to input from their normal habitat. The evolutionary value of curiosity, however, is that the limits of the habitat are forever being explored and, in time, may be breached. This process reaches its apogee in humans.

Hominid Anatomical Adaptations

The distribution of mammals into various orders, families, and so forth involved the exploration of and adaptation to new environments. We are interested in those adaptations that led to humans and, since humans evolved from primates, we are especially interested in the differences between humans and other primates. These present us with clues to our history, and, to some extent, with clues to the origin of human behavior.

The most significant anatomical differences between humans and other primates involve the dental apparatus, the postcranial skeleton, and the brain. Dentally, humans are most notable for the absence in males of large, projecting canines. Hominid canines do not protrude beyond the level of the other teeth, and, in addition, rather than being conical, are chisel-like in form, resembling the incisors. Incisors, canines, and premolars form a continuing series, unbroken by gaps (diastemata). The molars and premolars are large relative to the incisors and canines and are covered with a thick enamel, an adaptation for the chewing of tough food. The dental arcade is parabolic rather than parallel.

Masticatory apparatus is clearly related to diet. The masticatory apparatus of the first hominids appears to have been selected for powerful chewing, having closely packed teeth and the capacity to grind food with rotation of the lower jaw. The absence of large, interlocking canines facilitates such movement. Philbeam (1972) concludes that the basic hominid dental adaptations occurred early in hominid evolution, the chief changes during the last few million years being a reduction in the size of molars and premolars, with little or no change in the size of incisors and canines.

If this appears to be making much over something that seems unrelated to behavior, it should be understood that one can tell a good deal about the behavior of an organism if one knows what it eats. In addition, the fact that the males of even the early hominids did not have large canines suggests that they could not have used them in defense from predators or in intraspecific aggressive encounters. They therefore must have had other means for solving these problems, especially by the time they began to live in more open country.

The lack of sexual dimorphism in the size of canine teeth, as opposed to what is found in most other primates, may suggest something about early hominid social organization, although data here are obviously sparse and all conclusions must be quite tentative. The lack of sexual dimorphism in canines is at least consistent with the relative monomorphism in other primary human anatomical features. Human females, for example, average 80 percent of males' weight and 94 percent of their height (Crook, 1972). This is in contrast to some primates wherein the male is twice the size (or more) of the female. Gelada baboons, gorillas, and orangutans are remarkable in this regard. Hamadryas and savanna baboons also show considerable sexual dimorphism for size. Gibbons, on the other hand, show very little. In chimpanzees the difference is not nearly so striking as it is in gorillas or orangutans.

Given that sexual dimorphism is notable in a number of terrestrial forms, it might be thought that it evolved in response to a need for troop defense. Although this may be true in part, it does not hold in general, however. Orangutans are among the most dimorphic, and yet they are arboreal and do not live in groups. Gorillas are terrestrial but live in the dense forest, which is not frequented by the big African cats. Chimpanzees spend much time on the ground; indeed, they share with gorillas a unique adaptation for terrestrial locomotion—knuckle walking. Yet they are, as mentioned, not greatly dimorphic in size, nor do they live in large permanent groups.

Much of sexual dimorphism, rather than being an adaptation for defense against predators, may be a result of what Darwin originally called "sexual selection"; that is, selection for traits among individuals of the same species and sex on the basis, as always, of reproductive success. The tendency of primate males to form a dominance hierarchy is particularly apropos since the dominant males, although they do not do all of the copulating, do in some forms appear to sire most of the offspring. Large size, large canines, and so on are thus (along with behavioral disposition) a means to reproductive status, which is acquired chiefly by threat and intimidation.

We shall return to the subjects of sexual dimorphism and sexual selection later. For the moment the important point is that early hominids appear not to have been sexually dimorphic for size, either in body or dentition. There was not a great deal to undo in this regard for us to reach our present condition.

The masticatory apparatus of early hominids suggests then that they chewed tough food, and prepared it in a unique way, using the slicing action of the incisors and canines. We of course do not know what the diet actually was, but it probably consisted of roots, seeds, and perhaps also meat and even bone (Philbeam, 1972).

Postcranially, our most distinctive feature is our bipedalism, the ability to walk efficiently on our hindlimbs. Most other primates are quadrupedal; those that are not are arboreal (gibbons and orangutans) or are knuckle walkers. Bipedal walking in primates appears to be an adaptation for covering long distances economically, a feature necessary for an organism given increasingly to hunting on the open plain. In the striding gait achieved eventually by hominids, the body's center of gravity shifts very little in any plane, and the resulting restriction of movement means that a minimum of energy is expended.

Walking involves a series of flexions and extensions of specially adapted muscles and joints of the pelvic girdle, legs, and feet. However, such adaptations are for our purposes less interesting than other adaptations that occurred concurrently with the evolution of bipedalism. As a result of the upright posture, for instance, the ilium of the pelvis has been "bent" in humans (Campbell, 1974), with the result that there has been a tendency for the relative size of the pelvic canal to be reduced. But the pelvic canal is the birth canal, and its maximum size in the mature female limits the size of the fetus, especially its head. Since brain size is highly correlated with head size, and both are highly correlated with age, the size of the birth canal then limits the maturity of the fetus at birth. The outcome is that humans give birth to progeny when they are at an earlier stage of development than in any other primate (Campbell, 1974). This in turn means a longer period of dependence of the progeny, a longer period of parental care, and consequently a greater opportunity for learning. A greater opportunity for learning, in turn, is to be associated with an even larger brain and hence an even larger head size.

It should be noted also that a relatively "premature" infant could not, as helpless as it is, have become a successful part of the human scheme were it not for two other elements associated with the change to bipedalism. First, the mother had to have her hands free to hold her progeny. Nonhuman primate infants can cling to their mother's hair independently a few days after they are born. They can thus assure themselves of maternal contact. Human young, on the other hand, must quite literally be carried everywhere for at least a year, and effectively for two years or more. Only a bipedal primate has the capacity to do it.

At the same time, whereas the infants of other primates can quickly secure the satisfaction of their needs from their mothers through specific acts, the only behavioral avenue open to human neonates is crying. The human mother thus bears much more responsibility for her child's welfare than do the mothers of other species, and this had to have been selected for concomitantly with bipedalism and birth of the fetus at an early stage of development.

The change in the ilium resulting in more efficient bipedalism has proceeded further in human males than in females, evidently because the size of the female pelvic canal is about at its lower limit, given fetuses with such large heads. The structure of the pelvis in females has the effect of slightly diminishing the range through which the hip can move forward and back, with the result that for a given length of stride women must rotate their hips through a greater angle. This has not escaped the notice of men, for whom it has acquired some degree of erotic significance. (Not without reason, either, since hip rotation would be to the keen observer a reliable means of identifying a potential sexual partner at a distance.)

Just what began the evolution of bipedalism is not fully understood. According to Napier (1967), however, it probably did not occur in the open savanna, but in woodland-savanna areas, and was already advanced before hominids moved into the open. In the pure savanna the dangers are great, and food similar to that of the forest is not easily located. If the early hominids were in addition in the beginning stages of converting to bipedalism, they would have been unable to exploit the advantages of savanna life that do exist. The woodland areas, on the other hand, provide both forest foods and ready escape from predators while containing enough open areas to permit new locomotor adaptations. It is thus likely that the first steps were taken in the woodland, although striding was developed in the open.

Another adaptation concomitant with bipedalism involves the hand. Hands do not merely carry things; they are foremost exquisite devices for object manipulation. The capacity to detach objects from the rest of the environment and to hold and feel them, as well as to see them in three-dimensional clarity, represented a meaningful factor in human evolution (Campbell, 1974).

The basic difference between the human hand and that of other primates is the extent to which the thumb is opposable to the fingers. Although many primates have opposable thumbs, only in humans is the thumb long enough to carry a heavy masculature, which makes possible the "precision grip," such as is used in writing with a pen and in many other delicate operations wherein an object is held between thumb and fingers (Napier, 1967).

The most important application of the hominid hand to the environment lies in toolmaking, the extraction of one object from another with the purpose of its further use. Tools have been found that date from more than $2\frac{1}{2}$ million years ago. These are, however, rather simple, and tools of this sort apparently were not made by hominids having fully evolved hands. It was only much later, when the hand (and brains) of hominids were more advanced, that fine stone tools were manufactured.

This is interpreted as another instance of the reciprocal relationship between behavior and anatomy in evolution. It was not the case that a hand that could make tools evolved independently of toolmaking. Rather, toolmaking conferred advantages that led to anatomical changes that led to a hand that could better make tools, which in turn led to better tools, which in turn conferred more advantages, and so on. Simultaneously, a brain capable of conceiving more and better tools (as well as other things) was evolving, dependent upon and contributing to the changes in anatomy and behavior. The end product was of course the large-brained, bipedal, toolmaking, abstract-thinking, cultural creature we know.

The third, and for us the most significant, way in which human and nonhuman primates differ is in behavior. Complexity of behavior is in turn directly attributable to the size and organization of the brain. There are, alas, no fossilized hominid brains, so neuronal structures cannot be compared. The best data on hominid brain evolution come from the skulls, which provide information only about the grosser aspects of brain morphology. The brains of other living primates are therefore studied in an attempt to obtain some sense of the relationship between brain and behavior that may have existed at the inception of hominid evolution. This must be done with great care, since the current nonhuman primates are themselves products of independent evolution.

The human brain is approximately three times as large as that of the largest other primate. The average size of the brain of modern humans is 1400 cm³, whereas that of the gorilla is 500 cm³. The chimpanzee, which of course has a much smaller body than the gorilla, has a brain size of 400 cm³. The largest size difference between human and other brains is in the cerebral hemispheres, the structures which surround and essentially encapsulate the brain stem.

However, there was a change not only in hominid brain size, but, as would be expected, a change in brain organization. New parts having new functions were added, and the function of evolutionarily older parts were altered and integrated with the new. That the reorganization of the brain has been a major step apart from the increase in size is shown by the fact that human microcephalics, having brains much smaller than normal, behave in distinctly human ways.

The enlargement of the brains of hominids was not accompanied by an equivalent increase in the number of neurons. There are only 25 percent more neurons in the human cerebral cortex than in that of the chimpanzee (although this is an increase of approximately 1.4 billion cells). However, human neurons are larger, have more connections, and are spaced farther apart. In the spaces there are more of the supportive glial cells.

Noticeably expanded in humans are the sensory association areas of the cerebral cortex and their related subcortical regions. In particular, an area of the parietal lobe that has connections only with other association areas and not with primary sensory or motor regions has become enlarged. This "association cortex of association cortexes" is quite small in apes but immense in humans. Part of this cortex appears to correspond to the speech area, and it seems to be the place where integration of input from the different senses occurs.

The frontal and occipital lobes have also greatly expanded. The function of the frontal lobes is not clear, although evidence now indicates that they may be concerned with the initiation, maintenance, and inhibition of behavior. The occipital lobe is concerned especially with vision and, as one would expect from what is known of the importance of vision to humans, has increased in size and complexity.

Indeed, Jerison (1963) proposed that the major contribution to the hominid line lay in the increase in "adaptive" neurons, which were defined as those beyond the basic "housekeeping" neurons used, for example, for moving muscles, maintaining visceral functions, and so forth. The larger, reorganized hominid brain thus had more neuronal units and connections to devote to tool use and production, intricate social structure, and other functions. This proposal appears to have merit, although it is difficult to test.

Ramapithecus

The living apes are forest dwellers and, as far as is known, so were their ancestors. Between 15 and 10 million years ago a new habitat appeared in Africa, consisting of areas around lakes and rivers that flooded seasonally and thus did not support forests. In these places, next to woodlands but where there were open spaces, several primate lineages were "experimenting" with new behavior patterns. Only one, it is thought, survived: the first hominid.

In 1934 G. L. Lewis, then a graduate student, described the fossil specimen of what he believed to be a new primate. He considered it the most humanlike of the early apes, and so stated in his description. The following year a senior U.S. anthropologist, Ales Hrdlicka, wrote a polemical paper disagreeing with Lewis's conclusion. Hrdlicka's paper had the effect of discouraging inquiry into Lewis's find for more than 20 years, until Elwyn Simons restudied it and others resembling it. The result was that the fossil discovered by Lewis is now considered to be the earliest of hominid species, *Ramapithecus* (Philbeam, 1972).

Ramapithecus evolved about 15 million years ago, probably in Africa, and spread throughout what was then the African-Asian faunal zone. It was predominantly a forest animal, though probably a ground feeder that took advantage of foods in open grasslands by rivers and lakes and at the forest edge. The canines were small and were, or were coming to be, hominidlike, with broad cutting edges like incisors. The remaining dentition was adapted for chewing coarse food.

There is unfortunately little in the way of other evidence to tell us about *Ramapithecus*. No postcranial bones have been found, and nothing is known of brain size or organization. Comparison with modern apes nevertheless suggests that in the open foraging areas it was bipedal. This phase of hominid evolution lasted for millions of years, until bipedalism was established sufficiently for hominids to become habitual ground dwellers.

There is no evidence to suggest that *Ramapithecus* was a tool user or a hunter, at least not any more so than the chimpanzee. That is, the males may have used branches and rocks in displays and may occasionally have hunted small game, but they apparently had not taken up tool use or hunting as a way of life. As mentioned, *Ramapithecus* evolved not in the open savanna but in the woodlands, and the males had already lost their large canines.

Australopithecus

In 1924 a small skull, almost complete, with a natural brain case formed from limestone, was found during mining operations in South Africa. The skull came into the possession of Raymond Dart, a young professor of anatomy. Impressed with its combination of human and apelike features, he classified it as belonging to a species in the ancestry of humans. Once more, however, authority intervened, this time in the form of European, chiefly English, anthropologists. They objected to Dart's classification—first, because he was thought to be inexperienced; second, because their view was that early hominids were characterized by large brains and apelike jaws and teeth; and third, it now appears, because they were not consulted before Dart published his findings. (Much of this is, alas, a hopeless normal situation. The theories of the English anthropologists were in addition based upon the Piltdown skull, now known to be a fake.)

Dart, and others, persisted nevertheless, and in 1936 another skull was found. Soon thereafter the evidence became so overwhelming that Dart's man-ape, *Australopithecus*, was admitted to the paleontological lexicon as an ancestral hominid. Two paleospecies are in fact presently considered to have been uncovered in Africa, one ancestral to the other and both ancestral to humans.

Australopithecus africanus dates from at least five million years ago. It evolved in the open country and was bipedal, though not as efficiently as humans. It was short (about 4 feet tall) and light (weighing from 40 to 70 pounds). The brain volume averaged approximately 440 cm^3, midway between that of a modern chimpanzee and a gorilla, although the body was of course smaller. Most important, the brain was apparently undergoing a reorganization in the direction of later hominids, with a differential expansion of the parietal area. Hence, *A. africanus* may have been capable of rudimentary speech, and may have been a cultural animal in a way that other primates are not (Philbeam, 1972).

A. africanus used tools that can be dated to be almost three million years old, and it is quite possible that toolmaking is older still. Tools, associated animal remains, and locomotor and dental adaptations suggest that *A. africanus* was a cooperative hunter-gatherer, and not an individualistic forager. In all likelihood meat was eaten regularly, forming as it does among modern hunter-gatherers about one-third of the diet. The presence of the tools, grossly chipped stone implements, indicated that *A. africanus* had already begun one of the most significant events in human evolution.

On the basis of tooth wear, *A. africanus* individuals are considered to have lived no more than 40 years, with only 15 percent of them reaching the age of 30. The mean age at death was 20, and one-third of the individuals evidently lived beyond the age of reproduction. Maturation was apparently rather slow, allowing a long infant and juvenile period, and hence sufficient time to acquire the skills necessary to exist and to contribute to the group.

The *Australopithecus africanus* remains span some three million years. The earlier specimens resemble *Ramapithecus* whereas the later form a series approaching a second and later paleospecies, *Australopithecus habilis*.

Beginning in 1959 many different specimens, in most cases well dated, were found by Louis and Mary Leakey at Olduvai Gorge in Tanzania, by F. C. Howell at Omo in Ethiopia, and by Richard Leakey in the East Rudolf area of Kenya. For our purposes the most important finds have been of an *Australopithecus* form, which existed for about a million years, beginning two million years ago. This form showed a number of advances over, and is thought to have decended from *A. africanus* (Philbeam, 1972).

Australopithecus habilis was, like *A. africanus*, small (4½ feet tall) and light (50–70 pounds) and apparently lived in woodland areas around lakes and rivers. Like *A. africanus*, it probably walked in a slightly bent-kneed posture, being incapable of locking the knee so that the leg is straight, as do humans in their striding walk. In brain size, however, *A. habilis* exceeded *A. africanus* considerably, having a mean volume of more than 650 cm^3. The hand was also advanced, being capable of a powerful grip (necessary to manufacture stone tools). The fingers were still curved, however, and the thumb was relatively shorter than in modern humans.

Most interestingly, with *A. habilis* there is evidence of life in semi-permanent camps. The lake that was present near Olduvai, for instance, provided an attraction for all kinds of animals. A number of living sites have been discovered close to what was the lake's edge. From these, which were possibly seasonal camps, *A. habilis* hunted. It did not do badly, either. At one site the entire dismembered skeleton of an elephant was found.

Homo Erectus

The earlier hominids seem mainly to have been confined to Africa. (Other parts of the world have not been adequately surveyed, however, so the evidence for this is negative.) However, by about one million years ago hominids were assuredly distributed throughout much of the tropical and subtropical parts of the Old World. These hominids showed significant changes from *Australopithecus*, significant enough for them to be classified in the same genus as modern humans.

The first *Homo erectus* find was in Indonesia, in the nineteenth century. Fossils were also discovered in northern China in the 1920s and 1930s. Since such middle Pleistocene hominids had by then not been found in other parts of the world, the idea developed that *H. erectus* (which at the time went under many different names) was an Asiatic species. Later discoveries have demonstrated this not to be the case. *H. erectus* was widely spread, and was one species, although with considerable regional variation in morphology among populations (Philbeam, 1972).

It is thought now that hominids emerged from Africa at least two million years ago, at the *Australopithecus* stage. During this period, *A. habilis* was evolving into *H. erectus*. Whereas *A. habilis* was about

$4\frac{1}{2}$ feet tall, *H. erectus* was a foot taller. *H. erectus* was also considerably heavier, with much thicker bones. In addition, there was a notable increase in brain size. The average volume of *H. erectus* brain is estimated at 940 cm³, which begins to approach that of the modern human. (Some specimens in fact have been found with volumes in excess of 1500 cm³.) The expansion was, however, not general, being expressed chiefly in the parietal and temporal lobes, and possibly in the cerebellum. The overall shape of the skull became longer, broader, and flatter than in *A. habilis*. The causes of the brain changes are not known, but they were almost certainly associated with behavioral changes.

H. erectus was also a fully adapted biped, having none of the limitations of *Australopithecus*. Selection associated with increasing size of the home range is thought to account for this, which in turn is accounted for by more efficient (and possibly the beginning of big game) hunting.

A major innovation of *H. erectus* was in the use of fire. A very large fauna is found associated with *H. erectus* living sites and many of the bones are charred and splintered. Also, a number of hearths have been uncovered. Sites in France, Hungary, and China all show extensive use of fire as well as evidence of big game hunting.

Fire releases more nutriment from food, as in roasting meat or the boiling of tough vegetables, lessening the need for a heavy masticatory apparatus. This doubtless accounts for the reduction in jaw size seen between *H. erectus* and modern humans. But fire provides another service, one that was of far greater importance. Fire enabled *H. erectus* to live in the north temperate zone—and this, according to Campbell (1974), was the most significant step in the evolution of *H. sapiens*.

Unlike tropical environments, temperate regions are subject to extensive seasonal changes in temperature. There are usually two or three months of the year when plant growth ceases, and edible resources must be preserved. Fire may have contributed to this. It also seems highly likely that fire made cave dwelling possible, providing both warmth and protection from predators such as the cave bear, which in cool climates was a competitor for shelter. Thus the cave became a permanent home rather than a temporary base, reducing nomadism and contributing to population growth, since a larger number of children can be maintained in one place than is possible with a constantly moving band.

There is evidence of the use of fire in Europe about 350,000 years ago. One of the earlier occupations of a temperate zone is recorded in cave in France about 100,000 years prior to that. It was during this period that *H. erectus* began to master fire.

Homo Sapiens

Homo erectus existed for approximately 750,000 years and then gradually gave way to that paleospecies that we call "man." It was not quite man as we know him morphologically, but rather an archaic form, still having rather prominent brow ridges and large teeth. But the masticatory apparatus was by this time under the brow and a chin had developed; that is, humans had faces rather than muzzles—a trend since the earliest *Australopithecus* finds. Postcranially there is little difference between modern and archaic man. In brain volume, archaic *H. sapiens* had added another 300 cm³, reaching average values well within modern range.

The increase in brain size probably reflects changes in internal organization and in the morphology of neurons and glia, which are in turn related to an increase in the complexity of behavior. During the last 200,000 to 300,000 years or so humans had in fact participated in a cultural explosion that has changed their manner of living in ways we are only beginning to appreciate.

The most famous archaic humans are of course the Neanderthals of Europe, first discovered in the 1850s. Later discoveries of roughly similar humans in other parts of the world were initially described as if they were European migrants, but it is now thought that the various groups represent samples from a single, widely ranging polytypic species. This matter is an important one, for some earlier anthropologists had been led to postulate that Europe was the center of the evolution of *H. sapiens*, whereas the number of *H. sapiens* remains found in Europe relative to the rest of the world is more likely a function of the number of archaeologists working there. It is nevertheless probable, according to Philbeam (1972), that not all *H. erectus* populations evolved into archaic *H. sapiens*, and also that not all archaic *H. sapiens* populations evolved into modern humans.

The question of who evolved into whom is of scientific interest. Unhappily, it carries an enormous emotional burden as well: One need but peruse even casually any phase of the written human record to uncover exegetical passages on the topic of the behavioral inferiority of certain peoples, and normally such passages are concerned much less with performance than with origin. This may be something that cannot be helped; humans are not alone in distinguishing "we" from "thou" (although we are no doubt the most agile executors). Nevertheless, it is something the effects of which we must strive constantly to circumvent, not merely because such circumvention is compatible with current definitions of propriety, but because it is necessary for individual survival. We shall examine this in some detail later.

Homo Sapiens Sapiens

About 40,000 years ago there was a change in tool manufacture in Europe. New types of industries appeared quite abruptly, and it seems likely that they were brought at least to western and central Europe by migrants from eastern Europe, North Africa, and western Asia. The migrants were fully evolved modern humans, *H. s. sapiens*, complete with round, vaulted skulls, absence of heavy brow ridges, and facial skeletons almost completely under the brain case.

What happened to the Neanderthals, whence came the modern humans, and why did they evolve? Evidence is fragmentary and hypotheses tentative. The fate of the European Neanderthals is unknown. Some populations undoubtedly were absorbed by new ones; others died out. It is unlikely that any were directly destroyed by modern humans (Philbeam, 1972). It should be recalled that this was probably a time of great population movement, expansion, and decline, and changes came quickly.

In eastern Europe there is evidence that between 30,000 and 40,000 years ago populations existed that were either transitional groups or hybrids between Neanderthals and modern humans. The skulls are intermediate, and the postcranial bones also show a mixture of features, although generally they resemble *H. s. sapiens*. The hypothesis that the individuals came from a population that was part of a continuum from Neanderthals in western Europe to more modern types in eastern Europe and western Asia is favored by some. It thus appears probable that during the last 100,000 years the populations of Europe, North and East Africa, and contiguous parts of western Asia formed a mosaic of morphologies ranging between the specialized Neanderthals and forms differing little from today's humans. Thus, the more eastern and southern the populations examined from this period, the more modern they appear.

It is in Africa and adjacent areas of Asia that the earliest traces of indubitably modern humans are found. The oldest unquestionable specimen of *H. s. sapiens* specimens date from 45,000 to 55,000 years ago, and come from what is now Israel (a fact that may produce an eerie, if momentary, feeling in those who were raised in the Western religious tradition but are self-conscious atheists). What produced them? No one knows. It is proposed, however, that the origin and dispersal of modern humans was due to a shift in certain populations from generalized big-game hunting to the pursuit of large-herd animals, with a resultant change in social organization, and hence in behavior (Philbeam, 1972). Such a change in hunting methods is seen as requiring the formation of much larger groups than were previously necessary.

Are these recent social changes related to the changes in morphology? It is impossible to say, at present. There is evidence for relatively modern-looking populations, or at least for populations that contained some relatively modern-looking individuals, in Ethiopia 100,000 years ago. It therefore appears probable that between 100,000 and 50,000 years ago many populations in East Africa, Arabia, western Asia, and India did evolve in the direction of modern humans. Then between 50,000 and 30,000 years ago these newly adapted creatures swept forth, swamping and replacing other, archaic populations, until the earth was theirs.

HUMANS AND BEHAVIOR

Behavior, we noted earlier, has been subjected to selection throughout evolutionary history; in fact, changes in behaviors may have initiated the changes in the skeletal anatomy with which paleontologists were so long preoccupied. It follows then that an understanding of our present behavior requires an understanding of our behavioral past. But, as we also noted earlier, behavior does not fossilize. We are left then with bones, artifacts, a knowledge of ecological relationships, and the habits of other primates, as well as those of humans, to reconstruct human behavioral evolution. Any such enterprise must be speculative. Nevertheless, a good case can be made for the conclusions drawn by anthropologists in the last decade. It is to those conclusions that we now turn.

The major innovation in the evolution of early hominids was cooperative hunting. A number of adaptations were obviously necessary before hunting was possible, and many other human attributes evolved in conjunction with it, but fundamentally it was the acquisition of a hunting life by a species having the general behavioral features of primates that defined the evolution of *Homo*. What is unique, that is, is not cooperative hunting per se, which we share with dogs, wolves, and killer whales, nor basic primate behavioral tendencies which of course we share with other primates, but the combination. Certain adaptations must have occurred, other adaptations are likely to have occurred, in order for primates to hunt successfully. These adaptations, it is believed, made humans possible.

The most noteworthy fact about man the hunter, even as he is today, is that alone and without artifacts he is poorly equipped to hunt. Human adaptations have then chiefly been behavioral, since there were no special physical alterations related directly to the catching of prey. Humans began as primates living on the forest edge (perhaps pushed there by competition with cercopithecoids for food), increasingly ground feeders, going bipedal, having opposable thumbs, small canines, and binocular (and color) vision. Gradually they came to exploit a new environment. Meat doubtless did not make up much of the diet at first, although meat had always been a part of it, as it is with most modern primates. Later they came to depend upon it more and more, until adaptations to the exigencies of hunting had indelibly stamped them as a new species.

There is no evidence that *Ramapithecus* hunted regularly. The species is presumed to have lived near the forest, to have had a history of arm swinging, and is known to have had small canines. At this stage the social structure probably resembled that of the chimpanzee, with only temporary groups coming together at feeding areas. Similarly, there was probably little sexual dimorphism for size, the females being only slightly smaller than the males. *Ramapithecus* was probably an independent forager, but, as among chimpanzees, there may have been rudimentary cooperation among males for the occasional catching of small game, and some food sharing. Doubtless there was a dominance hierarchy among males and among classes of females.

It is certain that *Australopithecus* hunted. The fossil record between *Ramapithecus* and *Australopithecus* is, for the moment, lost to us. However, the living sites of *A. africanus* and *A. habilis* show both tools and the remains of kills. *Australopithecus* was a fairly efficient biped, tending more and more to live in the open where game was to be found.

H. erectus of course also hunted, adding very big game to his prey repertoire. Later in his stay he mastered fire, and with it a new environment that opened to him an even richer fauna. *H. sapiens* continued, exploiting the new environment, living, as did *H. erectus* in some cases, in permanent camps. Perhaps the last major change was the hunting of range animals, which would have involved cooperation of groups much larger than the males of one band. The harvest of migratory animals while the hominid population remained in place would be an important derivative, enabling the population to grow even more by reducing nomadism.

What are the implications of a hunting life, and how may they be fitted with what we know of human evolution and the behavior of modern humans and other primates? Cooperative hunting requires, first, obviously, cooperation among the hunters. It requires planning, foresight, communication, and, equally important, a commitment to the group. Animals lacking claws, fangs, and speed must depend upon intelligence to capture prey, and they must be able to share with others the knowledge involved. A premium was then put at the outset upon what was already the basic primate pattern: a social existence.

The emphasis was upon a different kind of social existence, however. Life apart from the group would have become not only uncomfortable, but almost impossible, as individuals became dependent upon one another for sustenance. Other primates are single foragers; each individual finds his own food daily, and none provides for another. Cooperative hunters, on the other hand, must rely upon one another if they are to survive. This deepens and expands the bonds usually present among primates.

It reduces, for instance, the direct effects of the dominance hierarchy. Not only the high-ranking males, but all males must participate if a small band is to provide for itself. The group cannot afford to alienate its young males through constant threat, intimidation, or physical attack, but must generate loyalty and a desire to contribute. This is hardly to say that status concerns were lost in human evolution; we know they were not. It is to say that the kind of rivalry typical of primates (and many other mammals), involving essentially a day-to-day, one-on-one confrontation among males for precedence in all affairs has been lessened and converted into new forms that better serve the needs of an interdependent and integrated group. By all means privilege remains, and by all means it is aspired to. It is the nature of privilege and the manner of obtaining it that (quite subtly it seems at times) has changed.

A second derivative of the hunting life is a division of labor. The males of most mammalian species are bigger and stronger than females (even when the amount of sexual dimorphism is relatively small). In addition, they are not burdened with (or capable of) the bearing and suckling of offspring. Given the long period of dependence of the young upon the mother among primates, and the increasing need for parental care among evolving hominids, it appears reasonable that the task of hunting should fall to males. It has, indeed, in all societies we know.

The role of females simultaneously expanded from that of mother to that of mother and cooperative gatherer. While the males were hunting meat, and were probably away for some periods of time, the females gathered the vegetable foods that provided perhaps two-thirds of the group's diet.

A third derivative of hunting is food sharing. This exists in incipient form among some primates, but not at a subsistence level. It is essential in a hunting society. Clearly, what is caught by all must be shared by all, else the system is bankrupt from the start. All of the males who were on the hunt must get a share of the kill. Similarly, males and females must share the products of their respective labors. The result is, once again, that a new sort of dependence is formed.

The picture that emerges therefore is one of cooperation and food sharing among males and of cooperation and food sharing between males and females. Both result in tendencies for increased interdependence, with one major result in the case of relation among males being a reduction in at least the more direct forms of intermale competition within the group, and a major result in the case of the relation between males and females being the inception of the human family.

Recall that a young primate relies almost exclusively on its mother. With the development of mutual dependence between the adult male and female, however, a new role emerges, that of "father." A major factor in this is the hominid's increasing creation of and need for tools and weapons. Juvenile males must learn to make the artifacts that will serve them as adults, and they must learn to hunt. Such things would not be the province of their mothers, who would be occupied with the very young, with training other females, and with acquiring vegetable foods. It became important then for males that the long learning period open to the young be used for acquiring hunting skills, and the one directing such acquistion in this case would be the father.

New kinds of social relationships, based broadly upon the logistics of food acquisition and distribution, thus developed in the evolution of humans. The foundation of these relationships was the already highly evolved social sensitivity that we know to exist in modern primates and we assume to have been present in early hominids. The effect of the change was to increase interdependence in such a way as to provide for the smooth functioning of a small hunting band.

Cooperation, as mentioned, requires foresight, a conscious prediction based upon knowledge, previously acquired, of the results of various acts. The kind of intelligence we think of as particular to humans would have as its basis the cooperation among males in the planning and execution of the hunt, and between males and females both in the division of labor and in the rearing of offspring. Successful hunting in animals not physically well equipped for it thus involved selection for certain behaviors, which in turn required and was reciprocally associated with selection for anatomical characters, primarily in this case the fundamental reorganization and later the exponential expansion of the brain.

Forethought, planning, is of little use to a group unless it can be communicated from one individual to another. There was therefore a necessity for selection for increased powers of communication simultaneous with and again reciprocally allied to the selection for the differing social relationships. This obviously also involved brain changes, leading eventually to the evolution of areas especially associated with speech.

A number of other features that mark the human circumstance are also thought to have evolved concomitantly with those mentioned above. The evolution of the family and the consequent development of special bonds between a male, a female, and their offspring meant that sexual perogatives would be restricted—as they generally are not among other primates. Human female sexual behavior, which is different from that of any other mammal, is believed to have developed as it did as a consequence.

Mammalian females at large are not receptive sexually except for several days surrounding the period of ovulation. This period is known as estrus (derived from the Greek for "gadfly"—hence, "sting" or "frenzy"—presumably because it reflects a dramatic change in behavior). In a mixed, promiscuous group, a number of females will come into estrus each month, and most if not all of the males will have access to them. If the family were to evolve, however, the situation would have to be different. Males show little periodicity in their interest in females; therefore the females would have to show little periodicity in their receptivity to males. This is in fact what occurred. Hominid females lost the estrous cycle common to other mammals, to the degree that there are, so far as is known, no physiologically preemptive time limits on female receptivity. Women are willing to engage in sexual activity at any stage of the menstrual cycle, as well as during pregnancy (although some evidence does indicate a heightened interest near the time of ovulation). This change in behavior was accompanied, as were the others, by an increased control by the "higher" (i.e., cortical) regions of the brain. Sexual behavior on the part of the female thus became much more a matter of conscious choice, as it would have to as it began to serve economic and political as well as reproductive ends.

We have noted that in a number of terrestrial primate forms the male is much larger than the female. This condition is thought to have developed for several reasons, intermale rivalry and troop defense being two. There may even have been a selection for smaller females as well as larger males in some cases, particularly when food was scarce. The effect of bipedalism upon the pelvic canal, the large fetal head, and the necessity for the mother to carry her progeny over long distances, however, meant that there was a lower limit upon the size of hominid females. This, in addition to the apparent lack of initial dimorphism, probably kept the size differences between the sexes relatively small.

The human female is, however, different from the male in secondary ways that are unique in her order. Among the more notable is the development in the adult of permanent breasts. These are not a reproductive necessity; the amount of actual mammary tissue in even a big-breasted woman may be no more than a spoonful (Harrison & Montagna, 1969). The rest is fat and connective tissue. What then is their function? Not surprisingly, to one reared in the Western tradition at least, it is erotic. The breasts of the human female are thought to act as more or less continuous signs of her sexual availability, and, in a species not otherwise greatly dimorphic, they appear to serve this purpose well.

The hypothesis is not as curious as it might otherwise seem. Females of other primates commonly generate physical and physiological, as well as behavioral, signals indicating that they are in estrus. The physical signs are swellings of the skin on the rear, near the vagina. They can be quite large, and quite colorful. The physiological signals occur in the form of pheromones (chemical communicators), which serve to alert and excite males. However, hominids apparently lost their capacity to communicate with pheromones. They also became bipedal, and the upright stance relocated the vagina and any signs around it to a far less conspicuous place. It seems plausible thus that evolution should seize upon some other anatomical feature to act as a sexual signal.

Permanent breasts are a good choice, for a number of reasons. First, no radical structural changes are involved. Second, the evolution of breasts could proceed as a part of a general secondary dimorphic phenomenon, the acquisition by females of the fatty tissue in various places (hips and buttocks being two others) that give them their special, ever-appreciated form. Third, and most interestingly, it is thought that, with the development of speech, hominids became essentially face-to-face creatures, and so a likely location for sexual signals would be ventral, on the front. Concomitant with this, and with the evolution of bipedalism, was the habit of copulating in the ventral position, a behavior uncommon except in humans. Ventral-ventral copulation would lead to ventrally located signs (seen also in pubic hair, no other function for which is known). Finally, something of this order is believed to have occurred in another primate. Gelada baboons spend most of their time sitting upright in order to feed on small morsels on the ground. Their rumps are, for primates, underexposed. As a consequence, in the female gelada there is sexual skin, which changes color during estrus, on the chest.

Another epigamic characteristic of the human female is her general hairlessness, particularly the acquisition of smooth facial skin and lips. This emphasis upon the face shows little cultural restriction, and is, obviously, a ventrally located feature. Relative hairlessness is a trait human females share with males, and is thought to have evolved along with sweat glands as a means of dissipating heat. That it has proceeded farther in females is, however, believed to be due to sexual selection. Males and females have the same number of hair follicles—indeed, they have the same number of hairs. It is in the length, pigmentation, and texture of hair on the face and body that the differences occur (Harrison & Montagna, 1969).

A related dimorphic feature is of course the male beard. Whereas in females there has been selection for facial hairs which are so small and light as to be almost invisible, in males the direction of selection has been quite the opposite. The male face is one of the two places where hair does not cease growth at a certain length; a beard is then capable of great dimension, the reason evidently being to increase the apparent size and ferociousness of the possessor. The beard is, thus, both a derivative of the primate past and an object of selection, due principally to intermale competition. It serves the same function as does the mane of the lion or of the baboon.

A face-to-face, speaking creature with a developing intelligence and self-consciousness is bound to be appreciative of the uniqueness of particular individuals. This would be true among potential sexual partners, for whom social compatibility would be an increasingly important factor. It would also be true of individuals who make notable contributions to or fulfill special roles within the group. What began as a division of labor between sexes thus continued into many divisions of labor, differentiated finally into particular roles. But the more roles, the more different ways of behaving, the more the group would be required to be aware and tolerant of behavioral differences. Society thus came to depend upon variety.

Variety presupposes the importance of the individual, for individuals differ in their competence to fulfill roles and in their inventiveness, and it is upon these that the maintenance of cultural standards, not to mention cultural progress, depend. Human behavior, it is thus argued, is to a fair extent directed by self-conscious intelligence; however, such intelligence is not the property of the group but of the individual. The group therefore comes to rely upon the intelligence of its individuals, and at the same time, because their behavior is so important to group survival, seeks to control it. At the cultural level, society thus looks to the individual as the source of novelty, not the reverse (Campbell, 1974).

Intelligence and intelligent activity are of course not necessarily correlated with whatever definition of good citizenship is current, nor with affability. There will always be social outliers who are sources of cultural invention. As a result, an important part of human evolution has involved selection for the processes that lead to the recognition and preservation of individuals. According to Campbell,

> Society must value an individual as a person, for upon individual people the evolution of society depends. In a cultural society the contribution of each individual to the population is no longer via the gene pool but may be direct and immediate, as a result of intelligent behavior. Cultural evolution finds its sources of variability, its "mutations," in individual behavior, and they can spread through the population, if they prove of value, within one generation. (1974, p. 367)

The human family is the simplest social unit with complete division of labor between adults. A larger group is required, however, if a primate hunting band is to survive. Even larger groups are desirable for capturing big game, but such groups must consist of males from several different bands, brought together for purposes of the hunt. It is upon such a basis that ever-widening social relationships are believed to have developed, and such social relationships are thought to have led to the common, if not universal, practice of exogamy.

Exogamy means "marrying out," in this case marrying out of the particular band, or later the tribe or the clan, into which one was born. It is a peculiar practice superficially; not at all primatelike, since most mating occurs within the group in monkeys and apes. Exogamy is explained in hominids by what was essentially the politicization of sex. When mating became not merely a matter of reproduction, but a matter of economics as well, it came to be governed by rules other than those of sexual attraction. Long-term alliances could be formed among bands that were required to cooperate for hunting by the simple and formidable practice of reciprocal provision of mates. Soon descent groups developed, often with rather rigid and detailed specifications as to who should be mated to whom.

Exogamy becomes significant in evolutionary terms when it is understood that its opposite, endogamy, results in inbreeding. Some of the direct consequences of inbreeding in mammals have been described in Chapter 3. In addition, endogamy would restrict gene flow across populations, so that new and more viable genes or "coadapted" sets of genes (Dobzhansky, 1962) would not readily increase in frequency outside of a small group. This has the effect of limiting the speed of evolution, for evolution can proceed only as fast as organisms better adapted to their environment produce viable offspring. It is a fact that hominids, in contrast to other primates, evolved very rapidly —indeed exponentially in the later stages. The changes were fundamental, especially involving behavior and the brain. It is thought now that the consequences of exogamy provided the genetic variance necessary for such rapid evolution (Campbell, 1974).

Exogamy may be seen as an extension of something found in virtually all human societies: the prohibition against sexual relations among members of the primary family. The universality of the incest taboo is remarkable, considering that few things of (apparently) cultural origin are shared by all societies. The prohibition against incest may thus be seen as having derived from basic requirements that must have been met if hominids were to form families and raise children. The family began, we have noted, as a reproductive, economic, and (later, no doubt) an affectional unit. There are thus reproductive, economic, and affectional roles that are appropriate for each family member. Individuals change their roles, of course, as the family matures, but they cannot *inter*change them with another family member if the structure is to remain intact. Particularly, economic and reproductive (sexual) roles must not be confused. A father and son cannot compete for sexual rights to a wife-mother, nor a mother and daughter for the sexual rights to a husband-father, and still maintain the economic and affectional separation required for familial existence. Sexual relations between brother and sister, while a less direct threat to the viability of the unit, still deprives the family of the capacity to make economic alliances through the exchange of mates. They thus probably came under the same prohibition as those involving the parents as soon as mate exchange was practiced. Shortly afterward, given our penchant (indeed our need) for ritualized fantasy, the myths surrounding incest no doubt began to form in the primitive consciousness of protohumans.

Finally, the "rise of the individual" (Campbell, 1974), appreciated as it is among those of democratic inclination, as necessary as it was and is to cultural change, and as greatly as it has affected and been affected by the course of human evolution, is not, it seems, an unmixed blessing. Individual self-consciousness in a societal context brings with it simultaneously a gift and a burden original to humans. It is held to be responsible for our most sublime ideas, and, it must be supposed, our most exquisite agonies. It is held to be responsible for the evolution of ethics.

This is, obviously, a topic of great speculation, and we have touched upon it before (pp. 137–138). Although the origin and effects of the family, the peerless human brain, bipedalism, hunting, the evolution of the individual, incest prohibition, exogamy, and so on are all established with often large, but never small, empirical support, ethics itself remains a subject but poorly understood, and its origins therefore even less so. Still, it is of interest to follow the thoughts of modern anthropologists here, for there are few others with either intrinsic appeal or explanatory power, and there are many which are unclear at best. The evolutionary viewpoint, begun by Darwin, brings at least some cogency to a troublesome, perplexing, and highly significant affair.

The problem turns upon the relations between the individual and society. The self-conscious individual is aware of his or her own wants and needs, and of the behaviors associated with their satisfaction. He or she is also aware that certain behaviors are restricted, while on the other hand certain other behaviors are required by society. These individuals must act, therefore, not only to satisfy themselves but the group.

The dilemma posed by such a circumstance derives of course from the fact that the behaviors desired by the group and those pleasing to the individual may be antithetical. Since the individual needs society (and is aware of it) he or she is, inevitably, forced to choose. Therein lies the doubt, and the pain.

How did this come to pass? "The key to man's nature," writes Campbell (1974), "lies in his evolution of social hunting. . . . In such circumstances, the actions of every individual affected others in the social group, and what is important is that it was of course the social group (as part of the population) that was evolving, not the individual" (p. 363). The message here is significant: it is the individual who behaves, but it is the population that evolves, and during the course of that evolution individuals were selected for who tended to behave *responsibly* to the population.

This proposition may at first seem inconsistent with others advanced, since we have noted in discussing the rise of the individual that self-conscious, intelligent creatures have been selected for, and these, it appears, must to some extent while existing within it come into conflict with society. It is now suggested that there was also selection for behavior that was not strictly self-serving to the individual but advanced the fortunes of the group. Can there have been selection for individuals capable of independent, even defiant, acts, and capable also of behavior loyal and responsible to their groups? The answer is yes. It is not unusual evolutionarily when two things are adaptive, but one seems contradictory to the other, for a balance to have been struck, and both to have been produced, neither in excess. Such balanced contradictions form the crucible of our distress.

Ethics is thus not a breach of the evolutionary process, as Thomas Huxley (see page 61) had maintained. Rather, it is a *product* of evolution, as supposed by Darwin (see p. 137). However, the socially virtuous behavior, developed through the evolution of cooperation, which he had thought to have become "instinctive," has not reached that level. It appears that it cannot if the will and intellect of the individual is to remain the creative force of society. There is therefore still a great deal of conscious choice involved, and, although as individuals we may have more or less natural leanings in one direction or the other, in most of us a substantial conflict remains.

What has this to do with hunting? Cooperative hunting requires the organization made possible only through *self-discipline;* the enterprise is doomed if individuals while rounding up the prey act in what appears to be their individual interest rather than the interest of the group. It is thus the capacity for self-discipline, the capacity to experience guilt if one fails to exercise it, self-consciousness, and the ability to anticipate the emotional consequences of one's acts that makes an ethical animal.

In this respect, Campbell (1974) argues, we have more in common with other social carnivores than with nonhuman primates. It is because of our mutual understanding of self-discipline that dogs and man, for instance, get along so well with one another, and indeed appear to understand one another. Dogs do seem to experience a kind of emotional malady when they disobey, judging from the appeasement signals they make when called to task, and this is interpretable as something akin to guilt. Do dogs then have a social conscience? Campbell suggests that they do. Are they ethical animals? No. They lack the self-awareness and intelligence that allow them to act on principle.

Kant described an ethical act as one that derived not from the pleasure of serving a good end, but from a sense of obligation to serve it. His conclusion seems, at least in part, applicable here. Were ethical conduct always natural and easy, it would not be a cardinal theme in the intellectual history of man, it could hardly be an issue that begins wars and brings down governments, and we would not require the ponderous, complex, sometimes peculiar, and very formal juridical system that consumes so much of our time, effort, and treasure. It appears, rather, that although moralizing comes naturally and easily, ethical conduct in one way or another costs us dearly. It is the human condition to be capable of choice. It is in choice that anguish lies.

For this reason Freedman and Roe (1958) speak of the emergence of *Homo sapiens* as the dawning of the "age of anxiety." Conflict, doubt, guilt, anticipation of results—all are part of the anxious animal. Humans not only share some facsimile of anxiety with other creatures, but are surely the most anxious of all. The striking thing, however, given this, is how well they manage. If we have doubts and fears, we also have the capability to overcome them, or at least not submit to them entirely, and this, perhaps more than anything, distinguishes us from all others. It is a common human statement: All societies seem to chronicle the exploits of protagonists, real or imagined, who have in effect stolen fire from the gods.

There is substance to such statements. The inner conflicts, the doubts, the frustrations, and the anxiety of which we in modern times have through the efforts of psychoanalysts and psychologists become so aware may be among the most important stimuli to cultural innovation. Anxiety is of course in some dishabilitating. In most of us, however, it produces the search for a better way, for the development of more adaptive behavior and more adaptive technology.

Biologically this may seem somewhat unusual, but it is not. It is the population that evolves, not the individual, and it is therefore the population with which evolution is (metaphorically) concerned. The roots of anxiety will be selected for if, even though the individuals in a population are less content, the population itself is more capable of survival. As Freedman and Roe (1958) note:

Man is not the only conflictual animal or anxious animal. He may not even be alone in his capacity to become neurotic or psychotic. It seems highly likely, however, that his present state of biological evolution makes him the most vulnerable of all animals to psychopathology, and the values of his social evolution make him the animal most likely to survive in spite of severe maladaptive behavior. (p. 461)

There is a small lesson in this. Evolution uses individuals opportunistically—not as unique and irreplaceable citizens of history, but as elements in a mindless, monumental process with no beginning and no end. One generation, having survived, is merely sustenance for the next. So as marvelous as the process is, and as much as its workings fill one with awe, do not love it. It does not love you. It takes another person, or a dog perhaps, to do that.

EVOLUTION AND CULTURE

Given the relationship between behavior and evolution now held to exist, it is not surprising that the philosophical foundations of some sciences other than psychology have recently been called to question. One such subdiscipline is a part of anthropology itself, the part that specialized in the behavioral or cultural aspects of the study of humans. This was known, appropriately enough, as cultural anthropology (or even "culturology"). It is of interest here because in many ways its development paralleled that of behaviorism, insofar at least as the two shared a number of assumptions.

One such assumption was the overwhelming importance of "learning," conceived, tacitly or otherwise, as something more or less added onto a common human nature, and as the thing that really made the difference among people. The remarks of cultural anthropologist Leslie White with regard to the "utter significance of biological factors as compared with culture" we have already noted. Another leader in the culturology movement was V. Gordon Childe. The effects of the approaches of both, however, were the same: to reduce interest in the evolutionary and biological factors associated with human behavior. As Mead wrote in 1958:

> During the past quarter-century, most anthropological research has neglected the wider evolutionary framework and has narrowed down to studies of the details of cultural sequences or of the dynamics of living cultural systems; implicitly or explicitly, these studies have insisted upon the independence of cultural phenomena from biological phenomena and often also from ecological phenomena. (p. 486)

This state of affairs is currently being redressed. Culture is believed now not to be an independent addendum to the evolutionary-biological nature of humans but rather to be an integral part of it. Culture and human biology evolved together, each reciprocally influencing the development of the other, to the extent that they are inseparable determinants of behavior. Just as we must have a human biology in order to be human, so, it is argued, must we have a human culture.

The difference between earlier and modern points of view is striking. Geertz (1973) notes that attempts to find "true" man amid his various customs have differed in tactics, but not in philosophy. Humans were seen as stratified, existing in biological, psychological, social and cultural levels, each superimposed upon the one beneath it and supporting the one above. Analysis consisted in removing the layers, each being complete and irreducible in itself. When that was completed the supreme importance of the cultural level, the only one not shared with other animals, would be revealed, for it would tell us what people really are.

The tactics employed to apply this strategy have of course been altered to suit the temperament of various times. At one end of the continuum, for instance, was Rousseau's noble savage, bedeviled only by the artifices and institutions of civilization, while at the other end was the uncultured brute, made less brutish only by those very artifices and institutions. Thus, "For the eighteenth century image of man as the naked reasoner that appeared when he took his cultural costumes off, anthropology of the late nineteenth and early twentieth centuries substituted the image of man as the transfigured animal that appeared when he put them on" (Geertz, 1973, p. 38).

A view consonant with that of startified man was the critical point theory of the emergence of culture. It postulated that evolution of the capacity to acquire culture proceeded on an all-or-none basis:

> At some specific moment in the now unrecoverable history of hominidization a portentous, but in genetic or anatomical terms probably quite minor, organic alteration took place—presumably in cortical structure—in which an animal which had not been disposed "to communicate, to learn and to teach, to generalize from the endless chain of discrete feelings and attitudes" was so disposed and "therewith he began to be able to act as a receiver and transmitter and begin the accumulation that was culture." With him culture was born, and, once born, set on its own course so as to grow wholly independently of the further organic evolution of man. (Geertz, 1973, p. 63)

The notion that so much could have come about in a single stroke stretches the mind, no matter how strongly one believes in the power of mutation. Even 30 years ago this would seem to have been difficult to imagine. Evidently it was not, but in any case it can no longer be supported, and all thoughts of culture being the last, separable, and definitive layer in the strata of human personality must be put to rest. All of the evidence indicates that behavior and biology evolved together.

But how does one think of humans, if not in layers? It is undeniable that there are internal neurophysiological processes, overt behavior, and bodies of tradition and ways of comprehending the world that we may call cultures. The error lies in reifying their separation, in treating what may be divided for purposes of analysis as separable in fact. Such reification usually results from a confusion between strategy and reality.

It is the commonest scientific strategy to parcel a subject into numerous smaller pieces, which can be studied more conveniently than the whole, in order better to understand how the whole is constituted. The confusion arises when one forgets or ignores the fact that the pieces really do not occur separately and have only been made to appear to do so for the analytic convenience of the investigator. One then begins to study a part of humans, or even a part of behavior, quite out of touch with other parts. Then once a discipline has been established (having become, in effect, a culture), its practitioners finally act as if their own level actually exists independently of the whole, an entity on its own terms.

Obviously, this tends to be a problem encountered more often at "higher" levels than at lower ones. With regard to humans, it is thus more likely to occur in psychology, sociology, and cultural anthropology than in biology. What indeed appears sometimes to have happened in these sciences is that investigators, wishing to study only certain things, have simply declared them to be unrelated to others. Such declarations often come in the form of scientific manifestos, one of which—behaviorism—we have treated at length. Another is the culturology of White.

It should be mentioned that the manifestos usually include a disclaimer, noting that the authors do understand that some time in the future the various fields will be united. However, their importance lies not in their words, but in their effect. The effect, we have seen, is to disjoin one enterprise into many.

Even this is not greatly harmful, provided that the enterprises can be reunited under the same aegis—that is, provided that the partitioning of effort produces useful results, regardless of the philosophical beliefs of the partitioners. But it is here precisely that the behavioral sciences have miscarried. There is little evidence that the analytic separation of humans from their biology has produced valuable new knowledge. On the contrary, the suspicion grows deeper and more formidable with time that the strategy in this case not only has produced barren results, but also has hindered thought on the kind of research that must be undertaken if viable answers are to be forthcoming. A glance at the fields that are growing out of the fractures of the old psychology (sociobiology, psychoneuroendocrinology, behavior genetics, psychopharmacology, and so on), each allied with one another and with biology, gives quick confirmation that the suspicion has merit.

This is not to say, obviously, that dividing up the behavioral universe into more manageable portions is necessarily wrong; it is to say that the behavioral universe must be divided along different lines than those proposed heretofore. What the new lines will be we are only beginning to glimpse, and they too may change. What does seem certain is that the study of human behavior will not again be separated from the study of human biology, but that they will merge, as indeed they are merging, into a united science of human study.

Where does this leave culture? To be considered again, in a different context. Man, notes Geertz (1973), "is an animal suspended in webs of significance he himself has spun" (p. 5). He has not spun them arbitrarily, however, but in relation to his evolving nervous structure, such that the two are inextricably bound. Thus, "As our central nervous system . . . grew up in great part in interaction with culture, it is incapable of directing our behavior or organizing our experience without the guidance provided by systems of significant symbols" (p. 49).

This is a most interesting conclusion. Our biology does not merely allow us to acquire culture, but requires that we do so if we are to function at all. A cultureless human, far from being an "intrinsically talented though unfulfilled ape" would be "a wholly mindless and consequently unworkable monstrosity" (p. 68). Humans are distinguished from other animals then not only by their behavioral plasticity and their sheer capacity to learn, but also by their dependence upon a certain kind of learning: "the apprehension and application of specific systems of symbolic meaning" (p. 49). To understand the special condition of humans we must discover the relationship between the essentials of those systems and the biology that predisposes humans to learn them.

EVOLUTION AND THE BEHAVIOR OF INFRAPRIMATES

We have thus far very briefly treated the evolution of primate behavior, particularly that of humans, because it is the most relevant to our considerations. However, it is by no means the case that scientific work in this field has concentrated exclusively upon primate behavior. On the contrary, much effort has been put into the study of infraprimates, in the hope of understanding the evolution of behavior in general.

Primate research does, of course, tend to dominate the present scene, since we have become far more aware than ever before of our kinship with at least certain species through its findings. Van Lawick-Goodall's (1968, 1971) reports of the behavior of wild chimpanzees have done much to demonstrate both the amount and the limits of that kinship. Impressive also has been the demonstration by other investigators (Gardner & Gardner, 1969; Premack & Premack, 1972) that chimpanzees can be taught to communicate with humans using sign language or plastic figures expressing words. The capacity for language and the concepts it embodies may thus not reside solely in us, making salient once again Darwin's point that animals, including humans, differ from one another in degree rather than kind.

The mainstream of modern work upon the evolution of behavior in infraprimates flows from ethology, to the extent that a new generation of texts providing ethological principles and experimentation has already been written (Altman, 1966; Barnett, 1963a; Hinde, 1969; Manning, 1967a, b; Marler & Hamilton, 1966). We have treated ethology in a very limited way in discussing the question of instinct (Chapter 2, pp. 133–143).

The subject is too vast and both too old and too new to be considered further here. It is sufficient to understand that relationships between physiological mechanisms, environmental influences, genes, and the principles of evolution are being sought in a variety of creatures ranging from man to insect. A quite small and unrepresentative sample of the topics investigated are: appeasement signals in black-headed gulls (Tinbergen, 1972), the importance of central nesting sites in kittewake colonies (Coulson, 1968), the function of color in coral reef fish (Lorenz, 1962), mating speed of fruit flies (Manning, 1967a), social behavior of honeybees (Rothenbuhler, 1967), and population control and social selection in Scottish red grouse (Wynne-Edwards, 1968). These occur as well as reports on larger issues, such as those on aggression (Lorenz, 1967) and sexual behavior (Wickler, 1973). Some of the latter will soon be discussed.

Chapter 6
The Nature-Nurture
Resolution

In the initial chapters of this book two different views on the provenance of behavior were presented. The first of these may be broadly identified as the "nature" position and the second as the "nurture" view. Neither alone was found to have been satisfactory. The early evolutionary perspective, concentrating upon inborn factors (nature), appeared to cast us adrift upon the sea of biological determinism, where we would be swept by the inexorable tides of natural selection to our genetic destiny, a journey of the unknowing into the unknown, beyond aid or intervention. In contrast, the approach of experimental psychology, fastened as it was to the relationship between behavior and the immediate effects of external events (nurture), would set us blank upon the flat, trackless plain of the environment, our force and direction subject only to the click of the conditioned response.

Each of these views has put a special ingredient into our assessment of ourselves. One can, indeed, understand a fair amount of political and social history in terms of the nature-nurture debate. That is, there have always been elitists and egalitarians and, throughout the revolutions, suppressions, reformations, reinstitutions, and so on with which the human sentence is so frequently punctuated, the former often seem to have espoused the biological and the latter the environmental cause. There is scant reason for this, except perhaps that the elitists were in power a good deal of the time, and sought reason in nature for their continued privilege. The egalitarians, on the other hand, seeing little correlation between biology and the right of political dominion, were more or less forced to dismiss the first in order to obtain the second. Science, we have seen, at levels highly refined and primitive, and at times old and modern, played its role in this by contributing, not always impartially, to both sides. In this sense it can be said that science subsidizes politics. (That may seem strange, given the purported ideals of science. It should not seem so, however, given the predilections of its companion. Politics is a strange bedfellow to nothing.)

Whether or not this is generally the case, of course, need not detain us. It *is* true that we can detect two basically disparate approaches to uncovering the determinants of behavior, and that whichever scientific view is dominant at the time is likely to have the greater influence upon our inward vision. We have also found, as mentioned, that these approaches, taken separately, are nonviable. Neither the environmentalism of Watson and Skinner nor the instinctivism of early evolutionists and some ethologists can explain behavior. It has been argued, rather, that a strategy emphasizing study of the combined effects of organismic and environmental variables has the best hope of success, and indeed is the logical successor to the previous two. No claim is made that this is new; very little is new in human thought. The claim is that it should be applied.

In later chapters we examined some facts and principles that are necessary to know if the organism is to be reintroduced at large into psychology. The transmission of genes, the chemical gene and its products, and the evolution of behavior (as exemplified by the evolution of humans) were discussed. However, the mere reintroduction of the organism does not fulfill the requirements of a better strategy. Rather, one must specify the way in which organismic differences are to be used in conjunction with environmental manipulations. That will be seen to be both simple and complex, and is the subject of the present chapter.

TYPOLOGICAL VERSUS POPULATIONAL THINKING

The strategy advocated here may seem complicated initially because a number of different, seemingly contradictory, precepts must be held in mind. This is a difficult task; it is always easier to play either-or than both-and. Nevertheless it must be done, and perhaps a useful place to begin is with an old problem in biological systematics, the question of types versus individual differences.

Classification is necessary to achieve order, and order is necessary if systematic relationships are to be ascertained. This applies not only to science, obviously, but to everyday life. It is in fact impossible to imagine human creatures evolutionarily making their way without classifying. An expanded consciousness discovers events and conditions that are far too many and too multifaceted to be treated on a one-to-one basis. Classification is, then, normal and helpful; we expect it and we use it.

The problem arises when we ask more of classification than it can give. We may group things together or separate them so as to help us take further steps in uncovering relationships, that is, but we must remember that it is we who have done the categorizing, not the universe. This is particularly true in the biological and behavioral sciences, where classification schemes are usually somewhat loose heuristic devices with many crossovers and exceptions. But we have a tendency to look upon them as ends in themselves, with the result that the categories become reified, as if they, and not the creatures composing them, were what exists.

This appears to have occurred in biology with respect to types. It began most recently and formally with Linnaeus, who brought order to biology by developing categories based upon the similarities and dissimilarities among organisms. Naturally others followed (for it was an impressive accomplishment), adding to and altering his scheme, and for a while most of biology was systematics, the business of biologists being the discovery and classification of new species.

Just as naturally, debate quickly centered upon the criteria for "good" species. Good species, it followed from the very idea of distinct categories, were ones within which there was small variation among individuals. However, even a small amount of variation left systematists with a problem. How could they concretely describe the characteristics of a species when those characteristics differed from individual to individual? It soon became clear that a convenient approach would be to describe the characteristics of a single, idealized form, as if it in some sense *represented* the species. This is indeed what happened, and the concept of the *species type*, a representative of the species, having its characteristics in ideal proportion, developed. Of equal importance, the differences among the organisms that actually constituted the species were thereafter (almost by default) considered to be *error deviations* from the ideal type. Thus the implicit notion was that a species type existed and that the individuals who failed to conform to it did so because of some sort of mistake of nature.

This is less otherworldly than it might seem. Recall that the belief before Darwin was in species fixity through divine creation. Since the Divinity was not perceived to be fond of variation among individuals (which must indeed seem a bit diabolical to a professional classifier), and since He could not err, individual differences had to be explained as nature's attempt to replicate the ideal type but missing. There are, thus, grounds for clemency for the older systematists. But they left a pernicious legacy nevertheless: One cannot think in terms of an ideal and take the approach that it is error that produces variation among individuals without coming to see variations through a pejorative eye.

The problem of biological unity and diversity is clearly one with which we still struggle. There are no final and secure answers; we only progress at all in the questions we ask. Darwin, for example, asked about the role played by individual differences when they were not conceived as error deviations and was led to the theory of natural selection. He also foresaw what should have been the effect of his theory upon the concept of the ideal type: "Systematists will be able to pursue their work as at present; but they will not be incessantly haunted by the shadowy doubt whether this or that form be a true species. The endless disputes whether or not some fifty species of British brambles are good species will cease" (1859, p. 447).

His point, as valuable now as it was then, was that one would not waste concern over ideal species types if he understood that variation among individuals of the same species is necessary for evolution, and that, indeed, species have evolved to be variable. Those that could not change did not last. Although some of the genetic variability may be "hidden" (Dobzhansky, 1955; Lerner, 1954), it is there nevertheless and can be demonstrated in programs of selective breeding.

Furthermore, in most cases the phenotype (what one sees or measures) must be a reasonably reliable guide to the genotype, because it is the phenotype that is selected for, not the genotype. That is, genes are not selected for directly, but rather evolution acts upon the morphological, physiological, and behavioral characteristics with which the genes are associated. For selection to be effective, then, individuals phenotypically better adapted to their environment must pass on a greater proportion of their genes than individuals less well adapted, and those genes must produce in the progeny generation individuals even more adapted. This cannot occur if the variation among individuals at the phenotypic level is chiefly error.

But, alas, Darwin was premature in his predictions. Systematists would not relinquish their representative types for another century or so because, one must suppose, they were used to them and still found them helpful and because the problem of unity (of the species) and diversity (among individuals of the species) is difficult even to state, much less resolve. Then too, there was so much in *Origin*, the distillation of 20 years of thought, that not all of it could be comprehended at once.

Somewhat later, type thinking was also aided by a seemingly unusual source: statistics. (This also considerably influenced the development of experimental psychology.) The statistical idea was simple and straightforward: when one is faced with a *distribution* of responses derived from a number of individuals, it is convenient to obtain a single value that in some way may be considered to represent that distribution. One can then deal with that single value rather than the many.

There are valid uses for such values, and we would be hard put to do without them (this is, after all, a form of classification). The most common of them is the arithmetic mean, the "average" we discussed previously (page 163). We generally use averages when reporting on large sets of data because we do not have the mental capacity to deal with more than a few values at once. The difficulty arises when, once having grasped the concept of average, we begin to wonder what to do about the distribution itself, what to do about the fact that not all of the individuals, probably very few of them, have a score that is the same as that of the average. If we wish to use a single value that characterizes a distribution, what do we do with the individual differences? Once more it is easy to think of them as caused by error. That is indeed what came to pass, and in statistical treatments of biological and psychological data, deviations from average values often go by the name of "error variance."

This stands reality on its head, for it is the mean value that is the fictional construct and the individual values that actually exist. Yet we tend to believe soon enough if we are not careful that nature intended to create the mean and that deviations from it are just so much flutter. And we may come to believe it without effort, first because the assumption is buried in the statistical methodology that we must use to analyze our results, and second because it appears to relieve a chronic problem (albeit by stating, in effect, that there is no problem).

Even so, there would be no difficulty if the procedure led to viable results. Assumptions count for nothing if one gets somewhere. More scientific right, indeed, may have been accomplished under assumptions that were scientifically wrong than otherwise. But we must suppose that in this case the assumption that individual differences may be considered error is toxic to the maturation of the discipline.

This is not to say that we cannot use averages. We still need them. It is to say that we must not reify them. It was Quetelet who invented the notion of "the average man," and that seemed to set people to think that such a creature not only occurred, but was the aim of nature (page 21). It is an ordinary sort of transposition. Statisticians dealt with curves where probabilities can be assigned systematically to intervals (pages 21–22). All they had to do, when moving from card games to biology, was to let chance, which generated the statistical distributions, become the culprit we often conceive it to be.

There was a great deal of overlap, then, between the representative species type of the biologists and the representative average of the statisticians, and between the notions of errors of nature and deviations due to the unpredictable, to chance. When biologists began to use statistical curves they thus found ready reason to continue their ways.

It was not until the 1940s that they turned around. Mayr, in his *Systematics and the Origin of Species* (1942; revised 1964), characterized the old view as "typological" and opposed to it the modern "populational" view derived from a renewed understanding and appreciation of individual differences. The old systematics, he noted first, was based almost exclusively upon certain morphological characters that, though useful to the taxonomist, were usually of no particular importance to the species. Second, all of the characters that had been described as good species differentiators were found to vary geographically. Clearly, a corrective was needed.

To this end, the concept of the static, monotypic, Linnaean species was replaced by the concept of the dynamic *polytypic species*. Members of the same species, rather than being considered ideally to be uniform, were seen to vary both within localities and across them. Thus natural selection produces not a representative species type but rather a series of populations, with variation within and between. It is variation, not uniformity, that is normal and necessary.

The difference between the typological and populational approaches is obviously of the utmost importance for any scientific concept of behavior. In addition, thoughts of ideal types and error deviations, of nature's successes and failures, lead inevitably to thoughts of superiority and inferiority. The concept of the polytypic species, on the other hand, leads to the expectation that differences will exist, both between populations and within them, that cannot reasonably be included in hierarchical evaluations. This does not destroy the species concept; instead it liberates the concept to include the individuals of which the species is composed and expands its concern to the differences among them. Respect for the uniqueness of the individual, long a social desideratum, may now be seen to have a foundation in biological fact.

INDIVIDUAL DIFFERENCES AND
GENERAL LAWS IN PSYCHOLOGY

The species type in biology had, and continues to have, its analogue in experimental psychology. Psychologists, both for reasons of statistical analysis and because of philosophical commitments, have attempted to find laws based upon average values taken on purportedly randomly selected groups, treating the individual differences within the groups as error. Such a procedure is not, per se, logically deficient; rather, it depends for its adequacy upon the nature of the "subjects" to which it is applied. The question, thus, is how meaningful are the differences among those subjects?

When the subjects are organisms, the chances are very good that the differences are meaningful. We have noted that a substantial portion of individual differences variance—that attributable to differences in genes—is the stuff of evolution. Treating genetic differences as error while attempting to write laws applicable to genetically different organisms hardly makes sense, and in fact has been not only nonproductive but counterproductive.

On the other hand, the "subjects" of the classical physical sciences may not be meaningfully different in the sense used here. Suppose, for instance, that one wished to write a law (a mathematical statement) relating the distance a body will fall toward the earth to the time it has been falling. Aerodynamics aside, it should make little difference what body is involved (and all bodies indeed fall at the same rate in a vacuum). Therefore, empirically, we may use the average distance traveled by a number of bodies in the equation, treating the slight deviations among them as measurement error. This is indeed the sort of thing done commonly in early physics. The equation relating distance traveled (s) to time (t) is: $s = \frac{1}{2}gt^2$, and it holds for all objects.

The same use of the concepts of the average and of error deviations is not applicable in a science of organisms, for even under the most stable of environments, organisms behave differently enough that a function rule written on their average behavior applies to very few individuals. This has been evident long enough to generate a humorous and frustrated summation known (to those at other universities at least) as the Harvard Law of Animal Behavior. Essentially the law states that the experimenter, given the best apparatus and controlling the conditions as carefully as possible, can expect the animal to do as it damn well pleases.

That is not an enticing prospect. An inability to write laws that describe and predict the behavior of the subjects of its interest brings to question a discipline's status as a science. And, needless to say, none of the behavioral sciences even approaches the standards of the physical sciences in this regard. There are no laws of behavior worth the name. There are a plethora of loosely bound assertions and assumptions, but they have fulfilled neither the formal nor the empirical requirements of lawful scientific statements.

In psychology the issue of the relationship between individual differences and general laws has appeared a number of times, but it has never been satisfactorily treated. Some personality theorists and clinicians did propose that we seek laws that are "idiographic"—that is, laws that describe the behavior, separately, of each individual. Individuals are so unique, they argued, that no general rule is adequate for them all. It followed that it would be more desirable to deal with individual cases than with functions based on some sort of average that is specifically applicable to few.

On the other side, experimental psychologists took the usual position that the business of science is general laws that are "nomothetic," relating behavior to its relevant variables in such a way as to predict the responses of a large class of organisms. The individual case is of use, in this view, only as it contributes to our understanding of general functions. One would thus no more expect to write a different law for the behavior of each organism than one would expect to write a different law for each falling body, for it is precisely in its generality, in its capacity to apply to a large class of objects, that the contribution of the law lies.

Insofar as the issue has been resolved, it has been resolved in favor of the nomothetic approach (Marx, 1963). Psychology as a science must be concerned with general functions, and the place of idiographic data in such a scheme is therefore limited to the role they play in the establishment of general propositions. The enduring question, however, revolves around exactly what that role should be.

Perhaps the best known answer came from Clark Hull (1945), in a paper entitled "The Place of Innate Individual and Species Differences in a Natural-Science Theory of Behavior." Hull saw two major tasks in psychology: (1) discovering general laws based upon the behavior of a "modal or average organism," and (2) dealing with the "problem of innate behavioral differences under identical conditions between different species and between the individuals within a given species." The first task derived from the cardinal assumption governing Hull's approach: namely, that the behavior of all organisms occurs according to the same set of primary laws. He thus proposed to postulate a set of primary laws and to deduce from them secondary laws that could be tested on an average organism.

If this sounds a little like the strategy employed by mechanical physics, it should. Hull took physics as his model; it is the basis for the "natural-science" reference in the title of his paper. Psychologists, he believed, should approach their discipline just as physicists had earlier. Since it was at that time believed that physics used the hypotheticodeductive method (pages 107–108), Hull in addition recommended the method for psychology.

But Hull also recognized the existence of "innate" individual and species differences and saw that psychologists are not free to treat the differing responses of their subjects as error in the same way as were physicists. What, then, was to be done with them? The solution was both ingenious and obvious: Treat them as parameters ("empirical constants") in the mathematical function rules describing the laws of behavior.

Mathematical rules stating the relationship between events in nature contain constants, parameters, and variables. Any specific rule pairs the value of the variable of interest with those of one or more other variables, using also the values of constants and parameters. The form which the rule takes defines it. The constants in the equations are, obviously, constant. Parameters are elements in the rule that can act as both constants and variables. They vary in value in different members of the same family of function rules, but for a particular member they are constant.

A simple example of such mathematical statements of relations in nature is the one given previously: $s = \frac{1}{2}gt^2$. Here s is the variable of interest, the distance that the body falls; t is another variable, the time it has been falling; the constant is $\frac{1}{2}$; g is a parameter because gravity is not the same at all points on the earth, and hence its numerical value varies from point to point, but for any given point its numerical value is the same. The *rule* is: $s = \frac{1}{2}gt^2$. But there will be a number of values one can get for s, given the same t, depending upon the value of g.

Hull proposed first to find general laws based upon the behavior of the average organism, and then to adjust those laws to suit particular instances by using species and individual values as parameters. The *form* of a particular rule would thus be the same for all organisms, making the law general, but it would contain a means of accommodating individual and species differences.

There are several difficulties with such an approach. First, values for the individual and species parameters would, a priori, seem hard to find, and indeed proved to be. Hull himself noted at the end of his article (1945) that "Despite the fact that a program of empirical research . . . has been in progress in the author's laboratory for more than a year, it must be confessed that not a single empirical constant of the twenty or so contained in his systematic approach to behavior has been satisfactorily determined." A year is a short time in these matters, but even after eight years the parametric values had proved just as elusive, both to Hull (1952) and to others, and the attempt to find them was for all practical purposes abandoned after his death.

Another difficulty results from the fact that no real being is the "average organism," and one must therefore substitute the average of the responses of a random group of organisms for such a hypothetical entity. This would be acceptable provided that individuals and species differ from one another essentially because of measurement error. They do not, however, a fact seen clearly by Hull, and indeed prompting his proposal. Unfortunately, this makes the proposal self-contradictory, for one cannot reasonably both assign great importance to individual and species differences and simultaneously seek laws based upon the average of their responses, holding the differences in abeyance in the form of parameters. Rather, such individual and species differences must be incorporated into the original formulation of the laws.

The contradiction, alas, went unnoticed and in the main goes unnoticed still. Hull's solution was accepted, not necessarily formally and in detail, but at least tacitly and at large, by most research psychologists. Partly this was due to knowledgeable endorsement and partly to vague but strong feelings that individual and species differences are something of a nuisance, or at least a concern which could be met after the "big" laws were uncovered. The elements of this outlook were finally fused as the principles of statistical analysis came increasingly to be applied to psychological data. Statistical analysis focused upon averages, particularly the arithmetic mean, and the result was that psychologists began to structure their experiments so as to enable them to study mean differences between random groups, which of course was perfectly in accord with current systematic ideas. It all seemed to make a good deal of sense.

Seeking laws based upon average responses and thereafter attempting to reconcile them with individual and species differences has, for obvious reasons, been called a strategy of "differences later." Opposed to it is the strategy emphasizing "differences now" (Vale & Vale, 1969), which explicitly requires that no assumptions be made with regard to primary behavioral laws that hold for all organisms, and that individual and species differences be incorporated initially into the search for the empirical relations that do exist. It is thus not preemptive with reference to the provenance of behavior; rather, both philosophically and practically it opens that provenance to organismic and environmental variables on an equal basis.

It is well to recall in relation to this latter point that in the "differences later" approach the essential determinants of behavior were invested in the environment. The "innate differences" of which

Hull wrote, the organismic or biological variables, entered into stimulus-response equations only as parameters, and thus enjoyed secondary status. The *laws* were to be between stimuli and responses. Organismic variables were seen as effectors of the determinant capacities of the environment. This both engendered and derived from positions we have discussed previously (pages 100–116). The result was to restrict thought and research upon the role of organismic variables in the whole of psychology.

Equally troublesome, though not quite so obvious, was another derivative of the environmental view. In order to demonstrate the efficacy of a treatment, the experimenter (in the simplest case) applies that treatment to one sample drawn randomly from a population while using another as a control and then compares the means of the two samples. When the difference between those values departs significantly from what is attributable to chance alone, he or she concludes that the treatment was effective.

If all of the individuals in the sample have responded in the same way to the treatment, there is little more to be said, and the system is probably operating just as it appears to be on the basis of mean differences. But suppose the individuals do not respond in the same way, or perhaps some do not respond at all, to the treatment? It is likely that a significant effect based upon dissimilar mean values will not be found. And what, in turn, is the likely response of the experimenter?

As a class, scientists do not thrive upon "negative" outcomes. Their contributions (indeed, often their livelihoods) depend upon reporting significant effects. To invest time, effort, and money in a program of research and find that it does not produce significant effects is in some sense to fail. One should then expect scientists to lessen the probability of failure. All else equal, the most direct route to this goal is, first, to use only treatments, no matter how far removed from what the organism encounters in nature, which produce obvious effects and, second, if an effect is still not found, to use more powerful treatments— more shock or whatever—until the point is reached where the differential responses of the individuals are simply overwhelmed by the manipulation.

In many respects something of this sort occurred, and the result was an experimental psychology of few and powerful treatments, a condition in which it was not only deprived of subtlety but precluded from discovering the actual processes it purported to find. An interesting example is provided by Estes (1960), in a paper entitled "Learning Theory and the New 'Mental Chemistry.'" He noted that some time ago experimenters were agreed that the concept of associative strength, representing the "distillation of centuries of theorizing about learning, not to speak of 70-odd years of experimentation in the tradition of functionalism, and, later, behaviorism" was the core of learning theory. This was seen in the concept of association of ideas proposed by the British philosophers, in the strength of stimulus-response bonds as developed by Thorndike, in habit strength as postulated by Hull, and of reinforcement as proposed by Skinner. In all of these theories the more often two events were paired, the stronger did the bond between them grow.

But Estes saw a fundamental problem: "The basic concepts and assumptions of learning theory are universally supposed to refer to states and processes in the individual organism. Yet the existing evidence for the assumption that associative strength is an increasing function of a number of reinforcements comes from performance curves representing average response measures over groups of learners." If the concepts and assumptions were sound, he continued, it should be possible to obtain direct evidence for them in individual organisms. To this end he designed a series of experiments wherein individual responses were not averaged.

The results suggested that the core of learning theory might require revision. Whereas it appeared from examination of group averages that learning occurs incrementally, the stimulus-response bond being strengthened slightly with each repetition, study of individual data indicated that learning occurs on an all-or-nothing basis. Individuals either learn the items on a trial or they do not (although they may forget them), with no "in-between." What, then, had happened during those centuries and that 70-odd years? Because an increasing number of individuals learned a given item as the number of trials increased, the interpretation was made after averaging that learning occurred by increments in each individual.

Doubtless many questions surrounding this issue remain. Nothing so important and so obscure is likely to be settled at a stroke. But a valid point is made here: namely, that research methods influence conceptual positions in direct (and one is almost tempted to say mysterious) ways. Methodology, we have noted before, can be a tunnel or a trap, depending upon the insight of the practitioners. In the case above, it was at least a burden, a part of the trained incapacity (page 115) especially associated with those prone to settle assumptive and methodological matters quickly, in the hope of "getting on with the business of empirically determining relations." Unfortunately, the chances of guessing right under such conditions are slim. "Getting on with the empirical business" then becomes in reality a continuous and rather single-minded commitment to futility, until something moves in to break the spell.

The methodological villain of this piece was the groups-by-trials design, in which groups of randomly chosen subjects received a series of trials and then a performance curve was obtained using their mean values. All but a handful of the studies in learning then extant were molded by this type of design. The problem, of course, lay in the fact that information was obtained only on group averages. And, as remarked by Estes, "No accumulation of experiments, however large, all conducted and all analyzed in accord with this same general method can provide a sufficient empirical check upon concepts and assumptions that refer to processes or events occurring in the individual learner." Estes's findings thus led him to write that "some of the most firmly entrenched concepts and principles of learning theory may be in a sense artifacts of a conventionalized methodology."

Another unhappy derivative of the average-organism mode of thought is its lack of concern with replicating results on a variety of organisms. This follows, of course, from the basic assumption that all organisms obey the same primary laws and from the judgment that individual and species differences can be accounted for after those laws are discovered. The influence of both is to lessen the conceptual importance of differences among organisms, with the result that the choice of which organism to use for what purpose may be made on the basis of convenience. We have touched upon this before (page 114); it is the principal reason why for a long period experimental psychology and rat psychology were almost synonymous.

The problem arises, once again, when organisms do not respond in the same way to a uniform stimulus. One would expect this to be true of different species, and of organisms of the same species that are in different ecological circumstances. But it is true also of organisms of the same species living in generally the same habitat. An interesting case is presented by Nakamura and Anderson (1962), who found in a study of avoidance conditioning that there may be an effect due even to the differences in vendors from whom one obtains animals of the same "strain" of rats (strain is in quotes because the animals involved were not genetically specified).

While attempting to extend the results of a previous study, Nakamura and Anderson discovered that Sprague-Dawley rats obtained from the medical school of their university learned in an avoidance conditioning situation so poorly that they could not be tested for the effect of shock on decrement of the conditioned response. This was unexpected, since their experimental conditions were similar to those used by the authors of the previous study, who also used rats. It was also disheartening because it is difficult to study effects on a particular kind of learning when the organisms are apparently incapable of that kind of learning.

The investigators therefore changed strains, obtaining Long-Evans animals from three different commercial colonies. These rats conditioned fairly well, but the group from one vendor conditioned at a rate different from those obtained from the other two. More animals were consequently acquired, and the "vendor effect" was confirmed. How well the animals conditioned was thus found to depend upon a number of variables not often considered in research in that area. The authors concluded that their results emphasized the "practical difficulty of interpreting the existing avoidance conditioning literature since the populations used have been poorly specified." They found further that "the existence of non-conditioners in the population represents an important problem in its own right."

This latter statement raises another issue, one often submerged beneath an ocean of method. To be valuable, a theory purporting to account for the occurrence of a behavior must account for its non-occurrence as well. Behavioral theories based on evidence obtained only from a certain portion of the population (usually the one that exhibits the phenomenon of interest most conspicuously) will be limited at best, and probably misleading. This is a common situation, particularly in areas of research where the only subjects used are those that meet certain criteria and other subjects are discarded. But it cannot continue. Psychology evidently is not in the happy position of some other sciences, among them genetics, that have progressed through the study of particular, most-suitable organisms at certain stages of their development. (This should not be surprising. The differences in complexity in that which is to be explained with regard to behavior, and, say, gene transmission or gene activity, make it reasonable a priori that behavioral principles would be much less accessible.)

It seems clear that the difficulties in the search for general relations between behavior and the variables that control it—generated by the fact that different organisms respond differentially to the same stimuli—can be overcome only by incorporating organismic differences into the law-building process at its inception. We cannot continue to put ourselves at the peculiar disadvantage of assuming that all organisms behave alike while we discard those that do not perform as we prefer; rather, we must avoid such limiting assumptions while systematically including organisms from all points of the distribution in our research. The law-building process then would be much slower than that foreseen by earlier psychologists, and the experimental base (and hence the training of experimenters) much broader. The outcome, however, should be far more profound and lasting. Certainly there would be fewer rude awakenings.

THE INTERACTION CONCEPT

In everyday parlance, to interact means merely to have an effect upon. Thus, when two people affect one another's behavior, they are said to be interacting. Not to interact means not to respond to the actions of another. This definition is often used by clinical and social psychologists and sociologists with reference to pairs of individuals, larger groups, classes, and so forth. It is also used by computer scientists under certain conditions with reference to their relationship with the instrument. They mean thereby that the machine is, there and then, responding to their input, and they in turn are responding to the actions of the machine.

A somewhat more complex definition of interaction is often found among biochemists and physiologists. Here two or more variables are seen to *combine* in their effects—for instance, in the case where enzyme activity is affected by temperature, pH, and the amount of substrate provided. The manner in which the variables combine is, however, usually left unspecified.

When the manner of combination is specified, a number of different kinds of interaction still may be distinguished. Basically, these divide into cases where effects are additive or nonadditive. We have noted this distinction before in discussing quantitative genetics (Chapter 3), wherein individual alleles are regarded as having certain values, and the effects of their combination are then measured as deviations from the sum of those values. Dominance variance was thus seen to derive from the deviation of values within loci from what would be expected if the separate values of the alleles were simply added. Epistatic variance was similarly seen to derive from the deviation of values across loci from what would be expected if the values of the alleles taken separately were added.

Exactly the same principle applies in other experimental situations that allow for adequate statistical analysis. When the experimental design is of such a nature that two or more manipulations or conditions are allowed to combine, then nonadditivity occurs if there is significant departure in the combinations from what would be expected if the values of the effects of the manipulations or conditions taken separately were summed. On the other hand, the effects are said to be additive if indeed their values in combinations are not significantly different from their sum taken separately. An example will make this distinction clear.

Suppose in an animal experiment we manipulated two variables, one organismic and one environmental, and we wished to study their combined effect upon a behavior. Assume we used three inbred strains of mice, S_1, S_2, and S_3. Genotype would then be the organismic variable. Assume also that we had both a treatment and a control condition, so that we can speak of two "treatments," T_1 and T_2. Samples of animals from each strain would then be subjected to treatments 1 or 2, and we could compare the scores of the various combinations to see whether the effects of the variables were additive.

· *Case I.* Suppose our experiments were without error and the scores were as follows:

	T_1	T_2
S_1	28	28
S_2	33	33
S_3	35	35

There is no treatment effect, because the scores in the two treatment groups are exactly the same within strains. However, there is a strain effect, since the strain means differ from one another. The effect of S_1 is found by computing its deviation from the overall mean, 32; that is, $28 - 32 = -4$. Similarly, the effect of S_2 is $33 - 32 = 1$, and the effect of S_3 is $35 - 32 = 3$. The score of any cell in the table may be determined if one knows the overall mean and the value for the appropriate strain effect.

· *Case II.* Assume the scores were as follows:

	T_1	T_2
S_1	30	34
S_2	30	34
S_3	30	34

In this case there would be a treatment effect and nothing more. The effect of T_1 is $30 - 32 = -2$, and that T_2 is $34 - 32 = 2$. Again, all scores are determined but by two factors, this time the overall mean and the appropriate treatment effects.

· *Case III.* Now suppose the scores were:

	T_1	T_2
S_1	30	26
S_2	35	31
S_3	37	33

There are both strain and treatment effects, since the scores within both strains and treatments are different. But notice that if the effect of a given strain is to raise a score relative to another strain under T_1, it has the same effect under T_2. It is always in the same direction. Similarly, if the effect of a treatment is to raise a score in one strain, it has the same effect in other strains. There are no scores that appear to be due uniquely to a specific strain-treatment combination.

Values for the effects are found by taking the mean for a strain or treatment as a deviation from the overall mean. Thus, the effect of S_1 is $(30 + 26)/2 - 32 = -4$. The effect of T_1 is $(30 + 35 + 37)/3$

$-32 = 2$, and so on. The important point is that the effect of each *combination* is merely the sum of the strain and treatment effects taken separately. Thus, since the effect of S_1 is -4 and the effect of T_1 is 2, their combined effect must be -2. The score for the S_1-T_1 combination should then be determined solely by the overall mean and the effects of S_1 and T_1; that is, $32 - 2 = 30$, and indeed it is. All other scores are determined in the same manner. One thus need know only the overall mean and the strain and treatment effects to know all of the scores. It is the fact that even though both variables influence the outcome of the experiment, they act in a way such that their separate effects need only be summed to give the effect of any combination that leads us to refer to their effects as additive.

· *Case IV.* Now suppose the scores were:

	T_1	T_2
S_1	28	28
S_2	41	25
S_3	33	37

The effect of S_1, found in the usual manner, is -4, that of S_2 is 1, that of S_3 is 3, that of T_1 is 2, and that of T_2 is -2. Therefore, both strain and treatment effects are operating. However, here one cannot obtain the effect of any combination merely by adding the effects of two variables. Under additive conditions, for instance, the score of the S_1-T_1 combination would be $32 - 4 + 2 = 30$, that of the S_3-T_2 combination would be $32 + 3 - 2 = 33$, and so on. That they, along with the other scores, do not agree with what should be obtained when effects are additive suggests that something unique to each or many of the combinations is influencing the scores. It is that combinatory uniqueness to which we refer as nonadditive interaction.

Simple graphs show the same differences in another way. Figure 6.1 presents the data of the examples in this form. Notice that in all but Case IV the slopes of the function lines are identical (in Case II all three functions are exactly the same). Parallel lines thus indicate additivity, whereas function lines that cross or diverge substantially indicate nonadditivity. In the latter instance here it is clear that the treatments affect the three strains differently: S_1 is unchanged by T_2 relative to T_1, whereas the score of S_2 is reduced and that of S_3 is increased.

The importance of nonadditivity is obvious from Figure 6.1. When only strain effects are obtained, one can make the same inference within all strains, regardless of treatment. When only treatment effects are obtained, one can make the same inference within all treatments

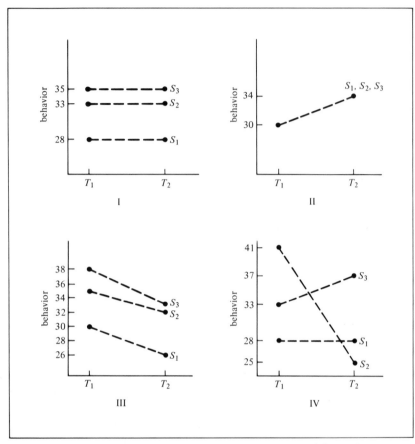

Figure 6.1   Graphs representing outcomes from a hypothetical experiment manipulating organismic and environmental variables.

regardless of strain. Even when strain and treatment effects are obtained, one may make the same inference with respect to the ways in which treatments are affecting strains, so long as the effects are additive.

When the effects are nonadditive, however, the interpretation changes. Nonadditivity suggests that one may not infer that the same process is occurring in all strains in response to the treatments. This in turn suggests that a more general process will have to be found in order to account for those responses. The presence of nonadditivity thus requires that a different inquiry be made. One may no longer be concerned with the effects of treatments alone (or of strains alone), but rather must be concerned with the more complex question of how treatments and strains combine to determine behavior.

We may for our purposes consider interaction to refer to instances wherein function lines are widely divergent, crossing either within the limits of the experiment or (it is assumed) beyond them. The lay definition of interaction as "to have effect upon," as well as the definition employed by some scientists, "to combine in having an effect upon," are thus found too broad to be useful.

The significance of interaction is, of course, statistically testable in a properly designed experiment. One can make a probabilistic statement, that is, as to whether the deviations from additivity are or are not due to chance. Significant organism-environment interaction may result from differences wherein the scores of organisms vary over treatments but maintain their rank order—that is, ordinal interaction (Lindquist, 1953)—or wherein rank order changes—that is, disordinal interaction. Disordinal interaction is of the most interest, for it clearly indicates the possibility that organisms are responding differentially to the treatments. Ordinal interaction showing a large divergence of function lines and a strong suggestion that they would cross if the experiment were extended is of interest for the same reason. Ordinal interaction generated by changes of response that are not in the same direction within groups are interpretively more troublesome. A straightforward inference in terms of process may not immediately be made.

This does not mean that these interactions are not of use in the search for general laws; exactly what will ultimately prove to be of most value in this respect remains as alluring and mysterious as ever. It means that we frankly do not know how to decide—that we do not have, and are unlikely for a very long time to have, a precise decision procedure. We must therefore make no proscriptions on the sorts of interactions which we discover. Indeed, we would make no proscriptions whatsoever, save against systems that dogmatically de-emphasize the study of interaction itself.

It should be mentioned, and is obvious with a little reflection, that organism-environment interaction is essential to evolution. If organisms did not respond differently to environmental changes, there would be nothing upon which natural selection could operate. There could not be selection for organisms making more appropriate responses, and thus there could not be the alteration in allelic frequency over generations that defines evolutionary action.

Erlenmeyer-Kimling (1972), who notes this, notes also that organism-environment interaction is responsible for the fact that organisms are differentially distributed across environments in nature. The most salient example is the covariance with which we are familiar among species and ecological niches. Within species, there is still further covariance between certain populations and microenvironments. The sickle-cell case is an example in our own. Such covariance is of course a product of evolution.

These facts bear directly upon the place of interaction in the strategy for uncovering behavioral laws. If evolution begins with and generates organism-environment interaction, if it requires and results in organisms that respond differently to environments, then such interaction is hardly the fuzzy nuisance that some psychologists have presented it to be. Rather, granting that evolution has itself been an event of sufficient magnitude in the determination of behavior to warrant our attention, so, it would seem, should be its basis and its consequences.

ORGANISM-ENVIRONMENT INTERACTION: A STRATEGY FOR BEHAVIORAL RESEARCH

It has been contended throughout this book that since we recognize two major classes of variables, one deriving from events inside the organism and the other from events outside it, and because we know that variables of the two classes are jointly responsible for behavior, then the proper concern of psychology is the manner in which organismic and environmental variables combine. The focus, that is, must be neither singularly upon the organism nor upon the environment; it must be upon both, and in such a way as to give us hope that general laws will be forthcoming.

This is no easy assignment. General laws must apply to a variety of organisms in a variety of circumstances, the greater the number of organisms and conditions the more general the law. The question then becomes: How does one study the manner in which organismic and environmental variables combine while using a variety of organisms and conditions? The answer is implied in the requirement that individual and species differences be incorporated in the law-building process from the start. What we must do then is to design experiments in which both organisms and treatment are systematically and simultaneously manipulated.

We wish to avoid situations wherein an environmental manipulation is made using two or more groups of randomly chosen subjects, and wherein (although this is much less common in psychology) an organismic manipulation is made under the same external circumstances. Rather, we seek to vary organisms and treatments together. This requires that the groups of organisms not be randomly chosen from the same population and that the treatments not be uniform across groups. This cannot be accomplished unless both organisms and treatments are varied simultaneously. It should also be noted, however, that whether or not such differential responses are forthcoming, the experimenter gains much at little cost by allowing organisms as well as treatments to vary. Fairly homogeneous (as opposed to random) groups are more sensitive to treatments because there are fewer differential reactions. Therefore at the least the experiment will be a finer instrument, and extremely powerful treatments will not be required.

Such experiments, if they became a part of the normal repertoire, would go a long way toward producing viable results. Knowing which organisms do what under which conditions is the foundation upon which a psychology of general laws could eventually be constructed. But perhaps equally important at our present stage as a science is the effect experiments of this sort would have upon those reciprocal banes of behavioral research: premature theories and narrow methodologies. Insistence upon the demonstration of effects in a number of different organisms, and not allowing discards, would in the future curtail the expansive tendencies so commonly (and understandably) found in theorists of the past. This benefit obviously derives not from dampened enthusiasm but from sober pause. We must, like everyone else, learn to conserve our resources and not commit them hastily to each and every cause. At the same time, a substantial debit should accrue against the constrictive tendencies of the kind of methodology that almost exclusively determines the nature of experimentation—that indeed rules in place of theory—and makes the form of the answer more important than the question. The excesses to which we were previously given (due not so much to our choosing poorly as to our knowing little) should thus diminish as data upon the nature and extent of differential reactions accumulate.

There is no logical restriction upon the number of groups and treatments used nor upon heterogeneity among the species of organisms. One could attempt to understand how organismic and environmental variables combine under any conditions. However, there are probably certain pragmatic restrictions both upon the size of the experiment and the interpretability of the data it generates. There also remains a pragmatic restriction, no matter how much we might work to broaden it, upon the range of our individual expertise. It does not follow, then, that if we should recognize the importance of individual and species differences or if we should accept the strategy advocated here, we would necessarily rush to acquire a huge variety of organisms and begin experiments using them all. That would, even if logistically practicable, produce an interpretative muddle.

Rather, each behavioral scientist should seek to work individual and species differences into his or her ongoing research. This requires not giant experiments but manageable ones, repeated on different sets of organisms, perhaps involving only genetically different groups of the same species or groups from closely related species, until the researcher begins to understand how the organismic and environmental variables combine to determine the behavior being investigated. In animal research on certain species such experiments are easily arranged, for groups exist that are genetically distinct from other groups of the same species but are genetically homogeneous within (i.e., inbred strains and their immediate crosses). Where such groups do not exist, the experimenter must study individual reactions and attempt to establish groups within which there is some degree of homogeneity and that come from different parts of the distribution. There are a number of techniques that may aid the experimenter in this respect, some of which are mentioned in Vale and Vale (1969).

What is envisaged then is not that all psychologists would experiment on all available organisms, but at best that fairly large number of psychologists would use at least a fairly large number of organisms, with a particular concern for the joint effects of organismic and environmental variables. This is in part only what comparative psychologists have requested all along (pages 135–142); if there is a unique feature it is the emphasis on the study of joint variation within each experiment.

That feature is, however, a substantial one, for it takes us beyond simple (or even not so simple) comparison, and moves us closer to the final goal of general laws. It is not enough, that is, just to compare the behavior of different organisms under different conditions; we must use the data produced by such experiments to find variables more generally related to behavior than those with which we began. This is always the task of science. The plaguing question inevitably is how.

It is proposed here that special attention be paid to those cases where organismic and environmental variables combine in such a way as to produce interaction. As we have seen, organism-environment interaction occurs when the responses of two or more groups of different organisms cannot be functionally related to a treatment variable but rather there is a different relationship between treatment and response for each group. A simple version of this condition is illustrated in Figure 6.2.

As is obvious, in Group I increasing values of the environmental variable are paired with increasing response values, whereas in Group II increasing environmental values are paired with decreasing response values. Overall (except where the two function lines cross), there are two response values for each value of the environmental variable. The behavior is thus a function of neither environmental or organismic variation alone but is a product of their joint effects.

(It might be argued that in the illustration there is technically but one function because there is but one function rule, that which defines a straight line, and the difference is only in sign of the slope. It should therefore be noted first that this *is* only an illustration, and linearity is used for clarity of presentation. Second, even if both rela-

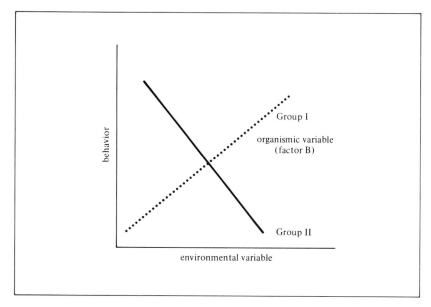

Figure 6.2   Hypothetical interactive relationship between an organismic and an environmental variable. (Source: J. R. Vale and C. A. Vale, Individual Differences and General Laws in Psychology: A Reconciliation. *American Psychologist,* **24,** 1093–1108, 1969. Copyright 1969 by the American Psychological Association. Reprinted by permission.)

tions are linear, it is by no means certain that, in terms of processes internal to the organisms, a decreasing response pattern can be considered merely the inverse of an increasing one. The differential reactions could be the result of natural selection, and thus involve processes in some organisms which are highly improbable in others. We shall then continue to use linear functions for illustrative purposes, keeping the above points in mind.)

Given the situation outlined in Figure 6.2, if we are to be left with anything more than the knowledge that interaction exists we must attempt somehow to "solve" it. What we would like, if we are to move toward a more general statement, is a function that holds for both groups across the environmental treatments used. To do this we need to find another variable, one that incorporates both the environmental and the organismic variation such that only one response value is paired with it at any point. This situation is illustrated in Figure 6.3.

Here the behavior is functionally related to one variable, X, in the same way for both groups. The behavioral effects of variables from both the organismic and environmental classes are now in equivalent terms; that is, they are both in units of X. Variation in X expresses the effect of variation on the organismic dimension across a given set of

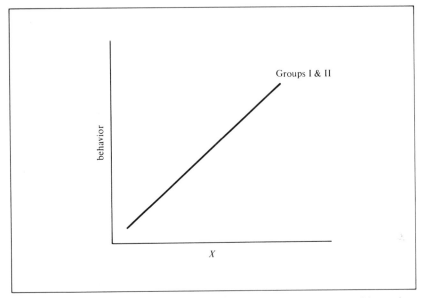

Figure 6.3   Hypothetical relationship between underlying variable and behavior. (Source: J. R. Vale and C. A. Vale, Individual Differences and General Laws in Psychology: A Reconciliation. *American Psychologist,* **24,** 1093–1108, 1969. Copyright 1969 by the American Psychological Association. Reprinted by permission.)

environments as well as variation on the environmental dimension across a given set of organisms. In terms of X, then, the effect of variation in environment is convertible to that of variation in organisms, and vice versa. A single relation exists between X and the behavior.

We may define X as a variable that *underlies* the interaction. Variables that underlie organism-environment interactions are those that, it is suggested, are to be uncovered in the search for general laws. It is thus proposed that we find variables that express the joint variation of organisms and environments and seek relations between those variables and behaviors. Clearly, the greater the range of organismic and environmental variation accounted for by such underlying variables, the more general will be the relation between the variables and the behaviors.

The question now, however, turns to how one uncovers underlying variables. There is no safe answer. No discipline has ever produced a procedure guaranteed to make it more profound. But there is some, albeit small, help in this: If a variable can be found whose relations to the environmental variable mimic the relations between the behavior and the environmental variable in the various groups, then it stands a chance of being a variable which underlies the interaction. This situation is illustrated in the following examples, one of which does not involve behavior, but each of which demonstrates the suggested paradigm in operation.

• *Example 1.* The Himalayan strain of domestic rabbits differs from other color varieties by a single gene of the color series. The four genes of this series are: A (full pigment), a^{chi} (chinchilla), a^n (Himalayan), and a (albino). Animals homozygous for the Himalayan gene (a^n/a^n) or heterozygous for the Himalayan gene with the albino gene (a^n/a) show a differential hair pigment response, depending upon the temperature of their skin. These animals have albino hair on the portions of their skin with a temperature of 34°C or above, but develop a melanic pigmentation in the hair on any portion of the skin subjected to temperature below 34°C, a temperature of 25°C being optimal. The pigmentation of other genotypes is unaffected by temperature. An idealized version of this situation is illustrated in Figure 6.4.

Thus, whereas the pigmentation of animals having either one A or a^{chi} gene or having both a genes is unaffected by temperature variation, the pigmentation of a^n/a^n or a^n/a animals is temperature-dependent. There is an interaction between genotype and environment.

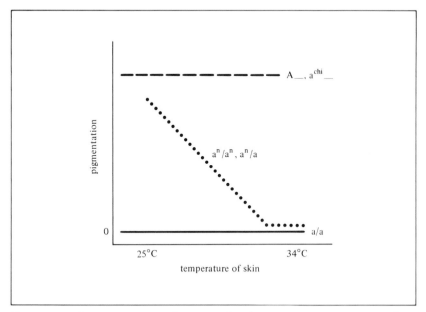

Figure 6.4 Representation of relationship between pigmentation and skin temperature for various genotypes. (Source: J. R. Vale and C. A. Vale, Individual Differences and General Laws in Psychology: A Reconciliation. *American Psychologist,* **24,** 1093–1108, 1969. Copyright 1969 by the American Psychological Association. Reprinted by permission.)

According to the paradigm a variable, X, whose relation to the environmental variable mimics the relations between the pigmentation response and temperature for all three classes of genotype should underlie the interaction. The X in this case is an enzyme in the skin, tyrosinase, that catalyzes the conversion of precursors into melanin. In Himalayan animals this enzyme is active only at temperatures below 34°C. Figure 6.5 illustrates this situation, and the relation in Figure 6.6 then follows. Pigmentation is lawfully related to the activity of the enzyme across all three genotypic classes.

Why the enzyme should respond differentially to temperature in different genotypes is another question. The paradigm could be used again, and another variable, Y, might be found that is related to enzyme activity across genotypes. An immediate suggestion comes from the fact that the a^{chi}, a^n, and a genes are mutations of the wild type, A, and the enzyme in these mutants may differ systematically in structure from the enzyme in A.

For instance, there are a number of nutritional mutants of the mold *Neurospora* that are sensitive to changes in temperature, in that they require the addition of growth factors to the minimum medium at certain temperatures but not at others. In these cases an enzyme in the mutants is active at some temperatures but not at others,

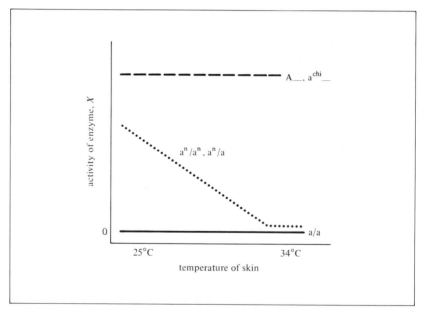

Figure 6.5   Hypothesized relationship between enzyme activity and skin temperature for various genotypes. (Source: J. R. Vale and C. A. Vale, Individual Differences and General Laws in Psychology: A Reconciliation. *American Psychologist,* **24,** 1093–1108, 1969. Copyright 1969 by the American Psychological Association. Reprinted by permission.)

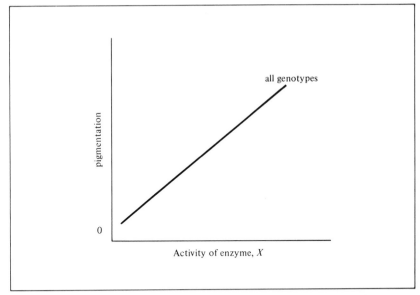

Figure 6.6   Hypothesized relationship between pigmentation and enzyme activity for all genotypes. (Source: J. R. Vale and C. A. Vale, Individual Differences and General Laws in Psychology: A Reconciliation. *American Psychologist,* **24,** 1093–1108, 1969. Copyright 1969 by the American Psychological Association. Reprinted by permission.)

whereas the enzyme in the wild type is unaffected. Wagner and Mitchell (1964), from whom the above example was taken, suggest that the changes in amino acid sequence of the enzyme in the nutritional mutants may affect its heat stability, leading to its denaturation at temperatures lower than that for the wild type enzyme. Enzyme activity obviously underlies the first interaction (with respect to growth ability), and a good candidate for the variable underlying the interaction with respect to enzyme activity would seem to be heat lability.

• *Example 2.* A physiological example involving behavior begins with an attempt to determine the relation between variation in population density and weight change of the adrenal glands in male mice. An increase in adrenal weight is thought to index physiological stress (Christian, 1959).

Bronson and Eleftheriou (1963) found that groups of the inbred mouse strain C57BL/10J showed an increase in mean adrenal weight (over isolated controls) with increasing population density, whereas the adrenal weight of colony-bred mice of a different species, *Peromyscus maniculatus bairdii,* stayed approximately the same as density increased. A simplified version of their data is presented in Figure 6.7.

This represents an organism-environment interaction wherein

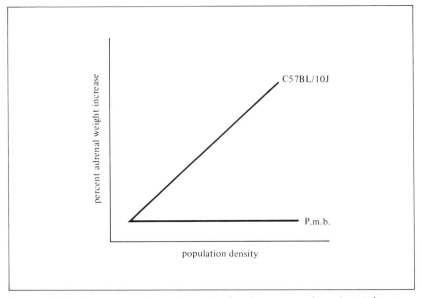

Figure 6.7   Representation of relationship between adrenal weight
increase and population density for two genotypes. (Source: J. R. Vale
and C. A. Vale, Individual Differences and General Laws in Psychology:
A Reconciliation. *American Psychologist,* **24,** 1093–1108, 1969. Copyright
1969 by the American Psychological Association. Reprinted by
permission.)

the relation between population density and change in adrenal weight
depends upon genotype. Without further exploration of the question,
one is forced to the limited conclusion that increased population density
may or may not be stressful, depending upon which genotype is
involved.

However, these two genotypes were not picked at random. King
(1957) had shown earlier that the C57BL/10J mice were far more
aggressive when put into groups than were the *P. m. bairdii.* A simplified version of King's data is presented in Figure 6.8.

Now with knowledge of an organismic variable on which the
genotypes differ and knowledge of the differential effect of the
environmental variable, a variable that may tentatively be said to
underlie the interaction between population density and genotype is
the degree to which the animals are subjected to aggression, as illustrated in Figure 6.9. Indeed, in a later part of the study, when mice
of both genotypes were subjected to attack by trained fighters, they
both showed substantial adrenal weight increases, with *P. m. bairdii*
even showing a slightly greater increase than C57BL/10. Thus, the
effect of aggression upon the adrenal weights of animals of both genotypes is approximately the same, whereas the effect of population
density increase is not.

These examples are reconstructions, and therefore do not convincingly demonstrate the efficacy of the paradigm. Ideally, one would want to present cases in which underlying variables were uncovered in experiments designed to do so. However, so little psychological research has even addressed the problem that such cases are infrequently encountered, if they exist at all. We must proceed then without explicit precedent.

But we need not proceed in blind faith. Organism-environment interaction is ubiquitous and recognized in concept (politely if without enthusiasm) by psychologists of most persuasions. The goal is to make it a focus for research. Two old, related objections against this have now been met. First, behavioral biologists have demonstrated without question the importance of organismic variables, cardinal among which are those contributing to individual and species differences, for understanding the origins of behavior. Therefore, neither the manifest claim that one can choose to ignore organismic variables while laws are sought between behavior and environmental manipulations nor the larger, covert assumption from which it derives, that organismic variables are of less importance, is defensible. Choice is no longer a refuge: However much the recognition of organismic variables increases the difficulty of behavioral research, the integration of such

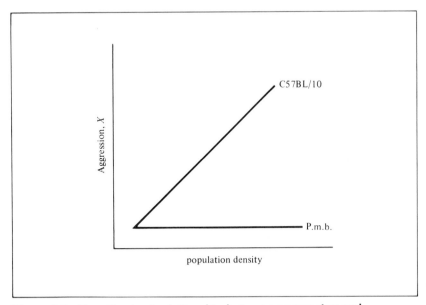

Figure 6.8   Hypothesized relationship between aggression and population density for two different genotypes. (Source: J. R. Vale and C. A. Vale, Individual Differences and General Laws in Psychology: A Reconciliation. *American Psychologist*, **24**, 1093–1108, 1969. Copyright 1969 by the American Psychological Association. Reprinted by permission.)

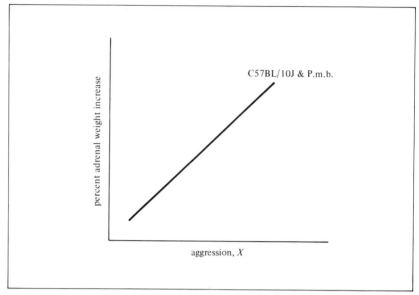

Figure 6.9   Hypothesized relationship between adrenal weight increase and aggression for all genotypes. (Source: J. R. Vale and C. A. Vale, Individual Differences and General Laws in Psychology: A Reconciliation. *American Psychologist,* **24,** 1093–1108, 1969. Copyright 1969 by the American Psychological Association. Reprinted by permission.)

variables on a large scale is necessary if a lasting contribution is to be made.

Second, it has been found that the simplest perspective on the study of behavior is not necessarily the most productive one. Quite the opposite appears to obtain. Far more data from experiments involving a far greater number of organisms under far more circumstances are required before we can hope to develop a science of behavior. It does not follow, however, that an increase in complexity need be considered synonymous with an increase in confusion, or that psychologists, even initially, need be inundated by the mass and diversity of their findings. On the contrary, careful work, respect for the proscriptive power of assumptions, and—perhaps most saliently— recognition of the fact that behavioral laws do not lie beneath the skin of nature like so many arteries of truth, ready to be opened at the first dissection, provide the philosophical structure from which the search for lawful relations may be conducted. We now have some small idea as to what sorts of relations should be found. It remains for us to begin the hunt in earnest.

Chapter 7
Unity, Diversity, and Society

A number of points presented heretofore may appear confounded—
if not actually conflicting. It would be well here to attempt to integrate
them. The points revolve around a theme that has always been central
to the human condition; it obviously is not new. We have been aware
of it at a variety of levels and dealt with it in a multiplicity of ways.
I refer to the relationship between unity and diversity in organic
nature.

We have treated this relationship from different perspectives in
previous chapters. However, we have not entirely come to grips with
it, and we must if we are to draw full philosophical benefit from the
lessons of the past and the questions of the present.

THE PRICE OF VARIABILITY

We are faced initially with the observation that nature appears to make
organisms different while at the same time making them alike. On the
one hand, that is, we have seen that variability is necessary for evolution—that indeed it is a part of evolution—such that several methods

of obtaining variety in sexual species have themselves been selected for. On the other hand, individuals of the same species must be similar enough, behaviorally and otherwise, to at least recognize one another and reproduce. In higher organisms much more obviously is required. There is thus both a need for and a limit on variability. It must be sufficient or extinction is certain because of inability to evolve; it must be restricted or provide the same danger because the integrity of a species may be lost, or never attained, unless an adequate number of individuals can contribute to the gene pool.

We thus find another of those balanced contradictions that seem so puzzling—and for which we pay so dearly. Such a contradiction was noted earlier with regard to the rise of the individual and evolution of the capacity to ethicize. The general case, discussed now, requires the same conclusion. Populations, not individuals, evolve; hence it is primarily the population, not the individual, that is served by the mechanisms of evolution. The result is that variation will be produced within populations despite its potential cost to individual members.

What sort of cost? Variability is produced by differences among individuals, and such differences in turn produce distributions of traits. In itself this is uneventful, unless a penalty accrues to those at certain places on the distributions. The sad fact is that frequently it does. We are distributed not only in weight, height, and other seemingly innocuous, visible characters, but also in susceptibility to physical and mental disorders and in intelligence—in fact in anything that can be measured—and a proportion of the individuals falling at certain places will inevitably be at a disadvantage relative to others. Further, such inequities may be expected to recur in each generation.

The most obvious instances involve deleterious single-gene mutations. We have noted a number of them, and there are many, many more. Seeing their effects, we might ask how they can continue to exist. Clearly they are quite the opposite of the gene changes with which we associate long-term adaptation.

The answer is, first, that mutation is the source of new genetic material, second, that such new material is necessary for evolution, and third, that there is apparently no way to restrict the direction of mutational alterations. Therefore, disadvantageous mutations will occur along with those that are advantageous. In order for there to be evolution at all, we must accept the bad gene changes with the good. The population as a whole can tolerate such a system or it would not be in use. However, a terrible penalty is imposed upon the individual recipients of some of the mutant genes.

Of course, selection operates against deleterious genes as well as in favor of beneficial ones. However, as mentioned in Chapter 3, when a gene is a rare recessive (as most of those in question are), then a very strong selection program over a very long period of time is required to alter its frequency substantially. This happens because with recessives the condition is usually expressed only in homozygotes, and only then are they subject to selection. Most deleterious recessives remain in the population in the heterozygous state.

The Sickle-Cell Allele

In other cases, strange as it may seem, an advantage may be conferred upon those heterozygous for a normal allele and one that is deleterious when homozygous. Perhaps the best studied of such instances is the sickle-cell allele, which we have already discussed in other regards (pages 189–190). Recall that the mode of inheritance of this allele is intermediate, so that the population is divided into three classes: (a) normals, (b) those heterozygous for one sickle-cell and one normal allele, and (c) those homozygous for the sickle-cell allele. The distribution of the sickle-cell allele thus generates a population that is polymorphous.

Those heterozygous for the sickle-cell allele are said to have the sickle-cell trait. Those homozygous for it have sickle-cell anemia. The advantage conferred upon heterozygotes is that they are much less susceptible to malarial infection than are normals. In a malarial environment, therefore, there will be selection for heterozygotes. This maintains the allele in the population, to its benefit, and of course to the benefit of those individuals who inherit one sickle-cell allele.

Evolution is said to be opportunistic—that is, it uses whatever genes, organs, and so on are available—for it cannot itself create. This is certainly the case here. A mutation, involving a change of one amino acid in 146 of the beta chain of human hemoglobin, was seized upon because of the protection it offered heterozygotes, and was propagated throughout the population until an equilibrium was reached such that stable proportions of normals, heterozygotes, and homozygotes were found (Dobzhansky, 1962).

However, this equilibrium exacts a double tariff: normals are likely to be encumbered by malaria, and homozygotes by an anemia that often kills them before the age of reproduction. Yet both are produced in mating between heterozygotes; in fact half their offspring will be one or the other. Matings between normals and heterozygotes produce half normals and half heterozygotes. In neither kind of mating, therefore, are all of the progeny protected. Some will be subject to infection, some will have anemia. Both thus pay a price, the first for a poor environment and the second as a part of the response to it.

If the evolutionary process seems, metaphorically, to be fighting fire with fire in this case, it is. Some individuals are genetically sacrificed in an already deadly environment in order that the population survive. This is harsh. Still, it surpasses a system wherein the population does not survive. The question that occurs now, however, is whether a better system might not have evolved.

The answer is yes, it might have, but probably not in the short period required. That is, a different genetically based resistance to malaria could have developed without a severe attendant detriment, but such a system would need considerably more time to evolve than a balanced polymorphism involving a single gene substitution. Both the environment and the genes would have to have changed slowly and in concert. Evidently this had little chance to occur.

Malaria (from the Italian *mala aria*, bad air) is produced by one of several species of the protozoan genus *Plasmodium*. It is transmitted by one of several species of *Anopholes* mosquito. In areas where a high incidence of malaria is endemic, the protozoan parasite is usually *P. falciparum*, and the transmitting mosquito *A. gambiae*. It is believed that a change in agriculture beginning no more than 2000 years ago produced an increased malarial environment in large parts of Africa, and is accountable for the high frequency of the sickle-cell allele there.

Specifically, it is hypothesized (Wiesenfeld, 1967) that the introduction of the Malasian agricultural complex led to the use of "slash and burn" techniques, which in turn altered the environment so as to favor *A. gambiae* and the spread of *P. falciparum*. *A. gambiae* breeds best in open, sunlit pools, of the sort that are left in a rainy climate when the forest has been cleared (by cutting and burning) and the land later deserted when it becomes barren. The Malaysian agricultural complex is one suitable to climates supporting tropical rain forest, being based upon root crops (principally yams), whereas other complexes, such as those based upon cereal crops, are not. It therefore opened up the forest to agricultural exploitation.

About 2000 years ago the Azonians on the east coast of Africa adopted parts of the Malayasian agricultural complex, and soon they spread through the forested regions to the west coast. Today a solid band, the "yam belt," runs across the sub-Saharan part of the continent. It correlates highly with the incidence of the sickle-cell allele in African populations.

The suggestion is, then, that a human cultural innovation altered the environment in a way that, in one important respect, affected a part of human evolution. The increase in the frequency of the sickle-cell allele was a direct evolutionary response to an environmental change for which humans were responsible. Its ramifications are still felt by those of African descent, long removed from danger of malarial infection, who have the misfortune to carry the allele.

There are other ramifications as well. For one, it may be that a large section of Africa suffered a lack of cultural exchange with other areas because the outsiders who would normally bring in new ideas could not survive there. Disease, in fact, could have created an unwanted fortification against the diffusion of ideas. We shall discuss this later. For another, the sickle-cell allele, although it is advantageous in single dose in a malarial environment, is not otherwise beneficial. Rather, even in heterozygotes it leads to a slight anemia, which must take its toll on the energy resources of the possessor. This cannot have helped the cultural development of the populations involved.

Malaria was not unknown in sub-Saharan Africa before the introduction of Malaysian agriculture; rather, a low level of infection is thought to have existed in some populations. The major vector then, however, was not A. gambiae but A. funestus. This is important because the behavior of A. gambiae is much more conducive to a hyperendemic level. A. gambiae bites humans frequently, has a high probability of survival during the period when the malarial parasite develops in its salivary gland, and has a long life span due to the stable and humid climate. Therefore, as the breeding space of A. gambiae was increased, the likelihood of a hyperendemic level of infection became greater.

A. funestus, on the other hand, prefers areas having heavy vegetation, and breeds in water not in direct sunlight. It is adapted to the forest, and there is less of a threat to humans as a carrier of the disease. Hence, although malaria was present among populations dwelling at the forest edge, it was not hyperendemic.

Populations living near the forest previously practiced Sudanic agriculture, which was dependent upon cereals and was unsuited to rain forest environment. Clearing the forest in the change to Malaysian agriculture opened new areas to A. gambiae, as noted. But equally important, the Malaysian agriculture could support more people because of a greater and more certain food supply. The result was an expanded population, and one that moved less than before. (The yam belt is now the most densely inhabited part of the sub-Saharan continent.) This further increased the probability of hyperendemic infection because infected individuals remained in close contact with uninfected ones, making it likely that the parasite would be transmitted through A. gambiae from the one to the other.

As the degree of infection increased, so did the frequency of the sickle-cell allele. At the same time, the allele protected a sufficient number of individuals for the population to continue to grow, and for the practices that led originally to high levels of infection to spread. The picture that finally emerges thus is one wherein a change in human behavior led to selection for a particular allele, which in turn became necessary for maintenance of the new behavioral activity, and this then resulted in a genetic equilibrium in which certain individuals are penalized for the benefit of the population as a whole.

Although the sickle-cell case best exhibits the principle in humans, it is not the only one in which selection for a heterozygous combination of a deleterious and a normal allele has been proposed. Fundamentally, the question is why severely deleterious alleles remain in the population at the same frequencies over long periods of time when there is strong selection against them. One answer, of course, is that the rate of mutation equals the rate of elimination. However, in many instances this would require a mutation rate far in excess of that found for other genes. Hence, it has been suggested (see Stern, 1973, p. 739) by a number of experimenters that selection for heterozygotes accounts for the relatively high frequencies. Cystic fibrosis and Tay-Sachs disease, both due to autosomal recessives, are two such candidates. So in fact are the genes associated with a predisposition to schizophrenia.

But this remains a poorly understood affair, and it may be that the sickle-cell case is too special to serve as a general model. Certainly it is rare to discover an agent as clearly defined as is the protozoan parasite that acts to modify the envronment so as to favor individuals heterozygous for an otherwise deleterious allele. The subject therefore requires a great deal more experimental attention.

Selection for Heterozygotes at Many Loci

Selection for heterozygotes where the fitness value of one allele is not greatly different from that of others is a different matter, and appears to be common. We discussed heterosis in Chapter 3. It is sometimes called hybrid vigor because the offspring of a mating between homozygotes for different alleles often grow faster, are less susceptible to disease and developmental malfunction, and produce more progeny of their own than do the parental lines. Heterosis is most easily demonstrated in crosses between inbred strains. Its biochemical basis appears straightforward: If different alleles code for different enzymes or enzyme forms, then an organism having a greater number of those enzymes or forms will be more fit merely because it has a larger biochemical armamentarium with which to confront the environment.

The same may hold with regard to genotypes in natural populations (Dobzhansky, 1962). In this case it is argued that selection in sexual, outbreeding species does not produce one optimal genotype but arrays of multiple heterozygous genotypes wherein the genes are *coadapted* to one another. The idea is that genes have evolved that act in heterozygous combination with a variety of others to produce a distribution of genotypes that are fitter in a wider number of environments normally open to the species than would be any single genotype, particularly a homozygous one.

The advantage of heterozygosity in this case is the same as in crosses between inbreds. The advantage of coadapted gene sets may seem less obvious but is equally important, for the existence of such sets would greatly increase the number of genes capable of acting in concert to produce an organism of high potential fitness. This in turn would result in the maintenance of many more genes in the population, providing genetic variability at relatively little cost to most of the individuals. The same outcome, a high degree of fitness, would thus be attainable through many genetic paths.

The Optimal Genotype and the Species Ideal

Rejection of the proposition that evolution produces a single optimal genotype in a population brings us closer to our everyday sense of reality and frees us from a pernicious old thought that at times has seemed to have scientific support. I refer of course to the notion of the ideal type. We discussed it before in relation to the general question of species and return to it now because of its relevance to the way we are likely to consider our own.

The concepts of the optimal genotype and the ideal species type clearly have much in common, and in addition they must have great intrinsic appeal, for despite bitter experiences we never quite tire of them. It is easy to think in terms of the ideal, first of all because of the order it seems to bring, and second probably because for each of us it represents, at least partially, an extension of self. It is also easy to believe, having conceived of some organismic ideal, that a single genetic substratum underlies it.

This can hardly be new to us; it lies beneath most theories of racial superiority and, even within the same basic population, has long been promulgated, sometimes quietly but often with great ceremony, both as an explanation of the status quo and as a reason for its continuance. However, just as the notion of the species ideal is invalidated by the modern concept of the polytypic species, so the notion of the optimal genotype is rendered nugatory by the concept of many heterozygous genotypes serving an array of adaptive norms. We are thus relieved, intellectually, of the insular burden of the one and should be prepared to accept the many.

That we often are not so prepared is another matter, which we shall discuss shortly. For the moment it should be noted that it is biology, long brought to the whipping post in this affair, that provides the convincing evidence for the necessity of variability, both among and within populations. The environmental perspective, that is, acknowledges variation but, even though claiming to account at least for individual differences through conditioning, has never been able to adduce any fundamentally sound reason for their existence. In contrast, biologists have shown that such differences provide the stuff of evolution. The diversity in which we (in good times) delight is thus more than a social desideratum; it is, in part, our potential for genetic change and hence, in the long run, our future.

Knowledge of this relationship results in a reversal of the outcomes of what sometimes are proffered as derivations from the environmental and biological positions on human variability. Variability has often been treated by environmentalists more as a moral than a scientific issue since, tacitly or otherwise, they subscribe to the underlying assumption that it is made and can be unmade by the alliance of stimulus and response. Though socially acceptable, even desirable, variability thus is not, from this viewpoint, a necessity. The biological position is profoundly opposite. Variability here is a requirement that, if unmet, would quickly have led to our demise as a species and is thus a requirement to which elaborate genetic mechanisms are devoted. It therefore acquires a deeply positive status in the whole of the human experience.

What does this mean? Scientifically, it means that individual differences must be considered far more than they have been in the search for behavioral laws. (We have spent some time with this assertion before and will return to it.) Otherwise, if one were to suggest a word, generated by the study of evolutionary genetics, which should describe the feeling of an individual vis-à-vis others, it would be respect. There is something to be gained when one examines the behavior of one's neighbors (and, fortunately, oneself) and discovers that all differences cannot be ascribed to toilet training, to mothering, to random associations of stimuli and responses, or even to opportunity or the extent of trying. We find a sense of harmony between our experience as individuals and the concepts of science. What we see in the eyes of other humans, what we see in the eyes in the mirror, is not merely a sum of momentary environmental fluctuations; it is both the history of the species, and in some measure, an expression of the diversity by which it has survived.

Distributions and Society

Much of this discussion, though self-evident, is, alas, for princes of the soul. That is, although we recognize differences, see what appears to be a genetic factor in them, and can respond to their evolutionary value, we also continue most to appreciate one or a few norms close to our own and reject, slightly or severely, those that do not correspond. The result is that individuals falling outside the limited confines of the familiar are, sometimes as much as are the possessors of major deleterious alleles, brought to the penalty block. Another price is therefore paid for variation, one that is completely within our behavioral jurisdiction and one that appears immune from appeal.

Skin color has recently been a favorite differentiator. However, anything that identifies someone as a member of another group, or even, as a last resort, as something other than oneself, will do. We know this well. We live now, indeed, under the hot skies of nuclear perdition because of it. Yet we do not relent, and there is no evidence that we have, deeply and en masse, ever relented. It is a lesson that evidently we cannot learn.

There may be ample reason for this, deriving from our mammalian heritage, and—although we may not like it any more than we like, say, typhoid fever—it is as foolish to refuse to recognize its presence as it would be to refuse to recognize the presence of the *Salmonella* bacterium. We succeed in our avowed intent only when we are completely open to understanding, and that includes, indeed it turns upon, our admission that the organic world exists as it does rather than the way we would prefer.

One fact of the organic world is variation, and another is its cost. We must expect both in each generation, for human life comes to us necessarily variable and predictably eager, like any primate form, to take on the animus of its group. Our task, which is no less because there is a biological component than it would be if there were not, is to maintain that variability while reducing its pejorative effects. To say that this will be difficult is laughably to understate the case. However, that does not diminish its import. Unless we succeed there may be no laughter at all.

THE EXPRESSION OF DIVERSITY

We have noted above that evolution appears to make organisms different while at the same time making them alike. One way in which they are made different is of course by speciation, wherein one or several sets of organisms that do not interbreed are derived from a common progenitor through the acquisition of differing gene pools. Another way in which they are made different is through operation of the mechanisms previously discussed (meiosis and others) that result in variation within populations. The first of these mechanisms is clearly dependent upon the second. Speciation requires genetic variance, for otherwise there could not be selection that would effectively differentiate groups.

Organisms are made alike in the same process. That is, differentiation not only results in gene pools among which there are differences in allelic frequencies, but also necessarily results in gene pools within which alleles are shared. Were this not true there obviously could not be species, units that change over time and differentiate into other units. Rather, there would be a single massive population, the individuals having to remain in reproductive contact with one another. The encumbrances upon evolution under such circumstances seem impossible to overcome.

Evolution, then, leads to diversity among sets of organisms (species) while it establishes similarities within them, generating reproductive units having integral anatomical, physiological, and behavioral characters. Individuals within species thus differ from one another, but not as much as they differ from other species. This is all relative of course and, though also obvious, serves to bring to light another of the balanced contradictions with which we live: Organic nature seems to assemble functional units that contain all the ingredients necessary for them to break apart. A species, that is, is something that appears to be striving simultaneously toward permanence and change.

Although a substantial amount must be, not all genetic variability within a population is expressed phenotypically. Suspension of the expression of variation is not incompatible with selection for heterozygous genotypes drawn from a pool of coadapted genes, but rather, according to Waddington (1957), is necessary because of the exigencies of development. The term he uses to describe this mechanism is "canalization." Waddington notes that many physical characters appear to develop as if through alternative pathways, only one of which is followed. The most common route is the one that would lead to a "normal" phenotype. This is most easily seen in traits such as the segments of legs in insects, in which there may be more or fewer joints than usual, but a joint either develops or it does not, there being little in between.

Two other, and for our purposes more applicable, characteristics of a canalized system are, first, that it may make use of a variety of different alleles to arrive at something of the same end point (hence the relationship with heterozygous, coadapted gene sets), and second, that development tends to return to the same path even after being diverted by major gene substitutions or environmental impress. The argument thus is that the development of a trait proceeds in a way that can be analogized by the notion of a ball traversing one of several canals, most of which will lead to normality, and that, just as the ball is not easily diverted from its path in the canal, so organic development is "buffered" against agents which would greatly alter its end.

The concept has merit, if only because it appears to make sense. Both different genes and different experiences are capable of producing great variance, usually far more than we would have predicted without experimentation. Yet the individual members of a species or a population often do not express these magnitudes of diversity: they usually in fact are distributed in such a way that the majority can be considered normal. This in turn is required if the species members are to remain anatomically, physiologically, and behaviorally close enough to one another to produce viable offspring. Thus, given the enormous diversity of which a species is capable, a mechanism must exist to keep individuals at least within reproductive distance of one another (and usually much more). In order for a breeding group to be established and maintained, therefore, not every genetic or experiential effect can be made manifest.

The result is that a great deal of variation is "hidden," and we remain unaware of it until altered circumstances demonstrate its existence. In this way species members can simultaneously be alike and different, the species as a whole having the integrity needed to exploit a given environment while also maintaining the flexibility required to deal with a new one. This at least seems certain for species at large. The question now is what particular relevance does the canalization concept have for our own? It presents another balanced contradiction: as members of a polytypic breeding group we are both more similar than we sometimes seem and more different than we actually appear.

Chapter 8
Organism, Environment,
and Behavior: Prospect, I

It has been our intention to discuss the role of scientific strategies in the study of behavior, and, in larger context, to note the relevance of their derivations and implications for man's understanding of himself. To this end, we very briefly examined the historical ground of the evolutionary and environmental movements, and, under the assumption that most readers would be more familiar with modern psychological than with biological thought, reviewed some pertinent biological concepts. However, both because our main interest *is* in broad strategy and because there are few reliable and meaningful data on behavior, we have not, except for occasional examples, dealt much with substantive issues. It is time to do so now.

We shall in the present chapter consider a number of behavioral problems. It follows from what has been said that we shall examine them neither from an exclusively organismic nor from an exclusively environmental position, but rather from the position of an interactionist. That is, we leave extensive documentation of the effects of genes upon behaviors to the texts (e.g., Fuller & Thompson, 1960; Hirsch, 1967; McClearn & DeFries, 1973) and journals (e.g., *Behavior Genetics*) devoted to them. Similarly, documentation of effects of environmental manipulations are left to the many books and journals in contemporary psychology. Our concern will be with those data that bear upon the combined effects of variables of organismic and environmental origin. In a recent discussion, Erlenmeyer-Kimling (1972) notes that inter- action data appear infrequently in behavioral review articles because researchers have produced so little to report. This is, as mentioned previously, all too true. Organism-environment interactions are some- times noted in studies addressed to questions on treatment effects, but rarely are they the focus of investigation. Such neglect doubtless de- rives from philosophical commitments on the part of the experimenter; interaction must have "psychological significance" for the investigator before its scientific significance can be pursued.

The number for whom such psychological significance has existed is few. A fairly comprehensive (although not exhaustive) list of those speaking to the interaction issue with immediate and serious concern in the past 20 years would include the following: Anastasi (1958); Erlenmeyer-Kimling (1972); Fulker, Wilcock, and Broadhurst (1972); Fuller (1960); Ginsburg (1967); Harrington (1969); Hebb (1953); Henderson (1968); Lindzey (1965); Manosevitz (1969); Sells (1963); Stanley (1960); Thiessen and Rogers (1967); Thompson (1967); and Vale & Vale (1969). Therefore we cannot expect to draw upon a large body of experiments or theoretical expositions in any presentation of the applicability of interaction thinking. The best available tack is to sift through some behavioral problems using the interactive perspective in the hope of showing its relevance both to a purely scientific and to a larger philosophic vision.

AGGRESSION

There can be no more important behavioral problem than human aggression. If we solved none other we could spend the rest of our time certainly in no worse circumstance, and probably in circum- stances far better than we have otherwise ever spent it, and face the eventual death of our species as an encounter with an alien. However, our wars, our other hostile acts, provide adamant testimony that there need be no foreigner involved in our demise. It is we who lie in wait; it is we out in the dark with glowing red eyes and murderous intent. The demon is not *among* us, it *is* us.

Why should this be? Why, after millenia of recorded talk of the benefits of peace and gracious hospitality, and we know not how much unrecorded talk before, is the balance still one of arms and terror rather than of harmony and gentle love? It is a sore, sore puzzle.

There are no final answers, nor should we expect much even in the way of specific scientific questions. Humans are poor experimental subjects when studied by other humans. Scientific questions, therefore, are often vitiated when translated into experimental manipulations, both by the limitations of the individual experimenter and the restrictions of his or her social group. The result is that there has been little, if any, solid, direct research on human aggression. Most psychological experiments purporting to study it are so far removed from real experience that the fact of their actually having been carried out and their data bent to theory is surprising.

We must therefore turn primarily to other means in order to uncover what hints at answers there are. These, especially the results of research using animals, can then be projected back, with caution, upon the human form, to see which fit. If some do, an outline may unfold by which we can begin to understand the phenomenon of aggression, in general if not in detail. To this we now turn.

First Considerations

Psychiatrist Franz Alexander (1941) remarked some years ago that "as far as the history of ancient and western civilization is concerned, periods of peace were nothing but preparations for a coming war." Although this doubtless is an overstatement, it is nevertheless true that ours is a violent past. But what is true also, and brings one up short upon first hearing, is that in spite of a large fund of available information we are for the most part unaware of just how violent it was. One reason, according to Tilly (1969), is that historians have tended to write "from the top," the only action garnering their concern being those that produced lasting rearrangements of power. Lacking that, violence has gone historically unnoticed. This analysis is confirmed by other historians (Graham & Gurr, 1969), who note that as of 1969 "violence" did not even rate an entry in the *Encyclopedia of the Social Sciences.*

Yet collective violence is normal (Tilly, 1969). It has accompanied us through the centuries, relentlessly at the center of political and economic life. It has been, indeed, an integral part of that life: those desiring to seize, hold, or realign power have continually used violence in their endeavors. Why, then, are we not historically impressed? Because "the collective memory machine has a tremendous capacity for destruction of the facts" (Tilly, 1969).

Sad but, alas, all too accurate. Knowledge of our violent past is not pleasant. However, it is essential. If we are to progress in understanding human aggression we must get the magic out, and this first consists of disabusing ourselves of the notion that there is a golden pattern of peaceful conduct from which we have somehow slipped or by external forces been led. Against this proposition history stands as a sobering corrective.

One need not be cynical to seek the truth, and the historical truth, so far as it can be determined, is that *there are no viable, righteous explanations for aggression.* We cannot, as is so popular, blame our cities (Lane, 1969) or even our machines (Tilly, 1969). We can find no historical reason to believe that aggression is an imposition that, but for the presence of some simple outside condition, would have long ago departed. On the contrary, it clings to us like the membranes of our brains.

The historical evidence does not support righteous explanations of aggression, and it therefore does not support the essentially moral approach to its study that has pervaded, and damaged, American sociology and social psychology for the last several decades. Any science that begins with assumptions about what ought to be leads only to useless research—and to results that are misleading at best. Rather, science must begin by asking what is. Only then do we discover the real possibilities; only then are we presented with the genuine options behind the "ought."

This may seem more than obvious; indeed it is presumably an article of faith among scientists. We should not be astonished, however, to find in areas that are both very important to the human condition and very difficult to investigate that moral preconceptions can determine the course of experimental and theoretical events. In matters governed by personal preference we disbelieve only what we are forced to, and the fact is that the horrors of war, particularly as related to Nazi doctrines of genetic inferiority, produced such strong negative reactions in an already environmentally committed American social science that it has not recovered to this day.

We do not like aggression. We wish it would go away. Apparently the most desirable thing to believe under these circumstances is that if it is not taught, it will. How does this translate into the language of experimentation? One prominent investigator put it clearly when he remarked that he chooses to see aggression as environmentally determined because he believes that is the only way he can do anything about it. Small wonder his experiments emphasize environmental effects, as does the theory endorsed and promulgated by him. The issue, of course, is whether we can continue to afford the luxury of this kind of choice.

All this is hardly to say that scientists should forswear vision. Far from it. The imagination of the best, indeed, sweeps through whole universes of discourse. Rather, it is to say that, regardless of one's sympathies, it is self-defeating to take scientific positions on moral grounds. Science is amoral. Scientists (we trust) are not. It is always bad for morality and bad for science when the two become confused.

We must therefore keep open minds, and be neither terrified nor disdainful of the creatures we may, when understood, turn out to be. We are better served, both immediately and in the long run, by accepting ourselves for what we are than pining for what we might have been. Indeed, it is only through such acceptance that we have any hope of maximizing the behaviors of which we approve, or at least minimizing those of which we do not. One does not control pestilence with prayers, but with medicine, and one does not control aggression with wishes, but with knowledge. We have only begun the real quest for that knowledge.

The Definition of Aggression

To deal with the subject of aggression, one would think, we must first define it. This, however, gets us instantly into trouble, for, as with many other charged and useful terms, we cannot say precisely what we mean. Aggression is not something that can be seen, like a tree, but is something we usually infer from behavior. It exists, in this way, just as love does, or friendship, or crudeness, or gentility. We have a general notion of what is meant, but in the specifics, single, global definitions begin to fail.

Aggression is commonly said to exist, for instance, when one organism acts to injure another. This seems straightforward enough, and does indeed capture something of the concept. But sometimes the injury is so indirect and remote that it is difficult to detect. Sometimes also it seems less a case of inflicting injury than of promoting or maintaining oneself. Does the definition help us in other ways? Turn the situation around and ask what behavior, at all times, under all conditions, is necessarily nonaggressive? It is difficult to think of any. Now consider which behaviors are, of themselves, at all times and under all conditions, necessarily aggressive. Probably there are some, but they are few.

The point is that it is often difficult to tell only on the basis of what is done whether an action is aggressive or not. What do we require for better judgment? Intent, perhaps. If the intentions of the perpetrators were available, we would know a great deal more about the situation and could proceed further in describing it as aggressive or nonaggressive. But trouble waits for us here also, because we have no really good way of measuring intent. Indeed, we must often infer intent from the act itself and the circumstances under which it was committed. Although this is a first principle in courts of criminal justice, the circularity obviously does not serve the purposes of science.

If we cannot separate intent from the act and conditions, but must do so in order meaningfully to define aggression, with what are we left? Perplexity. We must then, strange as it seems at first, relax our definition. We must admit to our analysis all that could reasonably be there, include more rather than less, trusting that if we can ever understand the congeries of phenomena we know as aggression we can define the concept, if required, at some later date.

"Aggression," then, is not a bad term; it is a loose one, with grave validity. We can leave it loose, being certain that so long as we can recognize particular instances of aggression, it can be studied. It is far better at this stage to be open, if somewhat uncertain, about precisely what aggression is than to be clean but narrow. We have too much to learn about the subject to begin by definitionally restricting it. Sparkling definitions, in fact, usually come after the work is completed, not before.

With this in mind we may eschew formal definitions, conceding only that aggression is usually correlated with conflict, physical or otherwise, however far removed from the immediate scene. Any behavior is potentially aggressive, depending upon intentions and circumstances. There is thus no simple category of aggressive or nonaggressive behaviors, but rather, aggression may be infused into almost anything, from dropping a hydrogen bomb to powdering a baby's bottom. Nor need any particular emotion accompany aggression; it may be done in "cold blood," as a matter of policy, impulsively, with great hatred, or reluctantly, after much reflection. All this, and more, can be aggressive behavior. What are we to make of it?

Phylogenetic Extent

One thing useful to know is how extensive aggression is. Excluding predation (killing for food), is aggression found over a wide variety of species or is it limited to a few? There is little disagreement among those who have examined the question. Aggression among animals, as among humans, is normal. Southwick (1970b) puts the case squarely: "Aggressive behavior is so widespread in the animal kingdom that it is virtually universal in some form or other in almost every animal which has the necessary motor apparatus to fight or inflict injury." Whatever species can fight, it appears, does. Aggression is absent only in those invertebrates that have no physical means of doing damage.

What does this tell us? First, aggression is generally not a cultural phenomenon in any meaningful sense (unless one would wish to argue that lobsters, seals, horses, and seagulls are cultural animals); it is not, in the usual sense, learned. This is not to say that the environment is unimportant with respect to the expression of aggression, but that aggression by and large is not acquired in the same way that, say, a knowledge of the multiplication table is.

Second, if living phylogenetically old forms maintain some physical character that can be found in the fossilized remains of their ancestors, and they share something of that character with phylogenetically newer forms, it is a reasonably safe assumption that the character evolved relatively early and the potential for it passed, with modifications, from one paleospecies to the next. Behavior does not fossilize, and so we have no direct evidence that aggression evolved long ago. However, the fact that it exists now in phyla with a long history suggests that it is not new to animal life, that in fact it has played an important role in organic evolution for some time, and that the potential for it has passed, like the potential for nonbehavioral traits, from earlier paleospecies to those of the present.

Third, if animals having highly different styles, living in very different environments, act aggressively, then aggression is not likely to be a lingering by-product of the course of some other particular evolutionary event. It is not, for instance, notably associated with carnivores; it is just as common in herbivores. Killing and eating animals of other species for a living thus does not make it more probable that one will aggress against one's own. Rather, if aggression is found among so many different kinds of organisms, one begins to suspect that it has a common use or set of uses that serves species at large.

Theoretical Issues

There are as many ways to study the subject of aggression as there are those who study it, but, as usual, investigators have tended to emphasize either environmental or biological factors. We shall not—indeed, cannot—examine all of the pertinent data, but rather we shall examine some current theory and research in the hope of gaining perspective upon the present state of the field. Later we shall attempt to show that it is within the interaction framework that aggression is best understood.

THE ENVIRONMENTAL VIEW

This view obviously ascribes to environmental variables the chief role in the origin and operation of aggressive behavior. However, there are differences among advocates of the environmental position with regard to the importance of organismic variables. Some consider aggression to be strictly a conditioned behavior that, like any other behavior, develops in response to outside stimuli in accordance with conditioning principles. Others admit organismic variables to their systems, but usually in a black box that carries various labels referring to organismic states necessary for the environmental manipulations to be effective. There is little or no attempt to integrate environmental and organismic variables. Others are concerned with the organism, and may even manipulate organismic variables, but still see the environment as the cardinal determiner of aggressive behavior.

Some environmental approaches attempt to explain aggression chiefly in terms of Skinnerian operant conditioning, in which the behavior is seen as shaped by the reinforcement contingencies of the environment. Other approaches attempt to explain aggression in terms of conditioning principles derived directly from Watson (ultimately Pavlov), who, it may be recalled (pages 98–99), believed that even though organisms are born with innate stimulus-response connections, such connections are immediately alterable by changes in environment. Thus much of what develops under the control of genes was seen as overlain by conditioned responses to the extent that it was irrelevant. Still other approaches stipulate the existence of internal dispositions and attempt to control for them but usually manipulate only the features of the environment that are considered important. We have noted earlier the problems likely to befall research programs of this sort (pages 111–113).

The analysis of aggression by experimental psychologists began, ironically enough, with a formulation which did not preclude organismic determinants. The frustration-aggression hypothesis, formally proposed more than 40 years ago (Dollard, Doob, Miller, Mowrer, & Sears, 1939), held that aggression is the primary response to frustration. However, the hypothesis did not require that the relationship between frustration and aggression be learned, in the sense that the organism acquires an aggressive response (as opposed to other kinds of responses) because it is found to be rewarding in the face of frustration. As stated later by Miller (1941) in a clarification, "no assumptions are made as to whether the frustration-aggression relationship is of innate or of learned origin."

Nevertheless, it was the "learned origin" that was seized upon, and subsequent research, usually undertaken by social psychologists, has continued in this direction. The frustration-aggression hypothesis itself fared about as well as one, in retrospect, would expect any such simple formulation to have fared. The circularity (aggression being the *measure* of frustration) soon became evident. It appears today that frustration may generate many things other than aggression and that other stimuli may elicit stronger aggressive responses than does frustration. As noted by Johnson (1972), "Whatever frustration leads to, it is neither the only nor the fastest path to aggression" (p. 136).

Few, then, endorse the hypothesis as originally proposed, although a fair amount of environmentalistic theory still derives from it.

A recent attempt to explain human aggression, which eschews biological factors, organismic dispositions, and so on, is social learning theory. As stated by Bandura (1973), "human aggression is a learned conduct that, like other forms of social behavior, is under stimulus, reinforcement, and cognitive control" (p. 44). Biology generates the neurophysiological-anatomical mechanisms that provide for aggression, but in this view it subsequently has little to do with whether or not aggression is expressed, because "the activation of these mechanisms depends upon appropriate stimulation and is subject to cortical control" (p. 29).

How is human aggression learned? By example. We learn to be aggressive by observing others behave aggressively, and furthermore we can acquire the appropriate emotional responses by the same process. Whom do we observe in all this? Models. Social experience is thus the origin of aggression, and it generates affective responses as well.

Except for the cognitive elements, this notion has a strong Skinnerian motif, one that reverbrates throughout its presentation. Thus, social learning theory rejects the usefulness of organismic dispositional concepts in favor of a strict S-R approach, taking, with Skinner, the position that "the causes of behavior are not found in the organism, but in environmental forces" (p. 41), and therefore that "behavior is environmentally determined" (p. 42). The results of research on infrahuman species are considered irrelevant to the question of human aggression because humans are thought to be unaffected by the organismic factors found in other animals. It is postulated that we learn aggression from one another in the same way in which we learn any behavior for which we are rewarded.

Upon what sort of data is this proposition based? Chiefly upon social psychological experiments undertaken in the laboratory. Clearly we cannot discuss them all, but let us look at one widely known experiment that may be considered representative. Bandura, Ross, and Ross (1963) allowed nursery school children to see adults behaving aggressively toward a plastic figure. They then frustrated the children, brought them individually into a room containing a variety of play materials, and recorded their behavior. Those children behaved more aggressively toward the play materials than did children, otherwise treated the same, who had either seen a nonaggressive model or no model at all. Further, the children who had seen a model tended to imitate the specific actions of the model.

Is this surprising? Not at all. In the first place, as Bandura admits, the actions of the models undoubtedly reduced the children's inhibitions against performing aggressive acts. Adults are authority figures, and for them to do something that a child can do clearly invites the child to do it. In addition, and this is important for any interpretation of the data, the models were not engaged in normal adult behavior (preparing a meal, washing a car) but were performing an unusual act. The novelty of this behavior for an adult—indeed, the fact that it is more likely to occur, and to have been witnessed, in children rather than adults—must also be considered a part of its drawing power, and to have contributed both to an increase in aggression and to the tendency for the children to imitate the specific actions of the models.

It is thus difficult to determine what has been demonstrated by this study. It is of interest, for example, that children exposed to no model at all showed a substantial number of aggressive responses when allowed to play with the materials—more than half the number of aggressive responses shown by the children who were exposed. What does this tell us about the origin of aggression? Is it "caused" by the actions of the models, or merely enhanced? Is it aggression per se that is learned, or have the children learned when it is permissible to be aggressive, and toward what?

These questions are of primary interest, especially when investigators speak of the causes of aggression as if they were literally to be found in the environment (Bandura, 1973). Neither organism nor environment alone can be said to cause behavior; rather, both are required. Behavior is their joint product. Further, it is the organism, not the environment, that actually behaves. In what sense, then, does the environment "cause" behavior? Only in the sense that environmental and organismic variables combine to produce it.

Few would deny this, but many seem to forget it in their theories. Such is the case here. On the one hand the existence of genetically determined neurophysiological mechanisms that are necessary if an individual is to act aggressively is admitted, but on the other hand it is asserted that the environment causes aggression. The result, in the reader, is bewilderment. From where in the environment did those necessary neurophysiological mechanisms come? They did not come from the environment, of course. They are properties of the organism.

The confusion that results when one, in spite of both logical and experimental contradiction, asserts that the environment does cause behavior is well exemplified in an analogy that is used to suggest that the results of research on infrahuman animals has little bearing upon our understanding of human aggression. Bandura (1973) states that

> . . . human sexual arousal is exceedingly variable and relatively independent of hormonal conditions. Thus, to produce a rodent Don Juan would require repeated administration of testosterone, whereas showing

him lascivious pictures of well-endowed female mice would have no stimulating effects. One would, on the other hand, rely on sexually valenced displays rather than on hormonal injections to produce erotic arousal in human males.

He concludes that "human sexual responsiveness is, in large part, socially rather than hormonally determined" (pp. 20–21).

These remarks contain a number of factual inaccuracies: (1) It has been known for some time that the administration even of very high doses of androgens does not raise the level of sexual activity of healthy, normal male rodents. In mouse as in man, therefore, one does not produce a Don Juan with repeated injections of testosterone. (2) Although lascivious pictures are unlikely to stir a mouse as they might a man, the mouse can certainly be stirred by other sensory input, namely odors from an estrous female. Men are highly dependent upon visual cues and mice are highly dependent upon olfactory cues, and here this particular difference ends. Both respond to the sexually valenced displays appropriate to their species.

More important, Bandura's analogy contains a semantic-deductive error. It is true that immediate sexual motivation appears to be more directly influenced by hormones in infrahumans than in humans. Castration, for example, produces a predictable decline in the sexual behavior of adult rodents, to the point where it effectively ceases, and this appears not to be reliably true in humans (there being, however, for obvious reasons, no large amount of systematic data on the latter). But does it follow that human sexual responsiveness is not hormonally determined? In no way.

What could Bandura's assertion mean? To understand the import of the question, imagine the human circumstance in the absence of sex hormones: (1) Externally we would all be female. (2) We could not reproduce, because there would be no puberty and therefore no ovulation cycle in females or spermatogenesis in males (and of course no way to obtain sperm in males in any case, short of surgery, even if spermatogenesis were possible). (3) There would be little interest in sexual activity, since this is associated with events after puberty—that is, with the capacity to respond physiologically. It would seem therefore that sex hormones play an essential role in our lives, given that we would have no sexual life—or any other for very long—without them.

What Bandura apparently means by his statement is that what humans find erotically pleasing is not the same across cultures. This is notoriously true. It is not the same within cultures. That, however, is trivial when one addresses the issue of origins because there is *always* human sexual activity regardless of cultural differences. That is, although cultural differences influence human sexual activity to some extent, such differences are not its origin. Its origin is evolution through natural selection.

The advantages of sexual (as opposed to asexual) reproduction are obvious and we have discussed them. Sexual reproduction is a part of the system that produces the genetic variance necessary for evolution. It is thus easy to see why sexual behavior should itself have evolved. It is requisite for sexual reproduction. It is not easy to see why the influences exerted by cultural differences upon what amounts to bits and pieces of human sexual behavior should be taken as the origin of that behavior. In any explanation of human sexual response it is surely required that the capacity for such a response be included. That capacity evolved over millions of years, and it resides in the species, not in the immediate environment. The environment, to my knowledge, has never mated, much less produced a child. These are organismic perogatives and occur in all cultures. The basic fact of human sexual behavior, then, is that it *exists*—and for very sound and comprehensible biological reasons.

To suggest, therefore, that sex hormones are largely irrelevant to human sexual responsiveness because the immediate environment plays a role in its expression is to err profoundly. However, it is just this kind of error to which social learning theory is prone. Because it does not consider organismic variables, it is forced to ascribe everything to the environment, losing sight of the most fundamental distinctions. "Environmental forces," by some sort of supernatural osmosis, take on the power to cause the phenomena they are believed to control. Control is thus confused with origin, the latter eventually being absorbed by the former.

Is it possible in the case of human aggression, as in the case of human sexuality, that a degree of environmental influence has been mistaken for origin and cause? It certainly seems so. Human aggression is taken literally to be a learned conduct, a skill, in the theory of social learning. Bandura (1973) thus defends the artificiality of laboratory research on human aggression by suggesting that aggression is rarely learned under real conditions in everyday life; that is, "Behavior that has dangerous or costly consequences is typically acquired and perfected in simulated learning situations" (p. 82). He cites military training, boxing, and hunting.

Considerable disagreement may arise over this assertion, but this need not delay us. The question now is whether the phenomenon of aggression has not been conceptually misaligned with some of the methods of its application. The question is whether aggression is a skill. The answer seems clear. No. What does it take for a very young child, when annoyed by another young child, to strike out at him? Where is the skill in this, the training? Where, for that matter, is the model? Who has taught an angry 2-year-old, having a tantrum, to scream and hit and throw objects? Need the child even succeed to have acted aggressively? Who has taught husbands or wives to destroy whole families, or quarreling friends to kill one another? Most violent crimes are sloppy, explosively inefficient. Where is the skill? The examples are endless; indeed we spend most of our time attempting to teach people from an early age not to act aggressively rather than encouraging the development, or even allowing the development, of any sort of aggressive skills.

It is true that in order to behave aggressively using an acquired skill one must acquire it, but that in itself provides little insight into the phenomenon of aggression. The fundamental fact of aggression, as of sexual behavior, is not that it can be found in various forms, but that it can be found at all. Also, as with sexuality, a basic fact is that aggression exists in all cultures. There have been more or less violent cultures, without doubt, just as there have been more or less sensual cultures, but the totally nonaggressive people occurs about as frequently as the totally asexual one—not often, or at least not for very long. The next question then is whether there is a good reason for this. Is there, as with sexual behavior, a sound biological basis, a biological function, for aggression? We shall be occupied with this issue for some time.

For the moment, we may note that social learning theory makes use of two conclusions with both of which we can concur: (1) We are social creatures and (2) we can learn. It follows that we can learn from one another. Research using this approach has not, however, demonstrated that it is aggression, rather than the appropriate targets and means of aggression, that is learned.

In addition there is another observation relevant to the genesis of aggression that social learning theory fails to explain successfully. I refer to the difference in aggressiveness between males and females that is found in virtually all societies. Maccoby and Jacklin (1974), in a recent and massive review of psychological and cultural research, examined this issue and concluded that the sex difference is genuine and cannot be accounted for exclusively by differential socialization (learning) processes. Sex differences in aggressive response tendencies, they affirm, have a biological base.

Social learning theory argues that aggression is an operantly conditioned behavior with some cognitive input. Other learning analyses employ "classical" conditioning principles, deriving from Pavlov, through Watson, to explain aggression. In this instance there is postulated to be an unconditioned stimulus-response relationship that, in the presence of a conditioning stimulus, becomes transmuted to a conditioned stimulus-response relationship. In Pavlov's most famous example, the presence of food led naturally to salivation in dogs, and, after a number of pairings of the presentation of food and the sound of a bell, the sound of the bell alone was enough to lead to salivation. The unconditioned S-R relationship was thus food-salivation, and the conditioned (learned) association was bell-salivation.

How would aggression be learned according to this scheme? First, there would exist an unconditioned S-R relationship, say between the feeling of pain and the response of an aggressive act. Second, some conditioning stimulus would be paired with the presentation of the unconditioned stimulus until the new relationship developed. In this way, an organism might be conditioned to behave aggressively in response to stimuli not originally eliciting aggression.

Attempts have been made to demonstrate that aggression can be learned in this way in animals, particularly rats (Ulrich & Azrin, 1962). Rats have been found to produce what is purportedly reflexive aggressive behavior when subjected to pain; they "boxed" one another when administered electric shock. One problem with this interpretation, however, is that such boxing does not resemble attack behavior for either laboratory or wild rats. Fighting rats bite, wrestle, and kick, often rolling over in the process. The "boxing" is seen when an animal under attack stands on his rear paws and attempts to fend off the aggressor with his forepaws, and the object is not to strike the aggressor injuriously, but merely to push him away. Such boxing is thus not attack behavior at all; in fact it can hardly be considered aggression, even aggression of the mildest, defensive sort. In addition, when only one animal in the cage is shocked it does not attack the other, but attempts to avoid him, and even when both animals are shocked they may not box unless they happen to be close together. Many other problems have been found with the original interpretation (Johnson, 1972).

This is not to say that aggressive behavior is not a possible, even a likely, response to pain, but that the kind of pain, the manner of its delivery, the state of the organism prior to delivery, and so on are all variables that affect the response, and they are not incorporated into the conditioning scheme. The use of an unusual, if convenient, source of pain—electric shock—muddies the interpretation seriously, for no species tested experiences electric shock in nature. Therefore, few investigators know exactly what has been or what could be demonstrated by experiments such as these (Johnson, 1972).

An attempt to explain human violence in terms of classical conditioning is reported by Berkowitz (1970). It is his conclusion that

> . . . aggressive behavior occasionally seems to function like a conditioned response to situational stimuli. On these occasions the observer reacts impulsively to particular stimuli in his environment, not because his inhibitions have been weakened or because he anticipates the

pleasures of his actions, but because situational stimuli have evoked the responses he is predisposed to make or set to make in that setting.

What does this mean? Essentially that people who, for whatever reason, are more or less ready to commit an aggressive act can be led to do so given the proper stimulus conditions. There is no need to look further, in this argument, into the motivations or long-range goals of the aggressor: If one is in an aggressive state of mind, then aggression-eliciting stimuli will influence him or her to behave aggressively. The analysis is meant to apply particularly to impulsive acts of aggression and to relate to its spread throughout society.

It seems eminently reasonable that people disposed to aggress should be more likely to behave aggressively, given the proper stimulation to do so. Why not? Once more, however, we must pay close attention to those slips between experimental cup and theoretical lip. We must subsequently pay attention to those elements of theory that would find their way into social policy.

A well-known piece of research by Berkowitz and Le Page (1967) illustrates the first point. Male undergraduates enrolled in an introductory psychology course at a university volunteered for the experiment in order to earn points counting toward their final grades. On the pretext that their work (to list ideas a publicity agent could use on behalf of a client) was judged by another student to be good or bad, they received either one or seven electric shocks. They were immediately given a questionnaire to evaluate their mood and were taken into another room containing the shock apparatus, where they were to evaluate the other student's work. They would now (so they presumably thought) administer shock to the other student (who was not visible to them) on the basis of that evaluation. The number of shocks they administered was the test of their aggressive behavior.

Some students, upon entering the room containing the shock apparatus, found lying on a table beside it a 12-gauge shotgun and a .38 caliber revolver. The weapons were said either to belong to the other student or to have been left there by someone else. In another experimental condition, students found badminton racquets and shuttlecocks, and in still another condition the students found nothing besides the shock apparatus. All were given the mood questionnaire again, and were then asked to evaluate the other students' work by the number of shocks they chose to give.

Not surprisingly, the students who received seven shocks reported themselves to be angrier than those who received one shock, and they delivered more shocks in return. Curiously, however, in no case was the mean number of shocks administered by students in any of the "angered" conditions equal to the number of shocks they received.

That is, although all of the angered students received seven shocks, the mean number they administered in return was 5.20. There is no explanation for this in the discussion of the experiment.

What of the angered subjects who saw the weapons? They administered a mean of 5.87 shocks, against the seven they received. Even those who believed the weapons to belong to the other student, the one whom they (presumably) thought shocked them, responded with a mean number of 6.07 shocks, again fewer than they received.

In addition, the difference between the angered group that saw the guns but was not told they belonged to the other student and the angered group that saw no object next to the shock apparatus was not statistically significant. This does not appear to be strong evidence that the guns elicited a great deal of aggression. It made little difference whether the students saw guns or nothing, so long as the guns were not believed to belong to the other student.

What is to be concluded from these data? Berkowitz (1970), in accord with his hypothesis, believes that they support the notion that some acts of aggression are due to the presence of aggression-inducing stimuli, in this case the weapons. If so, however, the effect is hardly strong. Shocking the subjects a number of times produced by far a greater increase in aggressive responses than did showing them weapons. The mean number of shocks given by all students who saw weapons and who were not angered by receiving seven shocks, regardless of who they believed the owner of the weapons to be, was 2.40, whereas the mean number of shocks given by students who saw weapons and were angered, again regardless of who they believed the owner of the weapons to be, was 5.87. That is a much greater difference than any other in the experiment. The most pertinent finding, therefore, is that shocks elicit shocks. If you shock someone, that person will shock you back.

In addition, there is the question, plaguing all human psychological research, of just how credible the experiment was to the subjects. Imagine that you are a student. You agree to participate in an experiment, get shocked, are examined for mood, see some weapons, are examined for mood again, and then have the opportunity to shock the person who shocked you. What, from the very structure of the experiment, do you think the experimenter wants you to do?

It is not an idle question. How often would you expect to find a shotgun and a revolver lying around in a university laboratory? How often is electric shock used to evaluate your work? How often have you administered shock to others as a means of evaluating theirs? And what would it tell you when twice in the course of a brief experiment you are asked about your mood?

The artificiality of the entire situation and the revealing sequence of events must in such instances alert the suspicions of the subjects. This is well demonstrated in the present case, for of 39 students whose data were discarded for various reasons, 21 suspected the other student (who was of course a confederate of the experimenter) of not being what he pretended to be. Twenty-one of 139 students, then, admitted their reservations about the study. How many more had reservations but did not admit them? We cannot know. All subjects were asked, but that hardly guarantees their answers.

I do not intend to suggest by the last remark that subjects in experiments tend to be liars. Quite the opposite. Such subjects, particularly students in the first flush of scientific endeavor, only let their desire that the experiment succeed cloud their reports on the experiment itself. Subjects, that is, generally want to cooperate with the experimenter; they want the experiment to be successful, and they want the data they produce to count and be meaningful, to contribute to that success. They are reluctant to bring up anything that threatens it.

This and other observations about the ways in which the relationship between the subject and experimental situation can determine the outcome of the experiment, quite apart from the researcher's avowed manipulations, has been investigated by Orne (1970). He makes several highly relevant points, all of which coalesce into a single theme: We cannot at all be sure of what to conclude from most social psychological research. He notes that "Generalizations from laboratory findings appear to have intuitive merit so that both the investigator and his scientific public are inclined to make the inferential leap to domains of behavior and experience beyond the laboratory. Such a leap is a complicated one, and often not warranted by current evidence."

Why so? First, human subjects are aware that they are being studied, and may not act as they would ordinarily on this account. That is, there may be motivational and perceptual consequences, due to the very fact that the subject is in an experiment, which alter his behavior enough to make inferences in other contexts hazardous or misleading. We may change what we seek to measure by the very act of attempting to measure it.

Second, the psychological experiment is understood by all to be episodic. It is isolated from the rest of the subject's experience, and all concerned know that when it is concluded the subject will be basically unchanged. Therefore, "Fears expressed, action undertaken, emotions felt in the context of an experiment are experienced as specific to that situation and intended not to carry over beyond it." But it is just such carryover that is sought. Information obtained under episodic conditions is regularly used to make inferences about enduring characteristics of individuals or groups as they manifest themselves in lasting situations. In usual social psychological research, like that of Berkowitz and Le Page (1967), college sophomores who generally do not know one another are used as subjects by an investigator who does not know them and whom they do not know and will rarely meet again in an experiment which may last only a few minutes. What bearing indeed are experiments such as these likely to have upon life as it is lived?

Third, as mentioned, the subjects, regardless of the circumstances, "acquire a stake in the research." They want the experiment to succeed and for their data to be useful. To this end they will tolerate discomfort, and even pain, so long as it is viewed as essential. Indeed, the more stress and discomfort associated with the study, the more likely is the subject to believe that it is vital. The more the subjects are required to endure, clearly, the higher is their stake in the success of the experiment and the less willing they are to believe that their efforts have been wasted.

Fourth, and relatedly, subjects can always find some meaning, some reason by which they can justify what they have been asked to do, in an experiment. Orne (1970) reports as follows on an attempt to create a meaningless experiment: (1) Subjects were put (one at a time) into a room with a pile of several thousand pages of serial additions in front of them. The investigator told them to work through the sheets as rapidly and accurately as possible. The question, of course, was how long the subjects would continue working. It was not answered, because they did not stop, coming out after several hours only to use the bathroom. (2) The task was then modified, so that besides the addition sheets there was a stack of instruction cards, each of which read: "Tear up the sheet you have just completed into a minimum of 32 pieces and throw it into the wastebasket. Then you go to the next sheet and work as accurately and rapidly as you can. Once you have completed the next sheet, pick up the next card, which will give you instructions as to what to do next." All the cards in the stack were the same, but the subjects worked on and on. (3) The experimenters tried again. This time there was the usual pile of addition sheets, but only three instruction cards, worded as above, except that the subjects were told to place the card at the bottom of the stack and pick up the next card for further instructions. The three cards were of course identical, which would be evident to the subjects after they had been through them once. Still, they worked on.

Why? Not a single subject doubted that the investigators were, in ways unknown to him, able to judge the accuracy of his performance. Further, all subjects reported that they found the experiment interesting and that they believed there to be a very important reason for it. Some subjects thought it was a frustration-tolerance task, others thought it was intended to determine inner-directedness or the like, but none found it meaningless. Once they were involved in it, they had no difficulty supplying meaning.

Fifth, the structure and conditions of the experiment can communicate to the subject what hypothesis the investigator is testing and therefore what responses are most suitable. Subtle cues, not fully recognized by the subjects, are very likely to influence their behavior. A subject's compliance, furthermore, is rarely willful or conscious, but rather, if perceived by him or her at all, is seen as "right." That is, it is not so much that subjects deliberately and directly seek to please the investigator, but that they behave in ways they believe to be appropriate for the circumstances.

Orne and Schiebe (1964), for instance, replicated the other characteristics of sensory deprivation experiments without actually depriving the subjects of sensory input. They required them to undergo a physical examination, provide a medical history, and sign a release form, and "assured" their safety by the presence of an emergency tray containing medical drugs and implements. The subjects were then taken to a well-lighted room, provided with food and water, and given an optional task. After taking some tests, they were left alone and told to report anything strange that they experienced. They were also told that if they could no longer endure the conditions, or became discomfited, they had only to press a red "panic button" to be released. There was, of course, no sensory deprivation, but many of the findings of sensory deprivation studies were produced, obviously because the experimental structure and events had led the subjects to believe they should be.

Sixth, although most social psychological studies involve an intended deception of the subject, it is unclear how often it is the investigators who are themselves deceived. This is particularly appropriate for aggression studies, wherein subjects are supposed to believe that they are inflicting severe pain upon someone else. However, it is naive for the investigator to take the position that subjects are behaving as they would outside of the experimental confines, for subjects cannot help but be aware of the investigator's culpability should anything untoward happen. This latter fact conveys a sense of profound safety to the whole affair. Investigators cannot afford to put people in the danger they regularly pretend to, and the subjects know it. What occurs then is a great deal of role playing by both the subjects and the investigators.

It is not that experiments provide commands as to what to do, but that they provide expectations. Those expectations, in turn, implicitly but effectively communicate the fact that in spite of appearances it is safe to proceed. Orne and Evans (1965) were able to get subjects to handle a poisonous snake with their bare hands, remove a penny dissolving in a beaker of nitric acid with their fingers, and throw the acid at a research assistant merely by structuring the experiment so that such behavior was expected. Is this blind obedience? Not at all. The subjects did not really believe they were in danger. They assumed that proper precautions had been taken so that no one would be hurt, and of course they had been. The interesting thing is that if subjects are asked outright if they would do such dangerous things, they emphatically reply in the negative. In the safety of the experimental ruse, however, they undertake them with apparent equanimity.

Finally, many experiments in social psychology end with an interview which is supposed to reveal whether the subject was aware of the hypothesis being tested. But as noted by Orne (1970),

> . . . the needs of the investigator in this tend to parallel those of the subject: he is no more anxious to learn that the subject has seen through his deception than the subject is to tell him. The peculiar interdigitation of the needs of the experimenter and the subject tends to result in what may best be described as a "pact of ignorance"—an unspoken agreement between investigator and subject that they shall not look deeply into the subject's awareness of the situation.

The investigator and the subject thus collaborate, however honestly, in the interest of a successful experiment. For the subject, success is defined in terms of the usefulness of his or her data. For the investigator, success is defined as support for his or her hypothesis.

What then does one conclude about social psychological research of the sort discussed above, particularly research on aggression? It lacks, and has always lacked, what Orne (1970) calls "ecological validity." The ecology of the experiment and that of life outside the laboratory are too different for there to be much acceptable generalization from the one to the other. This position has been rejected (Bandura, 1973; Berkowitz, 1970) but not answered.

We have examined only a few experiments seeking to demonstrate that aggression is learned. There are of course many more, and they are for the most part vulnerable to the criticisms enunciated here. What are we to make of them? At the very least they should be interpreted with far more caution than is common. The problem, that is, derives from their continued prominence in certain discussions of aggression, in spite of the repeated and relevant general criticisms of Orne and others (e.g., Elms, 1975; Rosenthal, 1969). The result is that unsupported ideas are sometimes treated by psychologists as facts and, perhaps more important, make their way as facts to the public and to the chambers of policy.

This is doubly curious in light of the self-admitted crisis in confidence that social psychology at large has currently embraced. For instance, M. Brewster Smith (1973), a leader of the field, notes that

> Near the end of the '60's, doubt and self-criticism became increasingly evident among American social psychologists—about the lack of cumulative gains with effort expended and of consensual paradigms to define the growing edge of scientific advance, about the artificiality and human irrelevance of some of the problems that had been pursued with great sophistication, about the instability of laboratory findings insofar as they often turned out to depend upon unexamined interpretations of the experimental situation by the human subjects, and about the questionable ethics involved in the deceptive manipulations that were typically required to attain control over these interpretations. The field is still in crisis, with no predominant new directions clearly apparent . . .

How could this have happened? Why should a scientific enterprise become so far removed from its own basic requirements that it disintegrated from within at first test? Smith provides a substantial clue: Social psychology, he suggests, was a "[once] smug little discipline with some of the features of a private club, . . . attentive only to a narrowly like-thinking audience." In the behavioral sciences at least, failure is the founding father of such clubs and a charter member of their audience.

But strangely the atmosphere of crisis—the realization that much of what has been undertaken in the name of social psychology is of dubious value—has precipitated no change in its manner of operation. As Elms (1975) remarks,

> The origins of the crisis—indeed even the existence of a crisis—do not readily appear in a survey of the research literature itself. The literature continues to grow at a fast rate; new theories are proposed, new research areas are investigated. The classical problems and theoretical approaches remain alive and reasonably well . . . without, apparently, much improvement in our grasp of human social behavior.

How could this be? One can only think of Kaplan's (page 115) notion of trained incapacity. It is not unusual that an area both difficult to research and important to the interpretation of human behavior should be endowed with false starts and strong opinions.

Our purpose in reviewing these matters, however, is only to assess the contribution of the environmental approach to the study of aggression as it is manifested in psychological thought. In so doing we find little support for the proposition that aggression is, in the usual sense, learned. Certainly the experiments undertaken to demonstrate that it is have not done so. The fact that some psychologists now recognize the tenuousness of all conclusions based upon laboratory studies of human social behavior only enhances our reservation.

THE BIOLOGICAL VIEW

In Chapter 2 we noted a certain uniformity among psychologists who advocate the environmental approach. That is, they may disagree about just what stimulus-response relationship accounts for most learning, but they operate with an inviolable consensus that behavior is learned. The methods, the machines, the very composition of the behavioristic enterprise reflect this philosophical commitment.

From the beginning a similar uniformity has not prevailed among those studying behavior who would, if forced to choose, place themselves in the biological camp. Early ethology, comparative psychology, the "traditional" biologically oriented behavioral disciplines, each developed a unique style, emphasizing different research levels and methodologies. The ethologists were descendents of the evolutionary instinctivists (pages 133–143) and thus tended to see behavior as innately determined. Their methods centered around the study of animals in natural and seminatural habitats. Comparative psychologists shared the evolutionary predilections of the ethologists but were on the whole less committed to innate determinants, and their methods were largely those of the laboratory psychologist. Physiological psychology derived from physiology, where there was neither an evolutionary nor a comparative interest. Rather, the interest was in the workings of the nervous and glandular systems. There was thus not great concern in physiological psychology with innate dispositions; much of the inquiry was given to the search for neuronal mechanisms of memory and learning. The methods were those of the animal psychologist of the time, augmented by some of those of the physiologist.

The biological view has thus been more inclusive over the past half-century than has the environmental view. (This is increasingly true today, as geneticists, neurochemists, neurophysiologists, physical anthropologists, and the like, as well as ethologists, comparative psychologists, and physiological psychologists bring combined efforts and techniques to bear upon behavioral problems.) It is therefore far more difficult to represent, particularly in a brief discussion. Such representation will not be attempted here. Instead we shall examine for the biological approach, as we did for the environmental, a recent well-known treatment which states its case with perhaps more vigor than prudence. In so doing, we shall attempt to separate the meritorious ideas from those that are less so.

The most famous recent comprehensive biological treatment of aggression is that of ethologist Konrad Lorenz (1967). It differs from environmental doctrine in a number of ways, and these are reflected in the ramifications of a primary question. The primary question for environmentalists is: How is aggression learned? For Lorenz and other ethologists the primary question is: What functions does aggression serve?

The thrust of a science cannot be understood without knowledge of its assumptions, and the assumptions obviously generate the initial questions. The environmentalist, we have seen, assumes that aggression is learned and then attempts to discover how. The ethologist assumes that no behavior so prevalent as aggression could have evolved without a use, and then attempts to discover that use and the mechanisms underlying it.

Although in principle these two positions are not antithetical (what is learned can certainly have a function, and what is functional can certainly require some learning or environmental involvement), both in execution and discussion they have often been put in deep contrast by practitioners on each side. In times past, the propositions were either-or rather than both-and. We shall return to this later. For the moment, let us concentrate upon the issue of the conceptual gain, if any, in the ethological question.

The gain is considerable. Aggression makes far more scientific sense as a phenomenon of evolutionary-biological use, with evolutionary-biological roots, than as a habit, of whatever kind, that has been learned somewhere and presumably can be unlearned or forgotten if ever there were an aggression-free generation. How does it make more sense? First, the ethological approach enables us to comprehend the existence of the phenomenon in a way not possible in the environmental frame. We must begin by asking: Why aggression? Why does it exist at all? Why this means of dealing with the world? Why is all of organic nature not at peace? The answer is that aggression works to certain ends, and those ends, by and large, enhance the survival and evolution of species. Therefore, just as sexual behavior cannot be understood at any fundamental level until the evolutionary importance of sexual reproduction is understood, so aggression cannot be comprehended until its relationship with the evolutionary process is known.

A second reason why the ethological approach is of value is that it allows us to consider aggression in broad perspective. We are not limited to circumstances that favor learning or to organisms having a culture; rather, we can look for rules that hold throughout animal phyla. If general rules are found, we may use them to aid in interpreting the role of aggression in particular circumstances and organisms. Finally, in addition to embracing a greater range of organisms, the ethological question leads us to examine a greater range of behaviors. Aggression within the framework of environmental research and theory is usually seen as negative, as aberrant and disruptive, and as due chiefly to a hostile milieu. The ethological approach allows us to consider the positive aspects of aggression, to study it within the evolutionary process, and thus to deal with the phenomenon as a whole. This cannot help but offer insights as to its operation among organisms at the higher levels.

Most aggression occurs among animals of the same species. (We shall not consider predation as aggression, and other direct interspecific conflicts are rare.) What, then, according to Lorenz, is intraspecific aggression "good for"? How does it serve? First, aggression among members of the same species tends to spread those members over the available habitat. This provides the most efficient exploitation of food resources and helps protect the many from a disaster that might befall the few.

Spacing is often associated with defense of a territory. The example with which we are perhaps most familiar is the songbird. Males of many bird species denote their breeding territories acoustically, warning off other males of the same species who might intrude. Those lyrical vocalizations thus are not expressions of love, but of occupancy. Less lyrical but equally expressive are the calls of some other birds, the most familiar being the barnyard cock. Crowing at first light, or just before, is a declaration of possession, not merely a high-spirited way to wake up. Should the declaration not be honored by another cock, there will almost certainly be a fight. (The willingness of cocks to fight, indeed, led to their being bred for the purpose.)

Mammals tend to explore by means of olfactory rather than through visual or acoustical channels, and again there are familiar examples. Dogs are notorious urinators, and they usually sniff their target first to find out who else has been there. When two dogs meet for the first time, they often stand head to tail, each sniffing the anogenital area of the other. The object of scent marking, which is common, is to make the presence of the marker known to other individuals or members of other groups. Failing to heed such a signal, depending upon where the animals find themselves, may lead to a fight.

The aggressiveness of organisms toward members of their own kind therefore results in their remaining farther apart than they might otherwise. The calls, the scent marks, and the visual cues in birds, mammals, and other animals are ways whereby contact is reduced among potential fighters (the actual damage produced by such aggressiveness being minimized) while the advantages of spacing are obtained. As contradictory as it may superficially seem, intraspecific aggression is thus a reliable and effective means of directly promoting animal welfare. We should expect that such aggression should have been selected for over a very long time.

A second service rendered by intraspecific aggression is related directly to reproduction. In many species most of the breeding is done by the strongest or most dominant males. In some species the males keep harems during the breeding season; in others they establish a nesting site or breeding territory. Even in groups such as savanna baboons, wherein a large percentage of the males may copulate with a female, during the ovulatory period the females form single-partner bonds with a dominant male. Generally, the more active the male in seizing and defending females and/or breeding sites and in acquiring status, the more likely it is that his genes will be passed on. This ensures that those in each generation best equipped behaviorally to reproduce will do so, and thus the reproductive vigor required for success will continue.

A third use of aggression, again related to reproduction, is to aid in the selection of males capable of defending their offspring or the females and young in their groups against predators. Thus a propensity for intraspecific aggression may be turned to "altruistic" ends.

Finally, in higher species it is frequently the case that the experience of older animals is an asset to their groups. The animals to whom most attention is paid and who tend to live the longest are those near the top of the dominance hierarchy. It is therefore important that those individuals capable of acquiring valuable experience also have the capacity to achieve status.

The above are direct, straightforward, and beneficial effects of aggression. Does aggression, according to Lorenz, function in other, less direct but positive ways? Yes. Indeed, it is the basis for the ritualized behavior, redirected activities, and social bonds, which are the very foundations of its control. This is an interesting proposition. Aggression, in this view, is ultimately responsible for what we recognize as the virtues of friendship and harmony. How could this be?

If the products of aggression are essential, if spacing, rivalry, and the achievement of status are evolutionarily desirable, then they will be maintained. Aggression will be selected for. However, there are certain effects of intraspecific aggression that are undesirable. Little is gained, that is, if organisms of the same species are constantly killing one another in territorial disputes, fights involving sexual rivalry, or in attempts to attain dominance. A species normally cannot afford to lose that many breeding adults. It is better if aggression is controlled, not so much to prevent its effects from being lost, but so that the individuals are not harmed and the effects remain the same. That, in the opinion of Lorenz, is the purpose of ritualized fighting and redirected aggression, and it is the source of friendship.

As cultural animals we are familiar with the role that signs and symbols play in our lives. Gestures, postures, and language convey meaning symbolically without our having to resort to, or sometimes even see, the real thing. Ceremonies or rituals wherein infrahuman animals engage in threats, displays, and perhaps some form of physically aggressive action serve the same purpose. It is possible, that is, for two animals to "fight" symbolically, for there to be a victor and a vanquished, for the issue to be decided in meaningful evolutionary terms, without the individuals having done much physical damage to one another at all.

To be meaningful evolutionarily such ritualized aggression must be subject to natural selection and thus must be related to the number of genes different individuals pass on. Since genes reside in organisms, the number of viable offspring produced by different members of a species is the measure of their success. There should be little wonder then at the connection between sex and violence: A large percentage of aggressive encounters revolves around breeding rights, breeding and nesting grounds, and the resources necessary to support progeny, for in many species the individuals who are defeated in ritual fights are deprived of the opportunity to breed. Their genes are as effectively removed from the species pool as if they had been killed. However, symbolic fighting has an enormous advantage over destructive combat in that the ritually defeated animal is still available if needed. Should the dominant or successful animals be lost, or for some reason no longer be dominant or successful, their erstwhile rivals can take their places. Evolution thus, metaphorically, achieves given ends while maintaining its options, once again at a price to certain individuals.

What is a rival fight? Van Lawick-Goodall (1971) describes the circumstances under which the chimpanzee Mike finally achieved dominance over Goliath:

> One day when Mike was sitting in camp, a series of distinctive, rather melodious pant-hoots with characteristic quavers at the close announced the return of Goliath, who for two weeks had been somewhere down in the southern part of the reserve. Mike responded immediately, hooting in turn and charging across the clearing. Then he climbed a tree and sat staring over the valley, every hair on end.
>
> A few minutes later Goliath appeared, and as he reached the outskirts of the camp he commenced one of his spectacular displays. He must have seen Mike, because he headed straight for him, dragging a huge branch. He leaped into a tree near Mike's and was motionless. For a moment Mike stared toward him and then he too began to display, swaying the branches of his tree, swinging to the ground, hurling a few rocks, and, finally, leaping into Goliath's tree and swaying branches there. When he stopped, Goliath immediately reciprocated, swinging about in the tree and rocking the branches. Presently, as one of his wild leaps took him quite close to Mike, Mike also displayed, and for a few unbelievable moments both of the splendid male chimpanzees were swaying branches within a few feet of each other until I thought the whole tree must crash to the ground. An instant later both chimps were on the ground, displaying in the under-

growth. Eventually they both stopped and sat staring at each other. It was Goliath who moved next, standing upright as he rocked a sapling; when he paused Mike charged past him, hurling a rock and drumming with his feet on the trunk of a tree.

This went on for nearly half an hour; first one male and then the other displayed, and each performance seemed to be more vigorous, more spectacular than the one preceding it. Yet during all that time, apart from occasionally hitting one another with the ends of the branches they swayed, neither chimpanzee actually attacked the other. Unexpectedly, after an extra long pause, it looked as if Goliath's nerve had broken. He rushed up to Mike, crouched beside him with loud, nervous pant-grunts, and began to groom him with feverish intensity. This was the last duel between the two males. From then on it seemed that Goliath accepted Mike's superiority . . . (pp. 125–126)

Not all rival fights are, of course, either so free form or so intense. Often two animals will simply seize one another at specified places and pull or push until one of them tires and goes away. There may even be special organs or anatomical modifications for this, a stag's antlers being a notable case in point. There are also often special features adapted for display: colors, more feathers, or—in some mammals, among them men—more hair. Note that the fight between the chimpanzees did not deprive either of them of breeding potential. However, as mentioned, this may occur, as it does among wild turkeys of southeastern Texas.

Watts and Stokes (1971) report an exceptional but revealing case of the relationship between dominance achieved through rival fights and the attainment of breeding rights among wild turkeys on the Welder Refuge. Young males of the same clutch form sibling groups that stay together for life. When they are about six months old, these groups break away from the brood flocks and form flocks with other juveniles. It is at this point that a young male's status is decided. In the exclusively male winter flock he enters into two contests, one to establish his position within his sibling group and one to determine the position of his sibling group among other such groups. Each individual first fights his brothers, wrestling, spurring, and pecking until one or both are exhausted. The strongest fighter becomes the dominant bird of the sibling group, and he is seldom challenged by any of his brothers.

Thereafter, the sibling groups challenge one another to determine their relative ranking as units. The larger groups have an obvious advantage, and generally one of them becomes dominant. It remains so with remarkable stability, its rivals rarely challenging it thereafter unless the leader dies.

As the winter flocks break up, both males and females go to the mating grounds. Here the females become available for courting, and the males oblige. However, the females are not equally available to all males. At each mating ground there might be 50 or more hens, and 10 to 15 sibling groups totaling 30 males. The sibling group that had gained dominance over the others in the juvenile flock moves among the females, courting, while the other sibling groups are displaced to the periphery. Moreover, despite the fact that the individuals of a sibling group court together, only the dominant male mates. Thus, the overwhelming majority of mating is done by the dominant male of the dominant sibling group at each of the mating grounds. His sexual perogatives are unquestioned; other males literally fold their tails at the approach of his group, and he breaks up the attempts of other males to mate with impunity (often settling on the prepared hen himself). Thus, of the 170 males observed at four mating grounds, no more than six were ever seen to copulate. These six accounted for 59 matings.

It is important to note that neither the chimpanzee nor the turkeys did physical damage to one another. Male chimpanzees, indeed, are much more likely to attack females and juveniles, however ignoble that may seem. Even then severe physical damage is unlikely. The Welder turkeys fight heartily but actually are incapable of doing much harm. In animals having powerful weapons it is a different story, and Lorenz notes that in rival fights these weapons are rarely used. Rather, despite the roaring, clanging, and screeching, both combatants may emerge intact, the victor doing no more than ejecting his rival from the vicinity. In some instances there may be definite appeasement gestures on the part of the loser. We discussed a few of these found among chimpanzees (pages 236–239) in another context. Among animals that do not form individual breeding territories but remain together in groups, these gestures, along with threats and displays, are integral to social regulation.

Aggression is controlled a second way through redirected activities. Anyone put in foul humor by the actions of a superior, against whom retaliation is inadvisable, who has subsequently found himself or herself snapping at family, friends, or unwary strangers will find this notion to be immediately valid. The exigencies of life arrange circumstances for organisms within which they could aggress but are prevented from doing so by other factors, fear being the most prominent. It is common to redirect the aggression toward a more acceptable target, with, presumably, less harm than would result from a fight with the original antagonist. An example wherein all worked well is given by van Lawick-Goodall (1968). An adolescent male, unable to obtain bananas because adult males were present, would dash away, have a tantrum resembling an adult display, and then return to his group,

apparently assuaged. Redirected aggression in the service of territory defense is evidently found in a number of unrelated vertebrate
groups.

Finally, aggression in the perspective provided by Lorenz has led
explicitly to the evolution of the personal bond. The problem is simple.
Given intraspecific aggression, how is reproduction possible? How do
geese get together with ganders to make goslings? The solution is
through the ritualized redirection of aggression toward any species
members *but* the mating pair. Each of the two thus channels the aggression, which is elicited by the close presence of the other, toward
the outside. The effect is to create a bond between the two, who now
recognize one another individually. Ceremonies in which both the individuals make threats against their neighbors in the presence of the
other, as well as appeasement and greeting gestures to one another,
are the result. The same sort of principle applies among members of
groups that remain together and distinguish between their group and
others.

In this way aggression has entered into the evolution of the
mechanisms for its control. Ironically, it is in response to the continued
presence of aggression that those behaviors usually most highly valued
by the sensitive human judge have been made possible. The tendency
to aggress against one's own kind, so maligned as the perpetrator of
evil, becomes the seed of love.

Moreover, Lorenz seems to consider that aggression is a "true"
instinct, that is, a genetically controlled character formed with little
or no environmental input. As such, it is "spontaneous." It is its spontaneity, indeed, by which an instinct is often recognized. Behavior
may occur under rationally inappropriate conditions, as when dogs
sniff, track, chase, and kill prey when their feeding dishes are full. Cats
of course do the same thing, often bringing home their victims for a
few minutes' play. The important point in Lorenz's formulation, however, is that the energy driving the behavior must be discharged or
lead to something analogous to an electrical overload. In extreme cases
a particular behavior will simply "explode" upon the scene for, essentially, want of prior release. More commonly, the organism is put into
a state of unrest and begins to search for the objects upon which the
behavior may "legitimately" be executed. This analysis applies not
only to aggression but to those activities evolved in response to it—
that is, to the ritualized and redirected activities mentioned previously.
These activities have, in Lorenz's thinking, become "autonomous," the
energy driving them thus requiring discharge every so often. They too
have become instincts, "hereditarily fixed," such that the behaviors in
potential are rigidly encoded in the genes to be run off under the

proper circumstances as music is run off by the paper tape of a player piano.

Much of this is of course highly speculative. In some cases the evidence favors Lorenz's view, in others it does not. No aspect of his treatment, however, has evoked such a negative response (e.g., Montagu, 1973) as the proposition that aggression, particularly in humans, is instinctive. We have discussed the problem of instinct previously (pages 133–143), and the conclusions are relevant here. The evidence does not support the notion that any behavior is totally innate, in the sense that there is no input from the environment. The mistake of equating genetic potential with that which is "hereditarily fixed" was one made by the evolutionary instinctivists. This does not mean that "learning" is always involved, but rather that behavior requires environmental participation if it is to develop normally. There is thus developmental feedback between a template provided by the genes and relevant parts of the environment at various times. (An excellent example is provided by Hailman, 1969.)

In studying fully developed, coordinated behavior, Lorenz apparently failed to appreciate the extent of environmental participation. He thus emphasized what he took to be the innate determination of many behaviors, and unnecessarily divided all behavior into two classes (Tinbergen, 1968). Had he looked further into ontogeny perhaps he would not have done so (Tinbergen, 1972).

Another criticized aspect of Lorenzian theory is the related notion that aggressive energy accumulates and must be discharged. Indeed, the issue of "spontaneity," by which Lorenz means internal motivation has deservedly been seized upon and made an important test in the organism-environment controversy. Why deservedly? First, in spite of defenses, there can be no doubt as to Lorenz's meaning. In the chapter entitled "The Spontaneity of Aggression," for example, he says, "In exceptional cases, the threshold lowering of eliciting stimuli can be said to sink to zero, since under certain conditions the particular instinct movement can 'explode' without demonstrable external stimulation" (Lorenz, 1967, p. 49). This states clearly that, although it may happen infrequently, behavior, including aggressive behavior, is sometimes derived entirely from within the organism. The proposal has not found support even among some other ethologists and has had the effect of making the concept of indwelling contributors seem less acceptable.

Second, the energy notation that Lorenz uses to frame his views on spontaneity appears anachronistic (even vitalistic, although he takes pains to say that it is not) and is likewise unsupported by research. It *is* true that the central nervous system is constantly active, not only capable of but continuously engaging in stimulation and inhibitory activities and is in no way the mere mass of circuits awaiting outside intervention that some psychologists have allowed it to be portrayed, but there is nevertheless no evidence for the accumulation of behavior-specific energy that must finally be discharged. That is, although certain parts of the central nervous system undoubtedly have more to do with some behaviors than with others, and although thresholds for stimuli that elicit those behaviors do vary as a function of organismic conditions, there is no good case that this has anything to do with excess energy or with energy specific to particular behaviors. The central nervous system, from what is known, simply does not operate that way.

The result of these criticisms is that the whole ethological approach has been made to seem more vulnerable than it actually is. In the dust surrounding the debate over instinct and spontaneity many have lost sight of the value of the original question: What is the function of aggression? Why did it evolve and what purpose does it serve? Lorenz may be wrong in many details, and is almost certainly wrong in his instinct-spontaneous energy formulation, but this should not distract us from examining his general analysis with care. One need by no means believe in an innate-learned dichotomy, in instinct (without the asterisk), or in spontaneous energy to believe in an evolutionary role for intraspecific aggression. Accepting the existence of such a role does not of course present us with the mechanisms governing aggressive behavior, and one would not expect it to. To understand these we will have to work further. What it does do, and this is of the utmost importance, is offer a direction for research neglected by the bulk of social scientists for the larger part of this century. It offers a return to organisms.

The abundant criticisms, by those of both biological and environmental persuasion, of some of Lorenz's views have, however justified, thus had the effect of creating a heated mist within which much of the accomplishment of ethology appears blurred. The real issue is not whether behavior is (rarely) motivated entirely from within, or even whether it can be, but the relevance of internal factors for behavior in general. That is, although Lorenz certainly stipulates the environment and the average environmentalist stipulates the organism, the crux of the matter is the extent beyond these points which the two sorts of theorists ordinarily want to go. Lorenz sees behavior chiefly as instinctive, genetically determined, spontaneous, responsive to the environment through stimuli that release specific, innate behavioral patterns. The average environmentalist sees behavior mainly as conditioned responses and therefore sees the organism as little more than something that provides the physiological apparatus making conditioning possible; for practical purposes the organism is a neutral, relatively passive learning machine. Each would concede to the other very limited portions, but it is the great body of behavior, not the extremes, that both sides wish to claim, and it is for this that they contend.

What, finally, can we then say of Lorenz's contribution? It is doubtless erroneous in some major particulars, but in its integration with evolutionary theory and with its broad, provocative hypotheses it has focused our attention as has no other recent treatise upon the biological aspects of aggressive behavior. Perhaps, given the progress of other biological disciplines, that attention will this time remain. In any event there will probably be fewer public assessments of the sort made by Bandura, who remarked in 1973 that "In professional circles, the works of Lorenz . . . were admired for their literary quality but severely criticized for their weak scholarship" (pp. 16–17). Lorenz (along with ethologists Tinbergen and von Frick) shortly thereafter was awarded the Nobel Prize.

THE INTERACTIONIST VIEW

The instinct controversy is an old one in Western thought (Diamond, 1974) and, we have seen, lives today. It is easy to understand why. We have never known enough to settle it permanently; each time we think we have done so, sufficient scientific knowledge has become available to refute one camp or the other. This has occurred because both camps have repeatedly sought to explain too much with basically limited theories. In addition, the provenance of behavior, particularly as it relates to humans, is always a political issue. That is, the human self-image plays a part in determining who does what to, or for, whom. This imposes a considerable encumbrance upon scientific parties and may influence the direction of research.

The scientific, political, and popular pendulum has thus swung from the inborn to the environmental to the inborn and back again, from a sometimes stultifying admiration of nature's handiwork to an equally oppressive fixation upon the prepotency of the environment. Evaluative and opposing stereotypes have emerged, usually: learning, culture, and the divine spark versus instinct, passions, and the dark flame. Such distinctions are, of course, more apparent than real; what one believes is strictly a matter of which part of the same animal one chooses to see. But the question now is what is to be done about it?

Tinbergen (1972) urges that we be aware of the profundity of our ignorance in these matters. He has a valid point, for preemptive philosophies have failed in the past. Indeed, not only have they not advanced knowledge, they have at times led it into regression by their conceptual limitations, by their inability to provide an adequate frame within which certain inquiries could be made. We cannot allow this to be repeated. The problems that face us are simply too large and important to be approached dogmatically again. Laying stress upon how much there actually is to be known rather than how much we should now know in principle is one way, perhaps the best way, of keeping track of the facts.

How, then, do we approach the subject of aggression? Humbly. We will find no final answers, only clues, for in truth the proper research is only beginning to be done. We can, however, come away with some sense of what is to be understood and how it can be understood in a way less prejudicial than some of those treated previously.

The proposition most congenial to the data, with regard to the ultimate "cause" of aggression, is that it evolved as an extension of the competition central to natural selection. Most aggression revolves around the attainment or defense of resources, status, and breeding rights. All of these are connected, and all have an enormous bearing upon the number of offspring an organism leaves. Genetic differences due to a disproportionate number of offspring from some parents as opposed to others is of course the basis of evolution.

For the most part, the more aggressive animals are more likely to leave viable progeny. They are also more likely simply to survive. An example is given in Marler (1976). During the winter among the red grouse of Scotland three distinct classes of birds emerge: (1) territorial cocks and hens, (2) nonterritorial residents, occasionally allowed into the territories, and (3) transients, living mostly in places not defended by the territorial birds. Among the territorials the death rate is 20 percent, among nonterritorial residents it is 99 percent, and almost none of the transients live. That is an obvious and powerful effect. Something of the sort may be expected in any species that experiences a seasonal reduction in food supply.

Aggression may thus be considered an important factor in the struggle to exist and to pass on genes. As such, however, it is different from the forms of competition we have heretofore considered. Most competition tends to be indirect, whereas intraspecific aggression usually involves a direct confrontation between at least two organisms, with the possibility that one might be hurt by the actions of the other. To preclude undue destruction of the breeding population, behaviors have evolved that ameliorate the physical effects of aggression upon individuals while allowing the evolutionary effects upon the population to continue. Ritual combat, with the attendant threats, displays, and submissive gestures, is one such means of amelioration. Avoidance is another. Some territorial animals that live alone through most of the year go to great lengths not to contact each other. If put into circumstances where they cannot retreat, however, they may fight vigorously, even to the death (Marler, 1976). Some groups of animals also regularly avoid one another (Marler, 1976). Scent marking, we have noted, is efficacious among groups that defend space. Redirected aggression—turning the threats and displays outside of the group—is also a useful means of preserving the ties among those within. A great deal of very complex behavior is thus required solely because of the destructive potential of intraspecific aggression.

The notion that aggression is an evolutionary extension of competition is hardly unique. Bigelow (1972), Hall (1970), Lorenz (1967), Tinbergen (1968), Marler (1976), Southwick (1970a, b), and Storr (1970), to name a few from several different modern fields, have all expressed it before. Indeed, it is to be found in Darwin, who noted in the *Origin of Species* that

> As the individuals of the same species come in all respects into the closest competition with each other, the struggle will generally be most severe among them. . . . With animals having separated sexes, there will be in most cases a struggle between males for possession of the females. The most vigorous males, or those which have most successfully struggled with their conditions of life, will generally leave the most progeny. But success will often depend on the males having special weapons, or means of defense, or charms; and a slight advantage will lead to victory. (p. 435)

Darwin did not miss much, and his proposition, in its present form, provides a raison d'être for the ubiquity of aggression, both phylogenetically and historically, which is quite unmatched in any learning analysis. Such analysis, in fact, when carried below the level of man and a few primates, become highly implausible. On the other hand, the evolutionary approach offers a tenable origin for aggression in all animals. Its workings, for better and for worse, become understandable.

Allowing that aggression is both a product of and a participant in evolution, how shall we further conceive of it? How does it operate within the organism? In truth we do not know, although we can say with certainty that this can be dangerous analogical ground. We have seen that the model proposed by Lorenz, based upon the accumulation of behavior-specific energy, succumbed for want of physiological support. There have been numerous other models. For instance, Scott (who, incidentally, believes that pain rather than competition is the evolutionary genesis of aggression) has outlined what appears to be a "rational-reactive" process (1971)—"rational" because the organism is seen as making the most logical response under the circumstances and "reactive" because external events are believed to be the principal factors in instigating aggression. According to this model, anger and fear are aroused by external stimuli, whereas "internal emotional proc-

esses have the function of magnifying and prolonging" the reactions
(p. 20). Internal reaction gradually ceases in the absence of external
stimulation. Most importantly, Scott concludes that

> The significance of the finding that such "spontaneous" stimulation
> either does not exist or at the most is an indirect tie-in with other
> motivational systems is that we can safely work toward an essentially
> non-violent world, without danger of harming the individual, and with
> confidence that, when we fail, it is our methods that are at fault rather
> than internal physiology inherited from remote ancestors. (p. 39)

It is the "tie-in with other motivational systems" that, we shall see,
breaks the back of this noble sentiment.

For the moment, however, let us inquire about the conditions that
could lend it plausibility. What could they be? They are none other
than the conditions under which the organism is content. That is, it
is probably safe to say that if the organism feels there is nothing about
which to fight, it will not fight. Scott gives as an example the case of
a pair of laboratory mice, one male and one female. Put together in a
separate cage with adequate food and water they will raise litter after
litter without conflict so long as the male offspring are removed at
weaning. Often they will live to a ripe old age, none the worse for
their peaceful conduct.

The assertion of course is that in a world of wants fulfilled the
expression of aggression is not necessary; it would indeed be counterproductive. But the question is: What is the probability of living in
such a world? Aggression is an adaptive behavior both for the species
and the individual. Naturally, outside of the conditions within which
it evolved, it stands a good chance of being nonadaptive. That is not
news. Neither is the fact that aggression is related to external events,
for that, after all, is the only way it could have originally evolved and
become adaptive.

We must then ask once again what is the likelihood of establishing a world that meets the criteria of no aggression-eliciting circumstances. It is essentially zero. To be sure, if we keep apart those who
ordinarily fight, the probability of fighting will drop. As it happens, in
laboratory mice males rarely attack females or juveniles, and convenient
logistical arrangements are possible. But this is not even true of other
infrahuman animals (Lorenz, 1967; Van Lawick-Goodall, 1971), much
less people, and in any case physical separation would be contrary to
our more cherished goals. It is unworkable, not to mention undesirable.

There will be wants in any foreseeable society, and we are therefore no better off to think that perfect conditions will eliminate aggression than to think otherwise. We have noted previously that in any
species there is no perfect, or even optimal, genotype. What is true of

organisms is equally true of circumstances. Perfection is a mirage, the false friend of more than one thirsting traveler.

A second problem with Scott's formulation is his proposition that, in order to be a factor in aggression, internal mechanisms must first be aroused by external stimuli and that they function to prolong and magnify such externally derived effects. This, quite simply, does not take into account a large amount of data (some of which will be discussed later). Much behavior, indeed, is now believed to result from a sequence of related intra- and extraorganismic events that leave the animal internally primed to react to specific elements in the environment. Fluctuations of sex hormone in seasonally breeding males is a case in point. The hormones may be excreted at a low level during most of the year but rise dramatically just before breeding begins. Males that have been tolerant of one another then not coincidentally turn to combat or territorial defense. When the season is over, the hormone level is once again reduced and the fighting stops. (Some males may quietly even reassemble into herds or flocks.)

Upon seeing a fight under these conditions, one might conclude that the cause was the proximity of another male. It is true that the animals do not attack just anything; their targets are very specific. If males never met, there would be no fights. But it is also true that the animals do not fight one another at times other than the breeding period, and that if castrated before the season begins they do not fight even then. Does it follow that internal mechanisms merely magnify and prolong the effects of external stimulation? In no way. In fact there are data that suggest that when the internal mechanisms are in the proper state the animals will actually seek one another out in order to fight. It is by no means a matter of random encounters. Loosely, a primed male looks for appropriate objects in the environment upon which he may act. That is some distance from the passive-reactive organism envisaged by Scott.

Shall we now conclude that for some incomprehensible reason having to do with an increase in sex steroid output the seasonal breeder has become aggressive? Of course not. Aggressive behavior in this case is adaptive, as we have noted, but it is adaptive in the usual way, in that it leads to the differential production of viable offspring. This in turn is made possible by differential access to natural resources (territories in some instances), to status, and of course to members of the opposite sex. Clearly then it is its "tie-in" with species-preserving activities that gives aggression its importance. If the environment is not normal, if the internally primed male is provided with natural resources and females, and has no requirement for status in his group, then he will leave many offspring without ever having fought, just like Scott's mouse. But in nature this would be most unique. Indeed, the evolutionary process itself is predicated upon an environment wherein demand exceeds supply, wherein advantage is obtained through competition. The mechanisms that serve evolution must therefore operate in such an environment. There is no evolution without natural selection, there is no natural selection without competition, and aggression is a form of competition. Alter the environment so as to eliminate aggression and the chances are good that evolution will have been eliminated as well.

This must be kept firmly in mind in considering the role of aggression under any conditions. To argue that aggression is superfluous because circumstances can be made such that individuals survive without it is to miss a fundamental point. Aggression is species-preserving in the manner described, not because it relieves the individual of excess energy but because it provides a survival and procreative advantage to those who express it. Since the environment is normally one that favors individuals expressing aggression, and since we know of no environmental manipulation applicable to animals outside the laboratory or presently acceptable to humans (save for those who deeply offend us) that can prevent aggression, the question of permanent control at this time is moot. That is, although aggression can be prevented, it can be prevented only in the usual ways: through separation, intimidation, and fear. There are no bargains.

An organism need not aggress to live in the same sense that it must take in nutriment to live. One does not die from lack of fighting with the certainty that one dies from lack of food. Sustenance is required both for individual and for species survival, and food-getting behavior has thus, obviously, been selected for. Reproduction is similarly required for species survival. However (no matter how we may sometimes feel), individuals can live without copulating. Indeed, in most populations, mating and the rearing of young serve the species at large, rarely the immediate parents. Yet the individuals, not the species, must mate and care for progeny, often at cost to themselves. Why do they do it? We can say broadly that anatomical, hormonal, reward, and behavioral systems have evolved to ensure that they will. Given sexual reproduction, sexual (and sometime parental) behavior is necessary. There is no way to have the former without the latter.

Aggression, like copulation, is not a biological necessity for the individual. But, like copulatory behavior, it has been strongly selected for and has come to be a part of the largely internally programmed behavioral repertoire of most species. It serves by ensuring that those individuals most successful in competition will leave the most progeny. However, unlike copulatory behavior, aggression per se is not absolutely essential for species survival. There is no equivalent of the relationship between sexual reproduction and sexual behavior in the case of aggression; aggression is not demanded by some prior and necessary biological circumstance. Life *could* have evolved even to the highest point exclusively on the basis of indirect competition; all direct competition does is intensify the selective process.

That, however, is enough. Evolution is not a moral force. It is a name given to a series of related events in nature, which in summed effect appear opportunistic and expedient, and, though also seemingly profligate, in the long run are most efficient. If something enhances the selective process at large, it will become a part of the evolutionary schema wherever applicable. This has happened again and again, attendant miseries to individuals notwithstanding. The relevant fact is that aggression works. It is therefore no coincidence that its expression is intimately related to two enterprises that *are* absolutely essential to the individual and to species: sustenance and progeny. Thus, although aggression is not itself a primary requisite for life, it has evolved to become so closely associated with primary requisites that for most purposes it might as well be one.

Given this association, we can take no succor from the assertion that "when we fail [to prevent aggression] it is our methods that are at fault rather than internal physiology inherited from our remote ancestors" (Scott, 1971). Why not? Because it is that very physiology that puts such limits upon our methods as to guarantee their failure. Further, although it is true that we are not the hostages of our ancestors, we are nevertheless their legitimate descendants. Alas, we know not what they were nor what we are. It is our obligation to find out. In the meantime, disowning our lineage does not free us; it merely casts us adrift.

Moyer (1968; 1971a, b; 1976) has presented a model more closely in accord with current thought. His first premise is that there are "in the brains of animal and man innately organized neural circuits, which when active in the presence of particular complexes of stimuli, result in a tendency for the organism to behave destructively toward certain stimulus complexes in the environment" (1971a, p. 228). A first corollary is that a particular aggressive behavior will not occur if its neural circuit is suppressed, interrupted, or in some way deactivated. A second corollary is that aggressive behavior will usually not occur even when the proper circuit is active unless an appropriate object or stimulus is present.

Moyer thus argues that collections of neurons have evolved that, when firing, predispose organisms to behave aggressively toward specific targets. Such neural networks are said to be innate, although that does not mean that environmental input cannot influence their development. It does suggest that whatever input is necessary will be found in normal environments.

Moyer further distinguishes among a number of different kinds of aggressive behavior, and suggests that the different behaviors are governed by separate but overlapping neuronal circuits. Such circuits are not closed; that is, they are subject to facilitating and inhibiting influences from other systems and have outputs to other systems. They are also located in different parts of the brain. There is no particular "aggression center."

The categories of aggressive behavior proposed by Moyer are: (1) predatory, (2) intermale, (3) fear-induced, (4) irritable, (5) territorial (defensive), (6) maternal, (7) instrumental, and (8) sex-related. We shall not consider them in detail, but rather we shall study a few as examples of the way in which biological mechanisms can operate in the determination of aggression.

What determines the activity level of neuronal circuits associated with aggressive behavior? Moyer proposes three states, which blend into a continuum. First, a circuit may be inactive and insensitive. Castration (or some other means of blocking the effects of androgens) in adult males usually produces this state, particularly with regard to intermale aggression, but with regard to a number of other aggressive activities as well. For millenia, breeders have indeed used it to produce more manageable animals. The interpretation is that the circuit(s) underlying certain forms of aggression require androgens in order to be sensitized.

In the second state, the circuit is sensitized but inactive. Although it is physiologically capable of firing, in the absence of the proper stimulus it does not. In the third state, the circuit is both sensitized and active; that is, it is firing when an appropriate stimulus is not present. (It is not, however, spontaneously active in the Lorenzian sense. There is no energy accumulation requiring discharge.) In this circumstance the organism may be restless, explore its surroundings, and respond very readily to aggression-eliciting stimuli. (Humans may report that they feel angry or hostile.)

The status of organisms vis-à-vis the continuum described above varies as a function of genes, hormones, other neuronal circuits, neurochemicals, and experience. We should therefore expect to find large differences among species, among individuals within species, and within individuals over time in the extent and type of aggression expressed. Indeed, this is the case. We have already discussed the fact that sex steroids play a large role in the expression of aggression among males. The same is true among females with reference to maternal aggression. It is also the case that intermale aggression responds to selective breeding (Lagerspetz, 1961) and that males of inbred strains have been shown to differ dramatically in aggression (Scott & Fredericson, 1951; Southwick & Clark, 1968; Vale, Vale, & Harley, 1971), providing clear evidence of the involvement of genes and of the importance of genetic differences.

It is equally plain that input from other neuronal systems is a strong factor in the expression of aggression, input from sensory modalities being most relevant in infrahuman animals. In humans, both sensory and cognitive input are obviously relevant. We know less about the role of neurochemicals, although there are strong indications that they affect aggression. Indeed, they may be the medium through which other effects (e.g., those of sex steroids) are produced. We shall discuss this later.

Finally, prior experience has a great deal to do with whether or not an organism engages in combat. This applies not only to the phylogenetically higher organisms with which we are more familiar, but to lower organisms as well. Lorenz (1964), for instance, makes the very significant point that for ritual combat to "work" the vanquished individual must be "as effectively and as permanently subdued as if it had suffered serious wounds." He notes further that "one is again and again surprised to observe how completely the loser of a ritualized fight is intimidated and how long he retains the memory of the victor's superiority. It is to be supposed that a very special mechanism must be necessary to make the experience of a lost battle so impressive, in spite of the lack of any bodily damage." It would not do if the victors and the vanquished were not properly aware of the outcomes of their struggles; otherwise there would be continuous, indecisive fighting, with little time for the survival and procreationally related functions of aggression to operate. Such affairs must thus be settled quickly. What most organisms have to learn then is not how to fight or even when or what to fight but who won or lost.

Moyer's proposal is one that includes the organismic and environmental elements necessary for any modern conception of behavior, and one that allows for a realistic variety of responses. He cites considerable data to support his basic premises. We shall not consider them all here. Nor shall we be too concerned with the categories of aggressive behavior, other than to note the obvious, that all aggression is not the same. Whether separate circuits exist for each category of aggression is also, depending upon the category, something of an open question. Certainly separate circuits for predation (which we have not considered to be aggression) and what Moyer would call irritable aggression have been found in the cat. We will discuss this shortly. For the moment, however, let us concentrate upon the general concept that Moyer presents. The notion of evolved neuronal connections that potentiate aggressive behaviors and are sensitive both to internally and externally derived input makes a good deal of sense. It is consonant with thought in a number of scientific fields, at large if not in detail, and as such represents things as we now imagine them most probably to be.

An example supporting the notion of neuronal circuits for aggressive behavior has, as mentioned, been provided in cats. Flynn (1972) found that if electrodes are implanted in certain brain regions and then an electric current is passed through them, behaviors are elicited that very much resemble those normal for the species. He examined two behaviors in particular. One he designated "quiet biting attack." In this case the cat stalks, seizes, and bites to death (at the neck) a rat placed in its cage. The second behavior he designated "affective attack." This is characterized by the animal arching its back, raising its hair, flattening its ears, baring its teeth, and hissing, growling, or spitting. It will strike a number of objects, including a rat or the experimenter, with its claws.

It is of great interest that in the laboratory the animals do not ordinarily exhibit these behaviors. The cats Flynn uses are friendly to the experimenters and do not prey upon rats. Yet artificial stimulation of certain brain sites (the lateral hypothalamus for quiet biting attack and the medial hypothalamus for affective attack produces complex, coordinated acts that are meaningful in the environment in which they occur. In neither case do the animals "mindlessly" attack empty space. If appropriate prey is unavailable, the laterally stimulated cat will only intently explore its cage; if an object is not present, the medially stimulated animal does not claw at the air.

Clemente and Chase (1973) have reviewed other studies that show the presence of brain circuits associated with aggression. They distinguish between predation and defensive aggression (Flynn's affective attack), and find evidence also that electrical stimulation in another hypothalamic region (dorsal) results in flight in the cat. They also note that behaviors analogous to quiet attack, affective (defensive) attack, and flight may be elicited in other species (monkey and opossum) by stimulation of homologous points in the hypothalamus. Karli, Vergnes, and Didiergeorges (1969) also report that mouse killings in rats is abolished by destruction of lateral hypothalamic tissue. In addition, aggressive responses have been found as a result of hypothalamic stimulation in ring doves, chickens, and pigeons (Clemente & Chase, 1973).

The hypothalamus is a phylogenetically old part of the brain, known to be involved in the regulation of very basic organic requirements, including feeding and reproduction. It is therefore not surprising that this structure is also involved in the expression of aggression. The exact neural mechanism, however, is not known. It is understood that the brain is not, as was once believed, a collection of "centers," each autonomously controlling some aspect of an organism's existence. Rather, although certain regions or sets of cells do seem to be chiefly responsible for given functions, it is still the case that the neurons of the brain are greatly interconnected, both anatomically and functionally, and behavior is the product of the integrated response of the nervous system as a whole. Memory, visual, auditory, olfactory, and other sensations as well as other situational and organismic factors involving many parts of the nervous system combine to determine behavior. The brain is not a vending machine having separate and distinct channels for each response.

We should therefore expect, and we find, that many different regions of the brain affect the expression of aggression. However, certain regions do play a larger role in its expression than others, and that also is to be expected. The brain has been evolving for a very long time. Older structures often retain much of their original function, although they may acquire different ones, and have superimposed upon them relations with newer structures. The amygdaloid complex, a mass of cells found in the temporal lobe, is a phylogenetically old region, which has connections with the hypothalamus as well as a number of other structures, from which it receives information from practically all sensory modalities. Stimulation of this region produces both defensive attack and flight reaction, and there appear to be different anatomical systems for each. Destruction of the amygdaloid complex, on the other hand, leads to docility (among other things).

The same kinds of responses may be found upon destruction of another structure, the septum. Lesioned animals "attack objects thrust toward them, are dangerous to handle, and show startle and flight reaction pattern" (Clemente & Chase, 1973). In contrast, stimulation of the septal area increases the latency for hypothalamically induced attack, and produces a marked suppression of motor activity. The septum may thus act as an inhibitor of hypothalamic aggression mechanisms. The aggressiveness resulting from septal lesions can be reversed by lesions in the amygdala.

The three brain regions discussed are by no means the only ones found experimentally to be involved with aggression. They are, however, the most studied, and we shall have reason to return to them later. The research cited above (and much other research) does then make a case for evolved brain circuits potentiating aggression, which normally develops in most organisms. How they operate is the next consideration. Clemente and Chase see a central role for the hypothalamus, noting that it is the focal structure in the elaboration of agonistic behavior patterns since its destruction disrupts both spontaneous and induced aggression. The amygdala exerts its influence on aggressive reactions principally by modulating the activity of the hypothalamus, while the septum appears to inhibit agonistic patterns by suppressing aggressive behavior of both amygdaloid and hypothalamic origin.

Results of this sort are not restricted to mammals and birds, or even to vertebrates (Brown, 1970). Most interestingly, Robinson (1971) reported that aggression could be elicited by electrical stimulation of the rhesus monkey hypothalamus and, further, that normally subordinate animals would under the influence of such stimulation readily attack dominants. It was possible thus to change the dominance order between a pair of monkeys by artifically activating an aggression circuit.

The electrophysiological evidence for internally programmed (i.e., genetically derived) neuronal circuits that potentiate aggressive behaviors in infrahumans appears at this point irrefutable. (We have mentioned but a fraction of it.) Supportive evidence comes in addition from neurochemical studies. Specifically, it is known that some chemicals found in neurons act as transmitters (across synapses), serving to excite or inhibit other neurons that may be associated with circuits involved in the expression of aggression. Of particular interest are monoamines (dopamine, norepinephrine, serotonin), which have been found to be differentially concentrated in brain regions, some of them known to be related to reproduction and aggression.

Reis (1972), for example, concluded that norepinephrine is implicated in defensive attack in the cat. When the behavior is induced by electrical stimulation, there is a concomitant release of norepinephrine from the synaptic vesicles. Valzelli and Garattini (1972) found a correlation between the tendency for isolated rats to kill mice and turnover (a measure of chemical activity) of monoamines in different brain regions. Hyperactivity, along with increased defensive fighting and sexual activity, was reported in rats (Benkert, Gluba, & Matussek, 1963) when the dopamine and norepinephrine precursor L-DOPA was administered. In contrast, chlorpromazine, a neuroleptic that blocks the action of neurotransmitters at the postsynaptic receptor site, inhibits defensive and attack behavior in both mice and cats (Hoffmeister & Wuttke, 1969). Lagerspetz, Tirri, and Lagerspetz (1968) have in addition found differences in brain monoamine levels between strains of mice selectively bred for aggressiveness and nonaggressiveness. These (and many other) results suggest then that the chemical constituents which one would expect if neural aggression circuits exist are in fact in place. It is not the case that a single substance "responsible" for aggression has been isolated (nor is it likely that one will ever be); however, the evidence is quite strong that certain substances acting at certain brain sites are involved, and these, as would be necessary, have been shown to be affected by relevant external and internal factors.

Genetic influences upon the expression of aggression have, as mentioned, been demonstrated in a number of ways. Males from different inbred mouse strains are radically different in this regard (Scott & Fredericson, 1951; Southwick & Clark, 1968; Vale, Vale, & Harley, 1971). In addition, selection beginning with a heterogeneous population has been most effective. Lagerspetz (1969), for instance, reports a significant difference in male aggressiveness by the second generation of selection and this continued through the seventh generation. Maternal effects were examined and found not to be an important factor. Karli et al. (1969) report that although breeding for mouse killing in rats did not increase the incidence of the response under normal circumstances, it raised the probability of killing after olfactory deafferentiation to almost 100 percent. Aggressiveness in wild female mice has also responded to selection (Ebert & Hyde, 1976), although the general level of aggression is not nearly so high as it is normally in males. This is reflected also by the fact that there was a lack of correlated response among the males of the selected female populations (Hyde & Ebert, 1976).

We have mentioned on a number of occasions that in infrahumans the expression of aggression is related to the presence of sex steroids. McKinney and Desjardins (1973), for example, studied the development of aggressive and sexual behavior, circulating androgens, and testis structure in laboratory descendants of wild house mice and found the expected correlation. Aggression did not appear until the animals were 35 days old, and sexual behavior reached adult levels by 55 days. At the same time, there was a 300 percent rise in the level of circulating androgens between 21 and 55 days. Growth of sex glands coincided with increases in androgen concentration. Luttge (1972) castrated adult CD-1 male mice, then injected them with androgen. Testosterone-injected males fought as much as did noncastrated controls, whereas castrated animals that received only an oil injection did not fight at all, indicating the dependence of this behavior upon testosterone. Barfield (1971) found that implantation of testosterone propionate pellets directly into the hypothalamic preoptic area of castrated male ring doves activated both sexual and aggressive behavior.

Aggressive behavior has in addition been found to correlate positively with plasma testosterone level and dominance rank in captive male rhesus monkeys (Rose, Holaday, & Bernstein, 1971). Frequent submissive responses to those above him in the dominance hierarchy by an animal that was aggressive to those below him was not correlated with a lowered testosterone level. This is interesting in view of another report (Kreuz, Rose, & Jennings, 1972) that psychological stress in young human males reduces testosterone levels. If the results of the two studies are comparable, it may be that submission among subordi-

nates in a highly structured situation is not the stressful thing we demo-
cratically believe it to be. The relationship between dominance and
testosterone level did not, however, hold in a troop of Japanese macaques (Eaton & Resko, 1974), although dominance and aggression
were highly correlated. One would not, in fact, expect so simple a
relationship to obtain generally. Clearly, certain levels of testosterone
are necessary for the expression of this sort of aggression under normal
circumstances but many other factors, including the genes of the individual and his experience, enter into the determination of the behavior.

Confirmatory evidence for the actions of sex steroids comes from
autoradiographical research, wherein uptake by nerve tissue is studied
using labeled (radioactive) hormones. Sar and Stumpf (1973) found
that testosterone was selectively retained and concentrated in specific
neurons of certain brain regions of male rats, particularly the preoptic
region, the hypothalamus, the hippocampus, and the amygdala. These,
it may be argued, are the principal sites of testosterone action. The hypothalamus and amygdala are strongly implicated in neurophysiological research on aggression (page 352). Similarly, McEwen, Pfaff, and
Zigmond (1970) found that labeled testosterone was concentrated by
the hypothalamus, preoptic area, and septum of male rats. Pfaff (1968)
found also that the female hormone estrogen (given as radioactive
estradiol) was concentrated in ovariectomized female rats by the same
structures. Most interestingly, McEwen and Pfaff (1970) reported that
the capacity for a female to concentrate the normal female hormone in
these structures is altered by an injection of male hormone (testosterone) at an early age. We shall discuss the effects of hormones in
neonates shortly. Stumpf (1971) reports also that specific neurons in
the hypothalamus, preoptic region, and amygdala concentrate estrogen. Finally, Inselman-Temkin and Flynn (1973) further found that
the latency for quiet biting attack stimulated electrically was altered
as a function of sex steroid level. This indicates that sex hormones influence the threshold at which the circuit is activated.

It seems clear then that sex hormones circulate to the brain, and
that their chief sites of action are those involved with reproduction and
aggression. A consistent, if incomplete, picture therefore emerges of
the relationship between aggression and hypothesized brain circuits
through neurophysiological, neurochemical, genetic, sex steroidal, and
brain autoradiographic studies. It can no longer be doubted that, whatever their final configuration and mechanisms, genetically derived
circuits that potentiate aggressive behavior exist in the brain.

Even more evidence has been forthcoming from the study of the
effects of sex steroids upon the ontology of aggression It has been
known for some time (e.g., Gorski, 1971) that the presence of sex
steroids during sensitive periods of the development of the mammalian

brain "organizes" it in a male fashion vis-à-vis the reproductive cycle. Thus, female rats given a single injection of the male hormone testosterone shortly after birth do not have the customary rise and fall of estrogen and progesterone levels as adults; rather, estrogen is secreted tonically, the way males secrete testosterone. Without a reproductive cycle there is of course no ovulation; such androgenized females are sterile. In normal female development no sex steroid is present during the sensitive period. In normal male development the testes secrete androgens during the sensitive period, become quiescent, and then begin secreting again at puberty, when the ovaries of the females first begin to secrete estrogens.

The picture that has emerged then is that the brain of the chromosomal female (XX) develops as female in the absence of sex steroids but can be altered to the tonic (male) form of hormone secretion in the presence of such steroids if they are encountered early in life. The brain of the chromosomal male (XY) develops as male in the presence of sex steroids (normally secreted by the testes) encountered early in life but will continue to develop as female in the absence of such steroids. It is thus entirely possible to have chromosomal females whose brains have in certain respects differentiated to resemble those of males and to have chromosomal males whose brains have differentiated to resemble those of females. The question now is what other effects, besides those upon reproductive physiology, do these manipulations have? Are there concomitant behavioral changes?

Yes. Further, they are in the expected directions. Chromosomal females "androgenized" early in life tend to behave like males; chromosomal males that are not androgenized (through removal of sex steroids by very early castration) tend to behave like females. A primary brain region affected by the presence or absence of sex steroids in young mammals is the hypothalamus. This structure controls the releasing factors that, when they reach the pituitary, govern the secretion of the trophic hormones stimulating the testes and ovaries, which in turn produce most of the androgens and estrogens in the system. The route by which hormones given early affect the reproductive cycle is thus obvious. But the hypothalamus is also deeply implicated in the expression of aggression. Should we expect that differentiation of the brain due to sex steroids affects the expression of aggression?

Once again, yes, and that is what has been found. Bronson and Desjardins (1968) discovered that female C57BL/6J mice (an inbred strain) injected on the third day of life with testosterone fought significantly more in standard tests as adults than did females given injections of oil. The incidence of aggressive behavior among female laboratory mice is usually negligible, whereas among males it can be high and severe enough to lead to death. It was also found in this study that injections of estrogen had much the same effect. This corroborates the neuroendocrinological data (Gorski, 1971) indicating that sex steroids in general if present at the sensitive period will defeminize the brain. (The ovaries are not active at the sensitive time in normal females.) In a further study, Bronson and Desjardins (1970) found a relationship between the period of injection of neonates, the dose given in the injection, and later aggressiveness among female mice of the same strain. These animals were injected with either testosterone or oil at 1, 3, 6, 12, or 24 days of age. They were ovariectomized at 24–26 days of age, and were injected with testosterone again at 70–75 days, following which the aggression tests began. Neonatal injections were maximally effective when given on the day of birth and less effective thereafter, becoming ineffective between days 12 and 24. The period when the brains of these animals responded to the differentiating effects of male hormone is thus limited from a time before birth to 12 days later, with effects diminishing as day 12 is approached.

In a second part of the study, the authors found that very small doses of testosterone were less effective in producing aggression later than were larger doses. Also, neonatal androgenization was alone not effective in producing aggression in adults; it was required that hormones be circulating in the adult brains as well. This suggests, as had been indicated by studies on sex behavior (to be reviewed presently), that sex hormones have two effects: an organizational or differentiating effect in the very young and a motivational effect in adults. Gandelman (1972) in addition found that neonatal androgenization of female mice leads to an incidence of pup killing that is similar to that of males. Females are normally not prone to this behavior. Finally, Goy and Phoenix (1971) found that juvenile female rhesus monkeys rendered pseudohermaphroditic by subjection to testosterone at an early age were much more like normal males in the extent to which they threatened, initiated play, and engaged in rough-and-tumble and pursuit play. This suggests (as does other material) that the behavioral effects are not limited to rodents.

It thus seems obvious that exposure to sex steroids during a sensitive period early in life alters the mammalian brain so as to affect the reproductive physiology and the aggressive potential of adults. But are all animals equally affected? No. There are species differences, for instance. Goy and Resko (1972) found that pseudohermaphroditic rhesus females still have reproductive cycles, although their sexual and aggressive behavior is masculinized (or at least defeminized). Are there differences among organisms of the same species in their reactions to perinatal sex steroids?

One would surely predict that there are, and the data support the prediction. Indeed, a model emerges that, though incomplete with regard to actual mechanisms, offers a theoretical framework within which at least some forms (and evolutionarily the most important forms) of aggression may be understood. The model is an interactionist one, using organismic and environmental (treatment) variables. I must use my own work as an example while emphasizing once more that interaction studies are rare and that mine are hardly definitive. However, they do provide some substance for the notions advocated here.

We have noted that there are differences in the extent to which male mice of some strains fight, some fighting readily, whereas the males of others can barely be induced to attack. The females of inbred strains, regardless of how the males of their strains behave, almost never fight. Recall now that there are differences in genotype among the various strains, but that within strains all genotypes are (with a very high probability) the same, with the exception of the genetic material on the Y chromosome. That is, males and females of a strain will carry the same genes except that females have two identical sex chromosomes (XX), and males have an X chromosome identical to those of the females and a Y. Assuming that all the genes that potentiate aggression are not on the male Y but occur on other chromosomes (an assumption consistent with other genetic research), then males and females of a strain carry much the same of genes involved in the expression of aggression. Why do the males and females of some strains differ so dramatically then in their aggressive tendencies?

Hormone research, we have seen, suggests that perinatal sex steroids are implicated. Females are not exposed to such steroids during the sensitive period, and therefore the neurophysiological groundwork is not laid for aggressive behavior later in life. One would expect then that females would be less aggressive than males, and that indeed is what is found at large throughout the mammalian class. But why the difference among males of various strains? Their brains have all been exposed to sex hormones. Should they not be equally aggressive under the same circumstances?

Not if genes and hormones interact. Suppose that genes respond to the presence of sex steroids during development. In the strains which carry the "proper" genes, the presence of perinatal sex steroids would lead to a high aggressive potential. In other strains, having fewer such genes, the presence of perinatal sex steroids would lead to a lower aggressive potential. It is suggested then that perinatal hormones have their effects through the activation of genes. In the developing organism the products of these genes differentiate the brain. The sequence is thus

$$\text{hormone} \xrightarrow{\text{activation}} \text{gene (DNA)} \longrightarrow \text{protein}$$

This is not as speculative as it sounds. Indeed, some time ago Tata (1966) would write that "Virtually every major process of growth and development, as well as metabolic activity, in higher plants and animals is initiated or regulated by hormones." Many studies have in fact shown that hormones increase RNA and protein synthesis in peripheral target tissues (Davidson, 1968; Dorfman, 1964; Hamilton, 1968; Tata, 1966). Some studies indicate that the same process occurs in brains. For instance, RNA metabolism in the amygdala and hypothalamus of newborn female rats was shown to respond differently to testosterone administration in one study (Clayton, Kogura, & Kraemer, 1970), and enzyme changes in the hypothalamus of adult female rats treated three days after birth with testosterone were found in another (Griffiths & Hooper, 1972). A new and characteristic RNA was even reported in the brains of adult female rats treated with testosterone when two days old (Shimada & Gorbman, 1970). Differences in activity of the enzyme sRNA methlyase were found between the sexes and among regions of the brain in a species of deermouse (Eleftheriou & Hancock, 1971).

In its simplest form, our hypothesis therefore is: If aggression-relevant genes are present, they will be activated by sex steroids during the sensitive period of the development of the organism, leaving it susceptible to hormone priming as an adult. If such genes are not present, they can of course not be activated, and the organism will be relatively nonsusceptible to subsequent hormone priming. It follows that females from inbred strains wherein the males are aggressive should respond to perinatal sex hormone treatment with an increase in aggression as adults, whereas females from inbred strains wherein the males are unaggressive should not. Sex-hormone-related aggressive behavior is thus a function neither of treatment (presence of hormone during the sensitive period) nor organismic (genotype) variables alone, but of both.

In one study (Vale, Ray, & Vale, 1972), females of three mouse strains were injected with testosterone when they were three days old, raised in isolation after weaning, and tested for aggression as adults. Table 8.1 shows the results for attacks against a passive male held (by the tail) in the cages of the females for two short periods over 3 days. The responses of females and males injected only with oil on day 3 are presented for comparison.

Clearly, the testosterone-treated females tended to behave like normal males of their own strains (statistical tests were run and of course showed a highly significant interaction between genotype and treatment). C57B1/6 and A males did not respond aggressively, and neither did the treated females of those strains. However, BALB/c males responded with a high incidence of aggression, and so did

Table 8.1 MEAN NUMBER OF ATTACKS BY TESTOSTERONE-TREATED FEMALES AND OIL-TREATED MALES AND FEMALES

	Treatment		
		OIL	
Strain	Testosterone (Females)	Females	Males
FIRST PERIOD			
A	0.00	0.00	3.08
BALB/c	11.98	0.00	29.25
C57BL/6	1.00	0.00	3.92
SECOND PERIOD			
A	0.07	0.00	1.75
BALB/c	10.07	0.00	20.62
C57BL/6	0.40	0.00	4.42

SOURCE: Adapted from J. R. Vale, D. Ray, and C. A. Vale, The Interaction of Genotype and Exogenous Neonatal Androgen: Agonistic Behavior in Female Mice. *Behavioral Biology*, 7, 321–333, 1972. By permission.

treated females of that strain, higher even than males of the other strains. Treatment and genotype interacted to produce the results.

This proposition makes a great deal of sense when extended to heterogeneous populations. If, all else being equal, aggression were merely the results of the actions of perinatal sex steroids (particularly testosterone), then males should be quite similar in the extent to which they show aggression. The fact is, they are not. If, on the other hand, the effects of perinatal sex steroids are gene-mediated, one would expect differences in aggressivity among males to the degree that there are relevant genetic differences among them. This appears to be the case, as it would have to be if aggression were to be subjected to natural selection.

The next step, if one were to seek to extend results such as these to a more general form, would be to find a variable that underlies the interaction between genotype and treatment. Alas, for sundry reasons this has not been done. There are, however, some fair guesses as to where to look. The monoamines mentioned earlier, thought to be associated with aggression, are affected by manipulations of sex steroids. For instance, the serotonin level in male and female rat brains differs over the course of early development, in that it rises in females on the twelfth day after birth. This elevation was prevented by an injection of testosterone on the day of birth (Lodosky & Gaziri, 1970). On the other hand, males castrated on the day of birth had serotonin levels comparable to those of untreated females. Norepinephrine and serotonin levels were found to differ between sexes and between strains, and as a result of perinatal testosterone treatment in females in another

study (Hardin, 1973). Gonadectony was shown to increase norepinephrine turnover in female (Anton-Tay, Anton, & Wurtman, 1970)
and male (Donoso, de Gutierrez, Moyano, & Santolaya, 1969) rat
brains. The monoamines are, incidentally, believed to play a part in
both reproductive physiology and sex behavior, so their involvement
with aggression and sex steroids is predictable and thus, to the scientist, comforting.

We may consider then that to a substantial degree the individual
differences in some forms of aggression found among infrahuman
mammals is due to the interaction of genotype and perinatal sex
steroids. Genes on the Y chromosome lead to the differentiation of the
gonads as testes, which in turn produce androgens, which normally
further differentiate reproductive physiology and anatomy as male and
alter brain development so as to make certain (broadly defined) neuronal circuits susceptible to sex steroid (particularly androgen) priming
in adults. A distribution of aggressive tendency among males may be
expected in any heterogeneous population. Highly aggressive males
should result from the combination of a large number of relevant genes,
sufficient sex steroid during the sensitive period, and sufficient sex steroid during adulthood. Fewer such genes or inadequate sex steroid
levels will lessen the probability of aggression. In normal (XX) females
the gonads differentiate into ovaries, but, unlike testes, they are quiescent during the sensitive period. In the absence of sex hormone, reproductive physiology, anatomy, and behavior potential further differentiate as female. Highly aggressive females may be produced by the
combination of a large number of relevant genes, sufficient sex steroid
during the sensitive period, and sufficient sex steroid in adulthood, just
as in males. This rarely occurs, which is a good thing, since these females are usually sterile and do not mate. The large differences in aggression among males generally are not seen among females, and they
do not form as structured and enforced dominance hierarchies.

Is there other evidence that aggression is best understood in terms
of genotype-environment interaction? Yes. Population number (that is,
the number of animals in a relatively confined space) has been hypothesized to be associated with the extent of aggression among males
of some species. As an extension of work mentioned earlier, we found
that of five inbred mouse strains studied, only one (again the BALB/c
strain) showed increased aggression with increasing population number (Vale, Vale, & Harley, 1971). The average of the attacks of BALB/c
animals with four mice per cage was twice as high as for any other
genotype, while with eight animals per cage it was almost 16 times as
high. Manipulating the environment by increasing the number of animals in an area may thus result in increased aggression among animals
having some genotypes but not others.

Other studies on genotype-environment interaction have been reported by Southwick (1970a), who found that mice from a passive strain raised by mothers from an aggressive strain showed an increase in aggression, whereas the incidence of aggression in mice from the aggressive strain raised by mothers of the passive strain was unaffected. Ginsburg (1970) also found differential responses of inbred mice to early experience. Of three strains (C3H, C57BL/10, and C/albino) subjected to pain stimulation during infancy, only the first responded after sexual maturity with an increase in aggressiveness. Similarly, when C57BL/10 mice were reared by rat mothers, their aggressiveness was reduced considerably, but the same was not true when another strain was used. This was confirmed in another study (Denenberg, 1970). Finally, Ginsburg (1970) found strong genotype-environment interaction in another experiment on early stimulation. In one strain there were essentially no effects. In another the effects were cumulative over the entire preweaning period. Still other strains were sensitive to stimulation early in preweaning life, in contrast with those which were affected only after the second week. It is notable that exactly the same stimulation during preweaning that leads to an increase in aggressiveness in male adults of one strain leads to a decrease in aggressiveness in males of another.

These studies, as well as those reported by Fulker et al. (1972) and reviewed by Erlenmeyer-Kimling (1972) offer insight into the magnitude and importance of genotype-environment interaction and, as mentioned previously, provide a model for thought about the determinants of behavior in heterogeneous populations. Students of behavior have in the past been inclined to accept measures of central tendency not merely as convenient single numbers characterizing populations, but as ontologically valid responses (pages 275–276). However, it is the average, we have seen, that is the fiction; the average is a numerical abstract of the individual responses of members of the population. Those responses, which are to the same stimulus, are often quite diverse. Summing them in an average may decrease the opportunity to trace the mechanisms that underlie them by obscuring that diversity. As Ginsburg (1970) remarks, "the central tendency of a genetically mixed population . . . represents an amalgam of styles, times and directions of responses." Continued failure to recognize this is an exercise in self-defeat. Some organisms will respond in certain ways and others will not, for reasons far removed from those usually associated with learning. This has, unfortunately, been a threadbare fact in psychology. Time and a change of mind will be required to give it substance.

Meanwhile, there are indications that humans are affected by exposure to sex steroids at an early age in some of the same ways as monkeys, rats, and mice. Money and Ehrhardt (1972) have described a number of syndromes in humans that approximate those produced by experimental sex hormone manipulations in infrahuman animals and have examined the behavioral results. The syndrome most closely resembling the planned androgenization of females is progestin-induced hermaphroditism.

Several decades ago a group of synthetic hormones that were known to have the biological action of the hormone that aids in maintaining pregnancy (progesterone) were administerd to women in order to prevent miscarriages. What was not known was that these "progestins" also have some of the biological action of androgens. A number of human female fetuses were therefore masculinized in utero, the anatomical symptoms ranging from an enlargement of the clitoris and some labial fusion to a complete penis and complete labial fusion (which formed an empty scrotum). These infants still had female internal reproductive organs because the progestins were administered too late to affect their development. Externally masculinized infants were without closer examination often regarded as boys with undescended testes. Infants with only incomplete masculinization were, however, examined further, found to be females, and subjected to corrective surgery.

A second condition, produced by endogenous rather than exogenous sex steroids, is described by Money and Ehrhardt as the female adrenogenital syndrome. This too results from androgenization during the fetal period (and possibly beyond), the problem arising because of a genetic defect in the fetus that causes its adrenal glands to release not the normal product, cortisol, but a precursor that is androgenic in action. If the fetus is female, the external genitalia are masculinized (but again not the internal organs, the effect registered too late for this), some completely, some less so. When the condition is detected early (which is likely because the malfunctioning adrenals lead to a salt loss that summons medical scrutiny), corrective surgery and cortisone therapy allow these females to develop as normal anatomical women. They menstruate (perhaps a year late) and can conceive and bear young. Without cortisone therapy (which may not begin if the salt loss is not too great and the condition is thus not diagnosed) the androgenic secretions of the adrenals continue throughout infancy and early childhood, precipitating the masculinization of secondary sex characteristics (voice, body and facial hair, and so on) and puberty eight to ten years too soon.

What are the behavioral effects of fetal androgenization found in these two syndromes? Considering only those who were discovered early, given surgical and hormonal care, and raised without question as girls, Money and Ehrhardt found the following: (1) Most of the girls described themselves, and were described by others, as tomboys.

There was evidence of dissatisfaction with being a girl (in comparison with normal controls) among those with the adrenogenital syndrome, although none wanted to change her sex at the time. (2) The androgenized girls tended to expend physical energy at a high level in vigorous outdoor play, games, and sports usually common to boys. (This may be an analogue of the rough and tumble, pursuit play that was increased among the androgenized female monkeys described by Goy and Phoenix, 1971.) There was definite interest in joining boys at play, particularly at a team game with a ball. Control girls preferred to play among themselves. (3) The androgenized girls were not, however, more aggressive in being prone to pick fights. In the opinion of Money and Ehrhardt, they probably avoided involvement with the dominance assertion and striving for position of the boys for fear of being denied access to their recreational groups, and they were not interested in the rivalries of other girls. (4) In keeping with their energetic activities, androgenized girls preferred utilitarian clothing significantly more than did controls.

Another syndrome, which we have encountered earlier as a chromosomal abnormality, is relevant to consideration here. Turner's syndrome (page 199) is generated by an XO genotype, which in turn leads to an external female phenotype, but ovaries have not been induced to form and the gonads remain primitive streaks that produce no sex steroids whatsoever. (Two X chromosomes are required for normal female development.) Turner babies therefore look like females and are raised as girls, with no evident difficulties until the time when puberty should occur. Lacking female hormones, such girls of course fail to menstruate and develop female secondary sex characteristics.

Money and Ehrhardt compared girls with Turner's syndrome to controls and found that, if anything, they were more "feminine." They did not differ from controls on four measures and were even less interested than normal girls in athletic endeavors; there were fewer incidences of fighting involving them; and they had a greater interest in personal adornment. These findings support the notion that perinatal sex steroids affect the ontogeny of human behavior, with the presence of such steroids tending to produce behavior normally associated with males.

What of chromosomal males who fail to receive perinatal hormones? They differentiate as females. The androgen insensitivity syndrome results from the inability of the cells of a chromosomal male to use the secretions of his own testes. This is due to a genetic defect, although the mode of inheritance is as yet unknown. The testes of an androgen-insensitive male produce androgens in amounts usual for males. They also produce much lesser amount of estrogens. (This is normal. The gonads of neither sex secrete only androgens or estrogens, although each occurs in much smaller amounts in the other sex. Indeed, the gonads are not the only source of sex steroids. The adrenal glands of both sexes normally produce small amounts of androgens, although not in the way associated with the andrenogenital syndrome.) The estrogens, to which the cells of the affected males are not insensitive, lead to a complete feminization of the bony structure and body contours, including the growth of breasts. The external genitalia appear female, although the vagina is just a shallow cavity that has no connection with a cervix or uterus. The internal structures are not male, but neither are they completely developed as female.

Problems of course arise when these androgen-insensitive males raised as females fail to menstruate, and it is at this point that a number of them have been detected. Money and Ehrhardt studied 14 such chromosomal males, and found them not to differ behaviorally from normal females. This once again supports the notion that genotype and sex steroid combine to influence behavior. Given the genetic defect, no amount of androgen can lead to masculinization.

An obvious concern in all human research of this sort is how much culturally derived expectations contribute to the conclusions. Dress, mannerisms, and the like are all part of shifting stereotypes of masculinity and femininity that may bear little relation to basic differences in behavior (McGuire, Ryan, & Omenn, 1975). There are, nevertheless, differences that hold over many societies and thus are unlikely to be culture-specific. Perhaps the most important difference, with regard to the Money and Ehrhardt data, is that boys tend to congregate and travel in larger groups, whereas girls tend toward smaller groups (often groups of two or three). It is just in these larger groups that boys expend a great deal of the energy for which they are noted, and it is within these groups that dominance and assertive behaviors are likely to appear and be challenged (Maccoby & Jacklin, 1974). Boys in groups "excite" one another in a way in which girls evidently do not, thus providing the context within which most aggression in young humans is experienced. The penchant of fetally androgenized girls to join boys in competitive play groups in Western society, even if they do not participate fully, thus suggests that the hormonal effect is a legitimate one, not dependent upon a cultural stereotype.

Other data support the proposition that aggression in humans is dependent upon evolved neural circuits that are subject to the same sorts of variables as in other animals. Both the environment and learning as it is usually conceived play a role in human aggression, with the latter influencing its expression considerably more than in other animals. The human case is, however, a far more difficult one to assess because of these learning effects and because the circuits, through the evolution of the extraordinarily complex neural system we know as the brain, are with our present methods relatively inaccessible.

There is nevertheless a fair amount of circumstantial evidence that such circuits exist. The responses of humans to perinatal hormone variations inconsistent with their chromosomal sex, mentioned above, is suggestive. Maccoby and Jacklin (1974), in their extensive treatment of psychological data on sex differences, conclude that males are more aggressive than females and that the difference is biologically based. This is consonant with the results of studies on infrahuman animals. Sex steroids also influence aggressive behavior in human adults. Lunde and Hamburg (1972) found substance in the common notion that there is often an alteration in mood among menstruating women, with aggressive feelings being highest on day 18 of the cycle, just after ovulation. They also concluded that, although the data are sparse, there is evidence for a connection between androgens and aggression in normal men, just as in other animals.

No human research has, however, aroused both scientific and public interest as much as that undertaken by neurosurgeons. Particularly, Sweet, Ervin, and Mark (1969), and Mark and Ervin (1970) have suggested that violent, impulsive actions may occur as a result of abnormal firing of specific neural circuits, and that this abnormality may be corrected either by lesions or electrical stimulations at given brain sites. The areas of chief concern for these investigators is the limbic system, which includes the hypothalamus, the amygdala, and the hippocampus—the latter two, being large structures in the temporal lobe, receiving the major attention.

The limbic system is involved in basic life-sustaining and species-preserving behaviors—noted, according to Valenstein (1973), by one blushing investigator as the four Fs—feeding, fighting, fleeing, and "sexual activity." What sort of evidence implicates it in violent human acts? First, it is seen that rabies, which in humans and other animals leads to irritability and assault without provocation, results from a virus that invades the hippocampus, presumably contributing to abnormal firing there. Second, it seems that after craniocerebral injuries young men in particular go through a phase of hyperactivity and belligerence. This is seen as due to abnormal brain activity as a consequence of the trauma, although the locus is too diffuse to be correlated with a given region. Third, tumors in the limbic system are associated with assaultiveness in some individuals. A famous instance is that of Charles Whitman, who killed 16 people and wounded 24 others in Austin, Texas, in August 1966. Finally, assaultiveness is correlated with temporal lobe epilepsy. Apparently the abnormal discharges leading to the one may also lead to the other.

Of even more interest are electrical recordings made at specific brain sites in those given to violent acts. One young woman had been a model child until she was 13, at which time she grew moody and uncommunicative. A few months later, after an argument with her aunt and uncle, she smashed up her room, tearing books and curtains and partially destroying the floor and the door with a heavy object. A year and a half after that, annoyed with the screaming of a 26-month-old child, she suffocated it, and then, frightened, told a girl friend what she had done. She was studied at several institutions, wherein she made a number of suicide attempts and an attack on a younger child, before electrodes were implanted. Grossly abnormal discharges were found in the hippocampus on both sides. These were shown to be increased when a tape recording of a baby crying was played.

Another young woman had psychomotor seizures of several types, including some in which she felt an urge to take long walks or runs (at any hour) and others in which she had the illusion that her upper limb and the left side of her face had become grotesquely ugly. During one of the latter episodes another girl accidently bumped her, and she stabbed the girl in the heart. Later, at a mental hospital, she stabbed a nurse who did not come immediately when she was asked for help. Electrodes were implanted, and an abnormal discharge was found in the amygdala of the right temporal lobe. Stimulation of this site (when the patient was unaware of it) led on one occasion to her smashing her guitar against the wall in the middle of a conversation with her psychiatrist.

Sweet, Ervin, and Mark stress, however, that the proper psychosocial setting is necessary for aggressive behavior to be manifested during or following brain stimulation. Anger is almost never mentioned by patients who are stimulated on the operating table, or thereafter even in the less constrained setting of a hospital laboratory. Why then are they aggressive under other circumstances? It is not precisely known. But such an observation is quite compatible with those from animal research. Flynn's cats, when stimulated, attacked rats but not their cages or the air, and animals involved in rival fights seek out appropriate targets. Thus, although the threshold for aggressive behavior may be much lowered when the firing circuits are stimulated by either an endogenous or exogenous source, there still must be some focus for it in the environment, even if it be so slight as an inadvertent touch.

We need not believe, as Valenstein (1973) remarks, that specific sites potentiating aggression in humans are as easily located and manipulated as some neurosurgeons suggest. Obviously these investigators present the cases which best corroborate their hypothesis. However, there are strong implications here that in the light of other data support the primary thesis that circuits which potentiate aggression in humans do exist. Such circuits are doubtless diffuse and overlap those potentiating other behaviors, but this is predictable given that aggression is both self- and species-preserving. We would therefore be surprised, given the probable evolution of aggression, if its expression were unrelated to events normally important to individual and/or to species survival.

The point is that we should neither search for nor expect to find reflexlike neural systems that mediate aggression. The whole trend of the evolution of the brain, of which we are the current final product, has been toward more interconnection among structures and toward the development of new structures (i.e., the cerebral cortex) that integrate, influence, and are influenced by older ones. Evidence that there is no "aggression center" that, when stimulated or disinhibited, produces aggression as a penny does a gumball in the proper machine is merely confirmation that aggression evolved as adaptive behavior. As such, it should facilitate the well-being of both individuals and populations, which means that it must be responsive not only to internal but to external conditions. This requires the processing of an enormous amount of information, which in turn requires the coordinated functioning of many neural complexes. We should not be surprised then that aggressive behavior in humans is difficult to produce by stimulation in individuals who find themselves in the unfamiliar (and sometimes completely dominating) environment of a hospital. Rather, we should consider that aggression under these circumstances is in fact what these individuals are internally programmed not to express. It is a bit of a wonder that any has been found at all.

We should also remember that the lesioning and brain stimulation procedures used, no matter how advanced over those of the past, are still crude, and it is entirely possible that in the future irrefutable data will be forthcoming. In the meanwhile, given all the other evidence, there is every reason to accept the proposition, espoused in one form or another by so many, that there are neural circuits that potentiate aggression in humans and that they are subject to much the same variables as in other animals.

We have then a broad but incomplete, though not inconsistent, picture of aggression. It may be conceived of as an extension to direct encounters of the competition that allows natural selection to operate, functioning to give advantage to some individuals because it results in inequitable access to resources and mates and to preserve species because the most capable individuals contribute maximally to the gene pool. Having been selected for primarily in conjunction with competition for resources and mates, it is naturally often expressed in situations involving one, the other, or both.

Genetic, neurophysiological, and neuroendocrinological research on infrahumans confirms the view that neural systems have evolved that potentiate aggression, and that they are sensitive to gene changes and alterations in physiological and hormonal status. Given that genotype, physiological condition, and so on vary among individuals, the organism-environment interaction model provides the most suitable framework for the analysis of aggressive behavior. Data specific to this topic are few, but they support the interaction approach.

Much that is required for an initial understanding of aggression is thus in place. Ethologists have provided its evolutionary "rationale," others have begun to uncover the mechanisms of its biology and the eliciting features of the environment, and still others are examining the ways in which biological and environmental variables combine to determine its expression. However, two important and related questions remain: (1) How precisely applicable is the ethological rationale? (2) Why is human aggression in some respects unique? We shall treat these in order.

The ethological explanation appears by and large to apply well to infrahuman behavior (from which, after all, it was abstracted). But at the limits even there the system sometimes fails, and the failures bear examination because their study provides some hint as to how tenuous the controls upon aggression can be in lower forms, permitting us to see with greater insight into aggression in the highest.

Marler (1976) thus remarks that "Although some animals are remarkably peaceful, I believe that the violence of animals has been underestimated. Most animals die before they reach old age (and) a great deal of animal mortality must have an element of social causation." The Hanuman langurs of India, for example, form two kinds of troops, one consisting only of males and one with a single adult male and a number of females and young. Occasionally an all-male troop will attack a mixed troop, and one individual of the all-male troop will replace the resident male therein, often doing him severe physical damage in the process. One of the first acts of the new male is to kill all the infants in the mixed troop and to drive out the juvenile males. The new leader then copulates with all of the females as soon as they come into estrus (Poirier, 1974). Similar behavior has been recorded in another species of langur, and in captive macaques (Marler, 1976).

Such acts are also found among birds. For example, herring gull chicks are attacked by nonparental adults if they stray from their nests. These attacks accounted for nearly half of the mortality of chicks in one colony. In a study of lions, 20 percent of the 53 cubs dying were found to have been killed by adults. An even higher percentage of cubs died of starvation because of the refusal of lionesses to share food. When there was only a small amount, the adults would aggressively repel the cubs. Finally, it was estimated that more cubs die because they are abandoned by their mothers than from any other cause. It is evident thus that "the relationship between adults and young, even that between mother and offspring, is not one of undiluted altruism" (Marler, 1976).

Fighting among adults, too, though not frequent, is not all that uncommon. Southwick (1970b) found reports of intraspecific violence among various species of lizards, muskrats, ground squirrels, tigers, hippopotami, musk oxen, grizzly bears, and of course rats and mice, to name a few. Elephants are notoriously social, but the males are known to engage in titanic battles that may continue until one of the contestants is killed. Injurious fighting can in fact be found in almost any species studied, including primates (Carpenter, 1974; Southwick, 1970b; Van Lawick-Goodall, 1971).

A particularly germane observation is that aggression is most reliably expressed to a stranger of like species and sex. This is consonant with everyday human experience and makes a great deal of sense if indeed aggression is basically a confrontational form of competition. The most serious competitor for resources and mates is the one most like oneself, a fact recognized by Darwin in *Origin*. With no social bonds to consider, the likeliest target for aggression then becomes the *stranger* like oneself.

In this regard, Southwick, Farooqui, Siddiqi, and Pal (1974) report that groups of free-ranging rhesus monkeys all attacked alien rhesus that were released nearby, with the exception of infants. In several cases the infants were adopted by females without offspring. The experimenters, like the strange adults and juvenile monkeys, however, were not so fortunate. They were once attacked by a troop even though they had been observing it without incident for three months. Southwick et al. (1974), having reviewed the data, conclude that such xenophobia is common in territorial animals and those with rather closed groups. It occurs in hamsters, mice, gulls, baboons, monkeys, wolves, lions, and many other animals. Even social insects are xenophobic: Members of one colony are unlikely to tolerate the presence of a member of another (Wallis, 1970).

It thus appears that the ethological program of sign and countersign wherein aggression is ritualized into harmless postures is not in life quite the success some texts would have us believe. There is more violence among animals than we were told there should be. However, this not so much suggests a defect in ethological observation or thought as it reminds us of the difficulty of stating principles by which all organisms rigidly live. More to the immediate point, it reminds us that a threat is only as good as is the probability of its execution, and therefore a system based upon purely sham aggression would not have long survived. The specter of real violence must hover over the ethological minuet, or else the system dances into biological invalidity. For selection to continue in favor of the ritualized behaviors that prevent aggression there must in each generation be a sufficient number of affirmative answers to the (figurative) question: Will you actually fight?

Given this, one would expect to find a certain amount of looseness, and for purposes of analysis it is reassuring to learn that it exists. It means that for aggression to function as it is believed to, and for the behaviors that minimize it to continue to be biologically valuable, the physical danger must be genuine. That it is genuine is shown in those cases we have briefly reviewed. This in turn, paradoxically it must seem at first, supports the ethological view.

Such looseness is reassuring, even more paradoxically, in the analysis of human aggression. Our behavior does not seem nearly so disparate from that of other species when it is understood that there is a fair amount of death and destruction among them too. This is an important point with regard to our image of ourselves at present for, in one of the curious turns taken in the intellectual and emotional history of our relationship with other animals, the social behavior of infrahumans has now to some become exemplary. The question seems to be this: Since animals settle their differences without violence, what is wrong with us?

We owe much of this to ethologists and an occasional psychoanalyst (e.g., Fromm, 1973). They have presented the view that animal societies are orderly and peaceful, with a few ritual bluffs here and there standing in for overt aggression. Humans, against this background, appear aberrant and bloody, ecological deviants out of touch with the past and out of control in the present. The search then begins for the environmental changes that will bring us back to nature.

A search is necessary, but not one motivated in this way. We find, first, that there is much more intraspecific aggression among other animals than originally reported, and we find further that there is good reason for this to be so. Second, it becomes clear that the very features of animal societies that serve to minimize aggression are those from which we have struggled to liberate ourselves for millenia. The fact is that the most efficacious means for the prevention of aggression among animals is recognition of the perogatives of rank. The system revolves around deference to those in power. Consider the following, from van Lawick-Goodall (1974):

> All at once a series of pant-hoots announced the arrival of more chimpanzees, and there was instant commotion in the group. Flint (a juvenile) pulled away from the game and hurried to jump on Flo's back as she moved for safety halfway up a palm tree. I saw Mike with

his hair on end beginning to hoot; I knew he was about to display. So did the other chimpanzees of his group—all were alert, prepared to dash out of the way or join in the displaying. All, that is, save Goblin (an infant). He seemed totally unconcerned and, incredibly, began to totter toward Mike. Melissa, squeaking with fear, was hurrying toward her son, but she was too late. Mike began his charge, and as he passed Goblin seized him up as though he were a branch and dragged him along the ground. And then the normally fearful, cautious Melissa, frantic for her child, hurled herself toward Mike. It was unprecedented behavior, and she got severely beaten up for her interference, but she did succeed in rescuing Goblin. . . . (p. 160)

Again:

After ten minutes Flo evidently could endure the tickling no longer and she moved away, leaving Mike with a benign expression on his face. And yet just two hours earlier this same male had attacked Flo savagely, dropping a huge pile of his own bananas and pounding the old female unmercifully—just because she had presumed to take a few fruits from a nearby box. (p. 127)

It is not our purpose to note the heroism or question the chivalry of chimpanzees, but to cite a feature common to their life. Not just Mike, but all males attacked females, juveniles, and other males lower than them in the hierarchy. Mike had paid his dues, incidentally. Before he began to move up in the way to be described he had been "threatened and actually attacked by almost every other adult male" (p. 121). Doubtless there would be much more of this were is not for the fact that the mechanisms described by ethologists do reduce the instances of violence.

The point, once again, however, is that violence is reduced under these conditions only by appeasement on the part of the subordinates. Fear, or its equivalent, keeps the peace. Further, a system dependent upon the use of ritualized behaviors to minimize aggression simultaneously maximizes the advantage of the dominants, for it reduces the probability of their being hurt. Although this in turn serves the population, it does so, as we have noted in so many other cases, at considerable cost to some individuals.

A human society based upon intimidation, fear, and privilege for the powerful is not what great and sensitive thinkers had in mind, and one must therefore wonder at least a little at the moral esteem in which infrahuman behavior is currently held. Certainly it has not always been so. On the contrary, we have heretofore found near ultimate fulfillment in human "civilization"—disparaging, with more or less indulgence, the antics of those other than ourselves. Indeed, it was the *animal in man*, a citizen of lower state, that was usually seen as responsible for our distress. Think only of all the demons having the form or parts of beasts.

However, we subsequently came to believe that animals are not so beastly after all, and the entire proposition was then upended, like an hourglass, to run the other way. Humans had gone wrong not because of the animal in them but because they had taken the animal out of its environment. It was deprived of the opportunity to engage in the purely symbolic forms of aggression and therefore turned to the real. "Civilization" was anything but a civilizing force.

Neither is quite the case. We are animals, yes, but of a different kind, and our behavior, including our use of aggression, is precisely what would be predicted given our evolutionary history. We cannot fault our ancestry—that is meaningless in any event, and besides it did not go awry—and we cannot fault our social environment, for it is just what such creatures would produce. We are left therefore with the rather startling conclusion that there is nothing to blame. There is no culpable entity within, separable somehow from our innermost and truest selves, nor any outside force that presses from our unwilling souls those lamentable acts that are our shame. In short, there is no villain. To some, this means the loss of an old friend.

The point, however, is that we are no more or less responsible for our behavior than any other species. Certainly there are valid reasons why we *should* act more often in accord with our professed ideals, but there are equally valid reasons why we do not. We therefore cannot attribute our failures to anything other than the whole creature who both professes and fails. To understand why, we must do two things: First, we must see that our ideals are often rationalizations that stand at variance with normal human behavior, even in the best of times; second, we must think always of distributions.

That ideals are rationalizations that continuously require enforcement is evident from just a quick look at the apparatus in any society, from hunting band to nation-state, that exists for this purpose. Humans, as we noted before (page 265), are ethicizing but not necessarily ethical beings. We are therefore never without some means of behavior control. Awareness of distributions is a particularly poignant necessity because, bluntly, it takes the actions of 20 of the good to offset the actions of one of the bad. It always has. And we must realize that a substantial enough number of the bad will be present in any generation to keep the others hard at work merely maintaining that very fragile compromise upon which a society desirous of high ideals is based. One might think of this as the germ model: Surrounding the societal corpus there will always be a sufficient number of those capable of turning it to their own ends to do so if vigilance is relaxed. We are never rid of them.

These considerations bring us to the second of the two questions asked a few pages ago; namely, why is human aggression sometimes unique? Why do we alone persist in the organized conflict that brings so much agony to our own kind? We consider now not the impulsive, isolated aggressive act of one individual against the next, which is common to human and infrahuman alike, but to the premeditated use of force by groups of men, large or small, that could properly be called war. The question thus is not why humans fight—all animals fight, and humans for the same reasons as the others. The question is why we fight wars.

The answer is disturbingly straightforward, but quite the opposite from that ordinarily predicted: the evolution of the human intellect freed us from the bondage of ritualized behaviors and at the same time increased our capacities for analysis, planning, and organization. Given that the association between aggression and the ends it serves did not change, war was the inevitable result. Thus, in contrast to Wallace Craig's (1921) early conclusion that animals fail to make peace because they are too stupid, it seems that men fail to stop warring because they are too smart.

A dehumanizing thought? No. Recall that for symbolic combat to be effective the defeated animal must *act* defeated; it must defer to its superior just as it would had blood been spilled. At a certain intellectual level, this simply becomes improbable. It is hardly that symbols hold less charge in general for humans—the opposite is the case; rather, it is that in humans the compelling character of specific ritual endeavors common to infrahumans has been deflected by the evolution of high intelligence and conscious individuality. Humans more than anything in the middle and late hunting stages became self-conscious problem solvers, using their brains rather than their always limited physiques to dominate the earth. The evolution of such intelligence and of striving individuality is inconsonant with continued recognition of the perogatives granted to ritually determined superiors.

We found then, for one, that the same capacities and strategies that allowed us to hunt and defeat larger, stronger, and faster animals could be turned upon other humans and, for another, that there were few internal inhibitions against doing it. Patience, forethought, organization, the very qualities that are attributed to the evolution of man-the-hunter, serve him equally as man-the-warrior. A psychological defeat need not have the near permanence it has among other creatures to one capable of analyzing the real strengths of an opponent and then plotting to overcome them. The acquisition and retention of power thus once more turned upon destructive struggle.

Hence we became, as it seems now, too smart for our own good. Aggression still brought its normal rewards, but intellectually we outgrew our infrahuman inhibitions against its application. And this adds enormously to the dilemma of the human condition, for we have evolved not only into creatures of great intelligence and self-consciousness, but also into creatures of normal aggressive penchants and few restraints against their use. The usual conflicts between individuals and society therefore assume the outlines of crisis. People unhappy with the power structure will eventually fight back; they will rebel.

The result is a damnably recognizable pattern in human affairs: Competition for resources and the disposition into dominance hierarchies among individuals, groups, and nations continues while at the same time the inequities they create are resented and resisted, often with great courage and at great cost. There is good cause, then, why collective violence is normal. Human social systems, in a way true of those of no other animal, contain ingredients in just the proportions necessary to blow them apart.

The Old Testament injunctions with which we are familiar (or perhaps all to unfamiliar as a historical point) as well as our knowledge of the rules by which human life has largely been lived thus take on, in this perspective, an awesome vitality. If people are not permanently subdued, they may mend and fight again. Thus it is terribly sensible when in conflict to destroy them altogether. For this reason our records are from the beginning crowded not only with the names of those who built, begat, and were begotten, but with those who destroyed, slew, and were slain. Total war is, therefore, in contrast to some current views (e.g., Dubos, 1973), not a perversion of modern times: it is not the product of long-distance killing, industrialization, or urban blight. Rather, the death of the enemy is and has always been a workable solution.

This perspective has been adumbrated, if not shared, by a number of investigators. For instance, in his famous exchange of letters with Einstein on the subject of war, Freud (1932; reprinted 1973) wrote:

> From the moment at which weapons were introduced, intellectual superiority already began to replace brute muscular strength; but the final purpose of the fight remained the same—one side or the other was to be compelled to abandon his claim or his objection by the damage inflicted upon him and by the crippling of his strength. The purpose was most completely achieved if the victor's violence eliminated his opponent permanently—that is to say, killed him. This had two advantages: he could not renew his opposition and his fate deterred others from following his example.

The slaying of the males of an opposing tribe, the enslavement of the women, the appropriation of grain and cattle, the destruction of dwelling places, the complete obliteration of enemies so that they exist no more except as references in boastful song does thus command an awful logic. The things we used to do, and do still, are therefore the acts neither of evolutionary freaks with an insatiable blood lust nor of ecological misfits unadapted to the stress of historical times. We have always killed our own, and the grim, sad fact is that it has always made sense. Usually those we killed were strangers, but invariably they were competitors. We simply put them out of business. The special circumstance that it was the business of living has ramifications that are chiefly ethical. Such ramifications hence are beyond our immediate consideration, for we know the contradictory ends in which they are regularly invested.

The particular human style, painful as it may be to admit, is thus not unnatural. Our friends the animals are much the same, and were they possessed of weapons and greater mind would doubtless join with us in bellicose crusades. Chimpanzees, for example, romantically the models of harmonic existence to some, are in fact persistent squabblers who do not desist from administering beatings to those whom they outrank and who offend them. (They are also persistent hunters (Teleki, 1973), snatching up and tearing apart young bushpigs, baboons, and—yes—young humans. Van Lawick-Goodall (1971) reports that while at Gombe Stream she was forced to keep her son in a cage at times for his protection.) We should therefore divest ourselves of the encumbering notion that we alone among the creatures of the earth have fallen to pathology, that we have been exponentially corrupted from some native happy state wherein all were one to a condition wherein all are at odds. The Golden Age, the Garden of Eden, Paradise—these are the fantasies with which we tempt ourselves. We know better, but the fantasies continue in diverse and subtle ways. In order to escape them we must begin to comprehend that it is all in us as humans. We are the antagonists, both the devils and the gods. If we require a grand vision to sustain us, and probably we do, that should be enough.

An example of the use of intelligence to promote oneself in the dominance hierarchy of chimps is recorded by Van Lawick-Goodall (1971). Mike, as mentioned previously, had not always been the dominant male. He learned, however, to bang empty kerosene cans in his charging displays, evidently frightening the other males with this unusual device until they acknowledged his supremacy—all that is but the previous leader, Goliath, with whom Mike settled accounts as described earlier (page 337). Van Lawick-Goodall describes one instance of the kerosene can affair as follows:

> A group of five males, including top-ranking Goliath, David Greybeard, and the huge Rudolf, were grooming each other. All at once Mike calmly walked over to our tent and took hold of an empty kerosene can by the handle. Then he picked up a second can, and, walking upright,

returned to the place where he had been sitting. After a few minutes he began to rock from side to side. Gradually he rocked more vigorously, his hair slowly began to stand erect, and then, softly at first, he started a series of pant-hoots. As he called, Mike got to his feet and suddenly he was off, charging toward the group of males, hitting the two cans ahead of him. The cans, together with Mike's crescendo of hooting, made the most appalling racket: no wonder the erstwhile peaceful males rushed out of the way. Mike and his cans vanished down a track, and after a few moments there was silence.

After a short interval that low-pitched hooting began again, followed immediately by the appearance of the two rackety cans with Mike close behind them. Straight for the other males he charged, and once more they fled. This time, even before the group could reassemble, Mike set off again: but he made for Goliath—and even he hastened out of the way like the others. Then Mike stopped and sat.

Rudolf was the first of the males to approach Mike, uttering soft pant-grunts of submission, crouching low and pressing his lips to Mike's thigh. Next he began to groom Mike, and two other males approached, pant-grunting, and also began to groom him. Finally David Greybeard went over to Mike, laid his hand on his groin, and joined in the grooming. Only Goliath stayed away, sitting alone and staring toward Mike. (pp. 122–124)

Van Lawick-Goodall then remarks that "Charging displays usually occur at a time of emotional excitement—when a chimpanzee arrives at a food source, joins with another group, or when he is frustrated. But it seemed that Mike actually *planned* his charging displays; almost, one might say, in cold blood" (pp. 123–124). Eventually the cans had to be hidden because Mike had learned to throw them, once hitting the observer on the head.

Mike's behavior did not of course approach the complexity and organization required for war, or even the planning needed for a successful act of individual violence designed to improve one's position in human society, and obviously it also took advantage of rather than circumvented the ritual aspects of dominance relationships. Nevertheless, it offers some indication of the role the intelligence and the novel use of objects (weapons) could have played in the evolution of agonistic contests among humans. Striving, self-conscious problem solvers will discover methods of achieving old goals in new ways.

Where does this leave us? Certainly with behavioral tendencies every bit as confounded as they have been portrayed to be. We are all things—not as individuals necessarily, but as a species—and we must form our expectations accordingly. We are not merely reconstituted apes; we are unique, but neither can we evade our primate, indeed our mammalian, heritage. We shall compete, generate dominance hierarchies, decry competition, deplore dominance, and so forth, not just for a few generations until we learn the true and final path, but effectively forever.

To say this gets us into trouble is to understate the case with authority. We live now in an age when miscalculation could destroy us all, and yet we continue in our ways. What is our choice? Those very penchants that create our deepest miseries are responsible also for our most sublime triumphs. It is the "indominable human spirit" that in all endeavors leads us past seemingly insurmountable obstacles, where we celebrate in glory until we generate the next disaster. It is high, melancholic irony; it is our hope and our bane.

These issues have, of course, been addressed before, if not through the same route. Freud (1932, reprinted 1973), for example, wrote to Einstein that "the attempt to replace actual force by the force of ideas seems at present doomed to failure. We shall be making a false calculation if we disregard the fact that law was originally brute violence and that even today it cannot do without the support of violence." Alas, it is so. Why else armies, police forces, jails? Freud, however, viewed aggression primarily as negative.

Psychotherapist Anthony Storr (1970), in his book *Human Aggression,* takes a more modern approach, noting in the introduction that "there is no clear dividing line between those forms of aggression which we all deplore and those which we must not disown if we are to survive. The desire for power has, in extreme form, disastrous aspects which we all acknowledge; but the drive to conquer difficulties, or to gain mastery over the external world underlies the greatest of human achievements." Much the same conclusion was reached by brain researcher Elliot Valenstein (1973). In addressing the question of the use of drugs or other brain manipulations as a means of pacifying political leaders and thereby reducing the probability of war (a proposal actually made) he notes that "there is no known way to eliminate aggression without reducing and distorting a number of desirable capacities that we would hope our leaders would possess" (p. 351). Historian Richard Brown (1969) offers much the same sobering analysis: "We must recognize that, despite our pious official disclaimers, we have always operated with a heavy dependence upon violence in even our highest and most idealistic endeavors. We must realize that violence has not been the action only of the roughnecks and racists among us, but has been the tactic of the most upright and respected of our people" (p. 76). Others have made the point differently, but these suffice.

It is important to realize, first, that aggressive tendencies will not go away merely because we do not like them, and, second, that we could not do without them anyway. They have been selected for in conjunction with life- and species-sustaining behaviors, with the result that they are a part of us. Given our evolutionary history, we are exactly what we should be.

It is therefore strange to read from gifted authors (Montagu, 1973) that the ethological view, as amended and applied to man, represents "The New Litany of Depravity, or Original Sin Revisted." The fact that aggression has a biological base does not prescribe its uses, and one must wonder therefore what depravity and sin have to do with it. Indeed, it is this very confusion between scientific propositions and moral proscriptions that has hindered our understanding from the beginning. We did not, for instance, wish to believe previously that we were descended from early apes. What, we wondered, would that make us? What would we then be? Nothing that we were not already. The knowledge produced no change in our behavior. We did not immediately fall upon all fours or eat from the ground. We did not *become* apes because we learned that we are related to them. Similarly, we shall not become more aggressive simply because we admit some degree of evolutionary-biological determination. (We should in fact inquire at this point as to the effectiveness in reducing aggression exclusively within the framework of the environmental view.) Rather, we shall seek as before to eliminate the detrimental expression of aggression, but now perhaps with some sense of the purpose and application of the phenomenon at large.

Our records show that we have killed our own. It has been argued, however, that this is due primarily to the increase in population associated with sedentary agricultural life. The members of the prehistoric hunting band were supposedly at one with one another and with nature. There is evidence to the contrary. Bones of ancient *Homo* have been found that are split, a procedure to extract the marrow, as have the remains of skulls with forced openings, a procedure to extract the brain. That is, there is evidence that we were cannibals. Helmuth (1973) notes that

> The finds of human bones and bony remains which were scattered in firesites or kitchen-middens are highly convincing evidence for the occurrence of cannibalism. These bones and skulls, as well as parts of the postcranial skeleton, show clear evidence of the application of force by men, and they indicate that the biological inhibition of killing an individual of the same species disappeared a very long time ago. (p. 229)

Bigelow (1972) makes much the same point, chiding in the process those who offer extraordinary interpretations in order to preserve the notion of prehistoric goodwill:

> Certain anthropologists, archaeologists, and historians assume that intergroup competition was inconsequential during prehistoric times. Their

works imply that war was more or less unknown until the dawn of history. [But] there is evidence which, though it cannot prove prehistoric violence absolutely, does suggest strongly that early humans did kill other humans. [Thus] prehistoric *nonviolence* cannot be taken for granted as a well-established scientific fact.

Faced with such evidence, we sometimes go to impressive lengths in our attempts to explain it away. A hole in a fossil skull may have been due to a rock that fell from the roof of a cave while the man was asleep; he may have stumbled and bumped his head against a sharp projection which happened to be about the size and shape of a prehistoric hand-axe; one anthropologist has even suggested to me that chimpanzees may have been throwing stones for fun and happened to hit prehistoric men by mistake. (pp. 8–9)

But Bigelow goes much further than this. He believes that warlike acts by prehistoric humans were not only likely, they were responsible for the rapid expansion of the human brain over the last million years. Intergroup competition, in his view, was thus a very strong selective factor in human evolution. Why would such groups compete? For the same reasons that most animals do—for access to resources. Bigelow proposes thus that it was for the more desirable hunting grounds that the groups fought, and that this put a premium upon the evolution of intelligence. Human brains had always to contend with other human brains. The more intelligent, organized groups would control more resources and would therefore reproduce and expand more rapidly.

It has been advocated here that it is the evolution of intelligence that has in humans led to the deritualization of aggressive encounters. Bigelow's suggestions are obviously not inconsistent with this. He emphasizes, in fact, that intergroup competition would not produce mindless aggression among individuals, but rather would lead to the evolution of capacities for group organization and individual self-control. A group does not effectively defend its properties, just as it does not effectively hunt, without cooperation. The fear and mistrust of the stranger, found among so many animals, and indeed known sadly to us all, would remain. However, against that a new premium would be put upon at least some cooperation among groups.

Is it likely, then, that through hunting *and* intergroup competition the particular human configuration has evolved? Certainly the proposition should be considered. All the elements are there, those balanced inconsistencies that provide the exaltation, tranquility, conflict, and dread in human life. Further, it is the only proposition that makes sense of long-term human behavior. As Bigelow astutely observes, if we were indeed physiochemically fashioned through millions of years of prehistoric peace, then our more immediate ancestors must have been hopelessly insane.

Granting that to be untrue, and granting the validity of the biological-evolutionary case, what is to be done? Where do the answers lie? First, it should be noted that we are in the same straits in which we have always been. Only the weapons are more horrifying. We can hardly conclude then that the conception and wide diffusion of noble sentiments has carried us beyond the threshold of war. Ideas neither fight nor make peace; people do. We must expect humans to continue to behave as humans. Further, we must expect them to behave as humans in *any* environment. Without doubt some environments are more conducive to hostility than others, but so long as we remain social, there can hardly be circumstances that will satisfy us all.

Second, it should be noted that in spite of the ravages of ceaseless conflict, we have multiplied and filled the earth—to the extent that population growth has itself become a potential new source of antagonism. We must face the unwanted fact that as a species our warlike actions have not hurt us that much; indeed, the requirement that one human brain contend with another may have resulted in increased capability for them all. It would appear then that there has been relatively little selection against the tendencies that lead to war. As a species it has seriously threatened us only in the past century.

We must thus predict that we will go on doing what we have always done. In the best of societies we now elect our political dominants rather than allow them to achieve power by force, but the machinery of government—the laws, the courts, the national and local police—through which they act is an agency of aggressive coercion as surely as is the hierarchy of a troop of savanna baboons. What we have achieved, that is, is a cultural check upon how we are dominated; we have undone neither the fact of nor the necessity for dominance.

Similarly, our recent efforts to share resources among individuals more equitably than they generally are or might otherwise be shared, while unquestionably helpful in reducing antagonism, require the full sanctions of that same coercive machinery. Brown's (1969) point (page 379) is most relevant here. No matter what the reigning ethic, we do not share willingly beyond our immediate groups; wider adherence to even the most reverentially held precepts must usually be enforced. The social progress we have made, which is genuine and does serve to lessen violence, nevertheless has itself depended upon and continues to depend upon acts or threats of aggression against specific individuals.

What does this mean? Are there answers? Yes, and they are surprisingly obvious. We should seek to identify with ever-larger groups (Freud, 1932; reprinted 1973), seek out and pursue just causes en masse (fight the "moral equivalent of war") (James, 1910; reprinted 1973), reduce frustrating conditions (Miller, 1941), come to know ourselves and others (Lorenz, 1967), and provide better models for our young (Bandura, 1973). All these will help. What should be recognized first, however, is that neither singly nor together do they generate permanent change since it is always possible *because* of the biological basis of aggression to fall back. Despite the moral munificence of any organization, there is that danger. Fighting is not something we forget.

Moreover, it should be realized that viable solutions will come only within the framework provided by our biological-evolutionary history. It is meaningless to imagine a world at peace because people have learned the error of their warlike ways. But such a world is at least conceivable if all find themselves in a single dominance hierarchy that allows them fair access and an economic system wherein resources are reasonably distributed. Finally, it should now be obvious that even a world society at peace would not under any conditions be a society that is aggression-free; on the contrary, aggression is needed to *prevent* war.

We arrive then at our accustomed paradox, at least when aggression and war are confusedly equated. The fact is, aggression is necessary, but not necessarily destructive. We must therefore channel its expression toward desirable ends; we must use it wisely to prevent its being used destructively. (This revives an ancient and insoluble dilemma: Who is channeling the channelers?) We shall not thereby rescue the energies of every mad dictator for high purpose; all we can hope is to lessen the probability of their being mad dictators.

But that is all we have ever, realistically, been able to hope. Past and future utopias notwithstanding, we must build upon what we are, not what we would prefer to be. We are striving, dominating, self-conscious, intelligent creatures. We have a chance, if we act within our capacities. Let us therefore not judge ourselves in haste. We have not done too badly. But let us also remember that we, as all species, are but one generation away from extinction.

SEXUAL BEHAVIOR

Sexual behavior is not nearly the analytical puzzle that aggression is. We have no difficulty comprehending its role in nature, and, although it may occasionally create problems for us, there can be little doubt that it must continue if we are to continue. Therefore we need not wonder "why sexual behavior" in the same sense that we wonder "why aggression." Sexual behavior is necessary if there is to be sexual reproduction. The smaller differences in sexual activity are thus of little account next to the fundamental fact that behavior leading to reproduction must, and does, occur.

The treatment of the topic here will be appropriately brief, the attempt being to present the case for biological (particularly genetic and hormonal) determinants of sexual behavior and their interaction with determinants of environmental origin. We wish merely to show that another important behavior is best understood within the interaction framework. We shall restrict our discussion initially to behavior actually involving reproduction.

It has been noted previously (see page 322) that in most mammals adult sexual behavior is dependent upon the presence of circulating sex steroids. Removal of such steroids (by castration or ovariectomy) leads to a radical decline in sexual activity. Conversely, the exogenous application of sex steroids leads to its restoration. This has been found in rats (Larsson, 1966; Powers & Valenstein, 1972; Whalen, Luttge, & Gorzalka, 1971), guinea pigs (Grunt & Young, 1952), mice (Champlin, Blight, and McGill, 1963), and monkeys (Michael, Zumpe, Keverne, & Bonsall, 1972; Resko & Phoenix, 1972) as well as other animals (Hart, 1974). Females at lower levels than primates are generally nonreceptive to males except during the ovulatory phase of their cycle; primate females copulate throughout a longer period of their cycle, although there may be an increase in sexual activity even in humans at ovulation (Luttge, 1971).

It has also been noted previously (page 359) that sex steroids act at an early stage during the development of the mammalian brain to influence adult physiology and behavior. In a genetic male (XY), the Y chromosome produces a substance that induces the primitive gonads to become testes, and the testes subsequently secrete the androgens that induce further anatomical, physiological, and behavioral changes. In a genetic female (XX) the primitive gonads differentiate into ovaries, which are by comparison quiescent until puberty. In the absence of sex steroids the genotypic female develops as a pheontypic female anatomically, physiologically, and behaviorally. Experimental (or other) alteration of this normal circumstance leads to genotypically dependent changes, as we have seen in the case of aggression.

What of sexual behavior? Predictably, it is influenced in much the same way. Summaries of observations and theoretical implications in this area were made particularly by Harris (1964) and Gorski (1973). The neuroendocrinological data suggest that a brain site also implicated in aggression, the hypothalamus, is the focus of alterations due to the effects of sex steroids present early in life. Basically, the sexual physiology of the organism is "masculinized" by sex steroids at a critical period, both genotypic males and females later secreting the hormones that regulate the secretion of gonadal steroids in a male (tonic) way.

The behavioral outcome in lower forms is reasonably straightforward. Gerall and Kenney (1970), for instance, found that testosterone propionate administered to five-day-old female rats led to a reduction in their receptivity to males as adults, and that the effect was dose-dependent—the animals receiving the highest dose (1250 micrograms) showing little sexual response to males even when given the normal female hormones estrogen and progesterone. Further, Södersten (1973) found an increased amount of malelike sex behavior (mounting) among female rats given testosterone propionate 24 hours after birth, even when these animals were ovariectomized and given estrogen. Similar results have been reported in mice. Manning and McGill (1974) showed that neonatally treated females from a cross of two inbred strains not only mounted normal females in estrus more than did controls, but they exhibited the full range of male responses, including the ejaculatory reflex. Goy, Bridson, and Young (1964) had earlier reported the same sort of outcome in guinea pigs. Thus it appears that the effect of neonatal sex steroids during the sensitive period is to masculinize or at least defeminize the brain in certain respects.

The data resulting from experiments on males confirm this view. Whalen, Luttge, and Gorzalka (1971), for instance, found that male rats castrated at birth (therefore deprived of androgens during the sensitive period) responded with female behavior (lordosis) to normal males when primed with estrogen. Grady, Phoenix, and Young (1965) had reported much the same thing in another study. On the other hand, males castrated at 30 days of age and treated with estrogen did not respond with a high degree of femalelike sex behavior (Arén-Engelbrektsson, Larsson, Södersten, & Wilhemsson, 1970). By 30 days the period when sex steroids can exert this sort of "organizational" effect upon the rat brain has evidently passed.

Shall we say, then, that sex hormones control the sexual responsiveness of animals? No. As Young, Goy, and Phoenix (1964) noted some time ago, "the same quantity of hormone elicits almost as many modes of response as there are individuals." This general notion is as cardinal to the understanding of the determination of sexual behavior as it was to the understanding of the determination of aggression. Differences among individuals, particularly those of genetic origin, must be considered in any proposition attempting to explain sexual activity.

Why? First, as was mentioned earlier (page 322), although certain suprathreshold amounts of circulating sex steroids are necessary for most animals to initiate and maintain sexual behavior, it is not the case that increased hormone beyond that point results in increased sexual behavior. Grunt and Young (1952) demonstrated this in guinea pigs. They measured the sexual activity of intact males, castrated them, and then measured the activity again under various hormone regimens. The animals were divided into high- and low-sex-drive groups on the basis of their performance with their testes in place. When they were castrated, the sexual responsiveness of both groups declined, and when androgen was provided exogenously, it increased. The important point, however, is that under androgen therapy the individuals returned to their former rank order. Even four- to eightfold increases of androgen above threshold failed to alter an individual's sexual responsiveness from what it had been prior to castration. The same results were obtained in a slightly different design by Whalen, Beach, and Kuehn (1961). They noticed that in most colonies of laboratory rats there are a number of noncopulating males—that is, males who fail to initiate sexual activity when provided with receptive females. They castrated a group of such noncopulating males, and tested them against a group of castrated males that had previously proved to be copulators, at a number of different testosterone replacement doses. All of the males from the copulation group ejaculated by the time the replacement dose reached 90 micrograms per 100 grams body weight, but not one of the noncopulators ejaculated even at 200 micrograms per 100 grams body weight. It was thus impossible to turn noncopulators into copulators by increasing the level of sex hormone. Other studies reporting individual differences in response to sex steroids have been conducted by Larsson (1966), Powers and Valenstein (1972), and Champlin, Blight, and McGill (1963).

The relevance of genetic differences for sexual activity has also been amply documented. Valenstein, Riss, and Young (1954) found that males of two inbred strains of guinea pigs displayed different patterns of sexual behavior, while Whalen (1961b) reported very much the same thing among inbred rat strains, and McGill (1968), McGill and Blight (1963), and Vale and Ray (1972) found gene-related differences in the sexual responses of mice. Relevant neuroendocrinological differences have also been reported. Everett, Holsinger, Zeilmaker, Redmond, and Quinn (1970) found, for instance, that females of the CD rat strain are far more refractory to brain stimulation (in the preoptic area of the hypothalamus) leading to ovulation than are females of the O-M strain. Most interestingly, this difference was maintained in females treated neonatally with androgens.

What can we then say of organism-environment interaction? Data on the topic are few, as usual, but the indication is that, as in the case of aggression, variables of both origins combine to determine behavior. Valenstein, Riss, and Young (1954), for instance, found that males from a genetically heterogeneous line of guinea pigs were unaffected by isolation during development, whereas males from two homogeneous lines (inbred strains) showed significantly less sexual activity when subjected to the same treatment. Riss, Valenstein, Sinks, and Young (1955) found similar results in a study that also involved castration and hormone replacement.

In direct interaction study we (Vale, Ray, & Vale, 1973) found that female mice of three inbred strains reacted differently to neonatal treatment with androgen, oil, or estrogen in tests of malelike sexual behavior. Similar results were obtained using the same animals in tests of female sexual behavior. Overall, of 19 comparisons showing effects, in 10 there was significant genotype-treatment interaction. This suggests strongly that when the experiment is such that interaction effects can be detected, they will be. Corroboration of this view is obtained in another study (Vale, Ray, & Vale, 1974), this time on males. Males of two strains (BALB/c and C57BL/6) showed increased sexual responsiveness as adults following neonatal treatment with testosterone, whereas males of a third strain (A) did not. Finally, Luttge and Hall (1973) tested the capacity of four naturally occurring androgens (testosterone, androstenedione, dihydrotestosterone, and androstanedione) to induce sex behavior in two lines of castrated adult mice. Testosterone was more effective in the SW strain than in the CD-1, whereas dihydrotestosterone was ineffective in the CD-1 but only slightly less potent than testosterone in the SW. Androstanedione had no effect in either strain. Androstenedione was slightly more effective in the SW. The two genotypes thus responded differently to three of the four androgens.

Data such as these support, although they are too scanty to confirm, the interaction paradigm with respect to sexual behavior in animals. There can be little question that individuals and strains differ in their responses to manipulations of sex steroids and that the differential responses are related to differences in genes. More work is of course required to uncover the chemical mechanisms whereby genes and sex hormones interact, but a viable hypothesis at this point is that the hormones activate genes that potentiate sexual activity and that gene products, rather than direct action of hormones upon the nervous system, generate the behavioral differences we see. This is of course exactly the conclusion derived from study of genotype, sex hormones, and aggression, and it helps to account for the fact that just as all individuals of the same species (having sufficient hormone) do not exhibit the same degree of aggression, neither do they exhibit the same degree of sexual behavior.

What of humans? As expected, the case is not nearly so clear, because we are cultural creatures, because there are ethical problems involved if we experiment upon ourselves, and because it is only recently that research upon human sexual behavior has been welcomed by the bulk of society. What data there are, however, indicate some consistency with results obtained upon other animals. Removal of the ovaries in adult humans for instance depresses female activity, but not nearly so much as in monkeys, rats, and so on. Female human sexual desire thus remains, and copulation occurs, without the presence of ovarian hormones (Luttge, 1971). However, removal of both the ovaries and the adrenals (which produce androgenic substances as well as other products) does appear to have a pronounced effect. In males the evidence is clearer, the loss of androgens leading in many instances to diminished desire, responsiveness, and often to an inability to achieve penile erections (Luttge, 1971). The magnitude of these effects is, nevertheless, also related to nonhormonal factors, such as the individual's attitude and expectations following castration.

The sexual activity of nonhumans can be restored by replacement of gonadal hormones. In humans, this appears to hold for males but not for females. An increase in sexual activity due to the supportive properties of estrogen to the vagina (expansion and lubrication) may enhance the likelihood of coitus in females, but there appears to be no direct connection between exogenous female hormone and sexual behavior in ovariectomized women. In males, however, the evidence does indicate that the effects of castration upon sexual behavior can be reversed by exogenous androgens (Luttge, 1971). It thus seems that androgens are critical to the human male for the maintenance of normal levels of sexual desire, potency, and activity. Consistent with the observation that deprivation of androgens (adrenal in origin) lessens female libido, there is good evidence that exogenous androgen heightens it (Luttge, 1971). Many researchers have therefore concluded that androgens rather than estrogens are responsible for the maintenance and stimulation of human female sexual eroticism.

Human females androgenized as fetuses (progestin or adrenogenital syndrome) but who received corrective surgical and/or chemical treatment at birth (page 364) do not show a difference from matched controls in manifest sexual activity, but they do differ in maternalism. According to Money and Ehrhardt (1972), fetally androgenized girls differ from normals in that the first signs of female gender identity, shown by doll play in childhood, does not appear. These girls preferred cars, trucks, and guns. Further, the lack of interest in dolls became later a lack of interest in infants, in contrast with girls of the normal sample. Indeed, a third of those having the adrenogenital syndrome preferred not to have children.

Neonatally androgenized girls also preferred careers over and above marriage and tended either not to date or to be late in dating boys during adolescence. There were, however, no signs of homosexuality in the erotic interests of these girls.

In contrast, among girls having the adrenogenital syndrome who did not receive corrective therapy at birth (but who received it some time later) and who were therefore subject to their own androgens for a longer period, there was a high incidence of homosexual (as well as heterosexual) imagery in dreams and fantasies. Four of the 23 had indeed already had homosexual experiences. The interpretation of this is not straightforward, however, because the virilization of their bodies may have influenced these women's sexual perceptions. Few of them reported a desire to have children or to be associated with small infants. On the other hand, a substantial percentage of the girls eventually married and did reproduce. Thus, neonatal and postnatal androgenization in human females, while it does influence behavior in a "masculine" direction, does not completely alter gender identity.

It has been noted previously (page 365) that women with Turner's syndrome (due most commonly to the lack of one of the X chromosomes) do not develop ovaries and thus do not produce ovarian hormones at any time (although when the syndrome is discovered, usually at puberty, they are given estrogens so that they will develop female secondary sex characteristics). How are they affected? According to Money and Ehrhardt (1972), they are more "feminine" than a matched sample of normal women, in that they show a lesser incidence of athletic interest and skill, a lesser incidence of childhood fighting, and a greater interest in personal adornment. All reported daydreams of pregnancy and caring for babies. Regardless of problems of sex role stereotype, one would thus conclude that the *absence* of gonadal hormones during the perinatal period and during early childhood in no way interferes with the development of female gender identity. Insofar as the identity concept applies to infrahumans, this is of course consonant with the results of studies on animals.

These cases, happily, constitute only a tiny proportion of the population. But, as is common in science, they teach us much (although there is still much to be learned). The same applies to another small sample of even more subtle etiological configuration; transsexuals. There are people of one chromosomal and anatomical sex who believe they should be the other and therefore consider that they are imprisoned in an inappropriate body. Their willingness to seek surgical and hormonal correction, and the physician's capacity to deliver it, has now made prominent an ancient, almost eerie human condition.

Green (1974) has studied this extensively, and brings to light a number of relevant facts. For one, transsexualism is at least as old as our records. Greek and Roman historians mention the practice abundantly (the more common circumstance then as today being male to female), referring for instance to the Scythian "No-men," who not only did women's work but behaved as women in other ways, and to the "Hypocritical Queens" (effeminate young patricians) of Rome. For another, it is extensive. Anthropologists find it in radically different cultures, there often being a place for affected individuals in special occupations or castes. In some North American Indian societies boys with girlish inclinations were made to choose their sex role early and were held to their choice thereafter. They might subsequently lead as full lives as women as physiologically and anatomically possible, even simulating menstruation, pregnancy, and childbirth. These practices were of course extirpated by Christian settlers (looking somewhat the other way, one would think, considering that the first colonial governor of New York, Lord Cornbury, executed his duties in full woman's attire).

Green concludes from historical data that although we can be certain of little with regard to the psychological or biological origins of transsexualism, two points do emerge clearly: (1) Cross-gender behavior begins early in the development of the individual, and (2) it is of apparent lifelong endurance. These points are confirmed by study of modern transsexuals.

What can it mean to be trapped in a sexually inappropriate body, to be one gender on the outside and another within? It means that the person thinks and has always thought of himself (herself) as the other sex, that he (she) has played the role of the other sex, and that he (she) very actively longs to *be* the other sex. One can feel the agony and confusion of this circumstance by listening to the accounts of transsexuals themselves. A particularly clear one was obtained from an anatomic male (unoperated) who had lived as a woman for ten years and had a permanent relationship with a man (Green, 1974).

> TRANSSEXUAL: I remember that as a little girl (sic) I used to lie in bed at night with my penis between my legs and my ankles crossed real tight and play a silly game and say if I did this, in the morning when I'd wake up it would be gone. This is very, very long ago.
>
> PSYCHIATRIST: How long ago?
>
> T: Definitely preschool. I don't know where I got this notion, but I just felt that it would go away by morning, and I was so disappointed because every morning I'd reach down and there it was. In kindergarten the kids used to make fun of me because I was girlish.
>
> P: How?
>
> T: I used to like to play with girls. I never did like to play with boys. I wanted to play jacks. I wanted to jump rope and all those things. I remember one day the teacher said, "if you play with the girls one more day, I am going to bring a dress to school and make you wear it all day long. How would you like that?" Well, I *would* have liked it.
>
> P: I am wondering if your parents were aware of anything unusual about you?
>
> T: They kept trying to make me a boy, and I resisted hard.
>
> P: What else do you remember from when you were little?
>
> T: My sister and I used to dress up in our mother's clothes.

p: How did that feel to you?

t: Natural. . . . if we played husband and wife, I'd always be the wife.

p: When did you become aware of sexual feelings toward men?

t: You know, the first time that it dawned on me that I was abnormal—it was really weird—it was my sophomore year in high school. I could remember the exact day because I was in class, and just being seated on the first day of school. Right next to me there was this kid standing there, and he was just the type I'm drawn to, very blond and very, very rugged, and I took a look at him and my whole heart seemed to go right in my throat. I felt like I must have blushed because I felt very warm and tingly. Every time he would look at me, it seemed, I'd just melt, And after school I went out to the beach. It was cold and foggy, and I sat there on the sand and these are the exact words I said to myself, "I started today. I've started something different. I don't know what it is, but I can't go back!"

p: Did you feel that you were homosexual?

t: Not *homosexual*. I felt that I was *drawn to men*. I could never think of myself as a man. I just couldn't.

p: What kind of sex have you found the most acceptable considering the limitations you have?

t: Well, I prefer rectally because its the closest to vaginal.

p: Is that enjoyable?

t: It's enjoyable to him and to me too. He prefers not to see or touch those organs in front. But, nothing really arouses me terribly at this point. Mostly I find satisfaction in pleasing him and knowing that I am satisfying him the best I can.

p: Do you have any kind of orgasm?

t: No.

p: How do you feel about not having an orgasm?

t: Delighted. I wouldn't want to have a male orgasm. (pp. 47–51)

A number of characteristics seen above are apparently common both to male and female transsexuals. First, they show an early preference for the company, clothes, games, and toys of the other sex. Second, they desire the anatomical insignia of the other sex. Third, they fantasize or dream later about normal sexual relations with members of their own anatomic gender. Fourth, when such relations are realized, they are not considered homosexual, the offending organs (when they are still present) playing no part.

What accounts for such a phenomenon? It would be tempting to suspect sex hormones, given the strong implications deriving from other research. However, there is at present no evidence of hormonal malfunction in transsexuals. They are not, as are the individuals studied by Money and Ehrhardt, physically or physiologically distinguishable from members of their anatomic or chromosomal sex, nor are the outcomes of the relations the same. Females having the adrenogenital syndrome do not, for instance, become transsexuals, and chromosomal males having the androgen-insensitivity syndrome have all but the internal organs of females, are raised as girls, and suffer no gender identity crisis. The biological etiology is thus, by current standards, quite subtle.

The same can be said of environmental determinants. There is no consistent pattern of familial or other external events that appears correlated with transsexualism. In studies of feminine boys, for example, Green found neither the absence of a father (male model) nor the presence of a domineering mother to be significant factors. Indeed, there were few differences at all between families having a feminine boy and those having none. Most important, the male siblings of male-to-female transsexuals are invariably masculine. This suggests that the operative environmental events, if any, are those to which only certain individuals are susceptible. If so, it is once again a matter of organism-environment interaction.

This is highly speculative, for data are few. It is possible, however, that genes related to cross-culturally identifiable masculine and feminine behaviors—that is, to characteristics normal for the two sexes throughout the species—are distributed in such a way (obviously on autosomes rather than on sex chromosomes) that individuals of one chromosomal sex (XX or XY) may nevertheless carry a strong genetic potential for the inclinations of the other. It is generally true of mammalian species that the two sexes behave differently, and there is every reason to believe that such differences should have been the result of natural selection. We should then expect that as the variance-generating mechanisms operate, there would be a small proportion of individuals with what is essentially a mismatch in their genetic input. Such individuals might be considered at risk for transsexualism.

An interesting point in this regard, mentioned earlier, is that males desire to be females more frequently than do females desire to be males. Green estimates that the ratio may be as high as six to one. Why such a disparity? At this point perinatal hormones may be reconsidered, because in the usual circumstance something must be added during a sensitive period in order that a male develop. Left otherwise undisturbed the fetus develops as female. Is it possible that sexual anomalies (homosexuality, fetishism, pedophilia, sadism, and so on, in addition to transsexualism) occur more often in males (which they do) because that which is added fails in some fashion to exert its normal influence? And, since that influence is known to occur through the activation or suppression of genes in other cases, is it possible that genes and hormones interact during a sensitive period (or perhaps one of many) to produce anatomic males with tendencies to behave as females? Is it similarly possible that females who wish to be males have a genetic constitution that is excessively responsive to the small amounts of androgenic substances that circulate normally early in development? Such a proposition makes a fair amount of sense, although the transsexual phenomenon is yet to be comprehended, even superficially.

Chapter 9
Organism, Environment, and Behavior: Prospect, II

SCHIZOPHRENIA

About 20 percent of all hospital beds in the United States are occupied by people diagnosed as schizophrenic (Rosenthal, 1970). This is not a small figure, and, as such, indicates in the most minimal way the scope of a major problem. Surround it by the suffering of the individuals directly or indirectly involved and it assumes critical dimensions. Further, the problem is intrinsic; it is not restricted, as some investigators once believed, by time, culture, or technological accomplishment, but rather it has endured the vagaries of circumstance throughout the history of man (Rosenthal, 1970). It is only our medical diagnoses that have been limited to the past few centuries; for much longer than that many have heard voices when no one spoke, have been possessed by unseen and unknown forces, and have been driven to apathy and withdrawal. Therefore we must come to grips with the frightening fact that minds go seriously awry, and go awry with substantial frequency.

The disorder was first systematically described by German psychiatrist Emil Kraeplin. He named it dementia praecox because he was

impressed by its early onset and because it appeared to be characterized by an inevitable progression toward a demented state. Eugen
Bleuler, a Swiss psychiatrist, later reformulated Kraeplin's ideas and
substituted "schizophrenia" for "dementia praecox" because he found
the more important symptom to be a cleavage of psychic functions, a
disorganization, rather than progression to dementia, although dementia was notable (Arieti, 1974).

Although schizophrenia is but one of a number of psychic disorders, it is doubtless the one that now generates the most concern. We
shall therefore briefly study it from the interactionist perspective in
the hope that some understanding of its presence among us may be
produced.

Common Characteristics of Schizophrenics

What does it mean to be schizophrenic? Many things, obviously, for
behavioral diagnostic boundaries are notoriously permeable. According
to Meehl (1962), however, four traits are commonly distinguished.
Initially, and centrally, there is a degree of "cognitive slippage"—
thought disorder wherein fragments of ideas are connected in illogical
ways, concepts are incomplete, and associations are disconnected and
strange. An example is given in a letter written by a patient with paranoid tendencies (Arieti, 1974).

Affidavit

LEGAL   The undersigned is being detained "here" Illegally under mis-
representation (criminal slander) The motive—we will mention
only the principal reason. Having specolated "on easy" possibilities of *criminal* exploitation public enemies have caused through
corruption a cloud to surround a life and by so doing try to *hide*
the *honorable talented upright* sane qualities, of a social *giant*
Life

MEDICAL   It beehove's you to know that I am sound and one hundreth per
cent Clean healthy ((Wasserman test)) blood test indicate 100
percent perfection quality purity.

URGENT   The writer has a *Sacred* Mission to perform, to give the american
Social structure a chance to *know* the Truth. I demand my release and go my way. It will save *you* the shame of being mixed
up with these public enemies parasites who are criminally . . .
fooling you I am an American able bodied capable of earning a
good and honest living I am within my own rights I believe in
Law and ORDER and in the sacred rights of man embodied in
Constitution and seth forth in the Bill of Rights of the U. States
of America Are you an American? then release an Innocent man

Signed   America (p. 402)

This patient's fears and desires are clear enough, and there is a certain logical structure to their expression. However, the bizarreness and weakening of thought structure is also evident.

It is more evident in the following letter, written to the staff physicians at Pilgrim State Hospital and reproduced also in Arieti:

Honerables
At the presents

past on hour future and the beginnin off the life every Doctor's on the capacity that have have to be sure off the business life and made it on confortables act on insuerance capacity acts at the moral mentaly on good Health. I'm my self I think any way for the true and give and all together the best onsuerds by ward for my inteligence to resolve a —problem that one plus one is one of them I myself think iff it is true or I mistake later I have to correct. Know at this presents moments I don't have to say anything and I don't have a word only on the same hared one and the same word have the ansuerd's (Liberty on free). Just only that I have an warred too. Anything more for my self I needed. I would like to have that consideration for finish and take a rest and to have a good time because many thing at my recuerments life. I made you at your self this propotion. To give me and oportunity to approff me a hoadcard for the few time that I won to stay here If you wanted to aproff me this allright and good too. Them when I resired I give you the thanks no for ever but for the presents mannern'ts allright and very good for all and pleasure with yours. (p. 407)

One senses that this person is concerned with having some say in the manner of his day-to-day existence, but his thoughts are so incoherent and disconnected that they have little specific meaning.

The inappropriate use of language is thus a diagnostic sign, the speech of many schizophrenics showing peculiar characteristics. Evasiveness in answers to questions is common. At times the individual may use abstract, impressive words, but in a meaningless way, and his or her communications appear warped by the introduction of extraneous elements. In more advanced cases, as in the second example above, it is difficult to comprehend what the individual it attempting to convey. Sentences are constructed of words that seem unrelated (resulting in what psychiatrists describe as a word salad), and at times the same word may be repeated in a seemingly stereotype manner (perseveration). There is also a tendency to invent new words by abridging or combining the usual ones (neologisms). Some patients indeed seem at times preoccupied by plays upon words, giving their written communication an unintended, strangely humorous content (Arieti, 1974).

The second trait for which schizophrenics are noted is described by Meehl as interpersonal aversiveness. There is thus a pronounced tendency toward "social fear, distrust, expectation of rejection, and conviction of his own unlovability which cannot be matched in its depth, pervasity, and resistance to corrective experience by any other diagnostic group." Such fear is exemplified in the first letter above.

The third trait found in schizophrenics is anhedonia—"a marked, widespread, and refractory defect in pleasure capacity. . . ." Affected individuals are thus evidently incapable of enjoying what others enjoy, or what they enjoyed before. Finally, Meehl describes a fourth trait, ambivalence, which is the tendency toward inner concerns and fantasy. This derives from the first three: cognitive slippage, aversiveness, and anhedonia. Thus, "if a person cannot think straight, gets little pleasure, and is afraid of everyone" he or she will learn to retreat to fantasy. Affected individuals are therefore withdrawn, apparently living in a world of their own, out of touch with reality. This is accompanied by a lack of physical activity, what activity there is being performed routinely at best. Often personal appearance is neglected, and there is deterioration of other habitual behaviors.

Not all psychiatrists, of course, would agree with Meehl's nosology, but it does seem to categorize at least the best known stage of schizophrenia. This is not to say that all schizophrenics manifest each symptom to the same degree or that there are not other, secondary characteristics that are commonly seen, but rather that most clinicians would agree that thought disorder, a sense of rejection, an inability to experience pleasure, and the tendency to live in fantasy are central to the schizophrenic condition.

Curiously (from the analyst's view) even with such disorganization, certain functions remain intact. Many schizophrenics appear to be aware of what goes on around them; however, they interpret it and react to it differently. As remarked by Arieti, "The sensorium and the intellectual functions are not seriously impaired. Orientation, memory, retention, attention, grasp of general information, calculation, and so forth may seem disturbed in many cases. The disturbance is actually the result of the other symptoms described and may disappear once these symptoms disappear" (pp. 33–34). This leaves, for the clinician, an awkward, confusing picture.

Stages of Deterioration

The picture is confused even more because some individuals show the progressive deterioration of faculties noted by Kraeplin, whereas others do not and may indeed improve to the point where they are judged to have recovered. Arieti distinguishes four stages of deterioration, the first dominated principally by confusion, fear, and disorientation, the second by delusions and possibly hallucinations, the third by the hoarding of objects and self-decoration, and the fourth by primitive oral habits and insensitivity to pain. In the last stages the patients are generally mute and nonreactive, except in the presence of food or certain very specific objects that can be placed in the mouth or other body cavities.

We have seen examples of cognitive slippage previously, one with a delusional persecutory flavor. The same delusion appears in the excerpt that follows (again written to a physician), but with a remarkable new element. Note, however, that in spite of the stereotyped phrases, this patient expresses himself quite clearly.

> I am a perfectly sane man who is imprisoned most unjustly and for no reason whatsoever, in the most detestable insane asylum. I am the victim of the worst injustice and most unbearable conditions, my being here is the worst sin, the worst mistake and the worst crime. nobody has the right to keep me here at all. Here is my story. The tenants who were living directly beneath us, where I used to live with my family (parents, brother and sisters), were talking to me telepathically, daily, continuously and invisible from their home and while I was at home. I was absolutely a slave and I was not able to avoid hearing or receiving mentally from them. They annoyed me extremely because they talked to me mentally. when I was at home, I did not teach telepathy to any member of my family. my mother did not believe that telepathy was true and consequently she thought that I was imagining that the tenants were talking to me telepathically and, because of that, she had me sent, suddenly, to King's park state hospital.
>
> The inmates, here, hate me extremely because I am sane and they always do their best to keep me here. They succeeded wonderfully and that is why I am here. In order to keep me here, (1) they talk to me telepathically, continuously and daily almost without cessation, day and night. (Inmates and employes talk to me telepathically, daily, and continuously without cessation, day and night) (2) By the power of their imagination, they force one another as well as other employes to mistreat me extremely, daily, and continuously and in every way to strike me, kick me punch me, choke me, knock at me, cause extreme pain in various parts of my body and to harm me in every way.
>
> By the power of their imagination and daily and continuously, they dilate my pupil and iris and they pull the veins, muscles and tissue of my eyes.
>
> By their imagination and telepathy and daily and continuously, they force one another to talk orally and to send their voices to my head, forehead, temples and heart. (5) By telepathy and imagination, they force me to say orally whatever they desire, whenever they desire and as long as they desire. I never said a word of my own. I never created an image of my own.
>
> What I am undergoing here is worse than the torments of hell itself. The situation is absolutely uncontrollable and irremediable. I must

surely get out of here at once or I will surely become insane or blind or I will surely die. please save me from insanity or blindness or death. please discharge me from Kings park state hospital as soon as possible during today or tomorrow and I swear by God that I will give or send you five hundred dollars, or any sum which you desire as high as one thousand dollars, as a reward for that and within three days after my discharge and I thank you most cordially. Gain my gratitude for life as well as that of my family. (Arieti, 1974, pp. 404–405)

In the third or preterminal stage some the "classical" symptoms are less in evidence—delusions and hallucinations being disorganized and deprived of emotional charge. There is severe disintegration of thought processes, and ideas are conveyed with great difficulty. Certain behaviors, such as picking skin, pulling out hair, and rhythmic movements appear, but what most distinguishes those affected at this point is the hoarding of objects and self-decoration.

The objects hoarded are usually of limited size and of no practical use. They may be old letters, toilet paper, pieces of wood, soap, strings, rags, hatpins—anything that can be put into a small place or carried on the person. In the beginning, the collection is kept in pockets, shoes, bags, or boxes, but later, as deterioration continues, cavities in the body may be used. Males often carry small things in their ear canals or noses. Females often use their vaginas. One such patient of Arieti's had stored in her vagina a metal cup, two teaspoon handles, several pieces of soap, and a little rag. Another patient hoarded objects in her mouth. During meals she would remove them and then would return them after having eaten.

The patients are by this time uncommunicative, and it is difficult to ascertain what meaning the objects have for them. Most do not answer inquiries, but only smile. One attempted a logical explanation: She had filled her mattress with toilet paper in order to be prepared in case she developed diarrhea. Another patient remarked only that she used what God had given her.

Self-decoration, although less common than hoarding, is also practiced by schizophrenics in the preterminal stage. Pieces of paper or rags may be cut into bands for bracelets, rings, necklaces, or belts. Small boxes, buttons, stamps, corks, or coins may be placed on the chest. Many patients (predominantly female) paint their faces in an uncommon manner.

Normal observers find no aesthetic value in such decoration, the result being to their eyes ridiculous and even disfiguring. Such is evidently not the case for the schizophrenic. Although apparently otherwise disinterested in their environment, participants appear interested in decorating their bodies. However, they often deny this. When asked about her paper bracelets, for instance, one patient said that she used them to cover her arms. Another said she used her paper necklaces to tie her neck.

The terminal stage of schizophrenia begins with an increase in activity among these notably inactive and withdrawn people. Their behavior then appears sharply reactive and impulsive, and they may become more destructive and violent. Delusions and hallucinations are evidently gone, and verbal expressions are absent or reduced to a few disconnected utterances. The most remarkable change, however, is found in relation to food. People who had to be tube-fed because they did not eat on their own now develop voracious appetites, to the extent that anything edible is promptly consumed. Often such "food grabbers" will eat their own and the meals of several other patients in a few minutes if not prevented from doing so.

Terminal schizophrenics may remain at the food-grabbing stage indefinitely, but the majority regress fairly rapidly to a level that is characterized by "placing into mouth." Here edibility is not a criterion; any small object is liable to be put into the mouth. If it cannot be chewed it may or may not be rejected. One case, which also indicates disintegration of behavior, is reported as follows:

> This patient entered Pilgrim State Hospital in October 1933 at the age of 32. On admission he had delusions and hallucinations but was fairly well preserved. The diagnosis of dementia praecox, paranoid type, was then made. Subsequently he showed a steady downhill progression in his mental condition. He became negativistic, mute, manneristic, and idle. He had the habit of wetting and soiling and required a great deal of supervision. On frequent occasions it was necessary to tube feed him. He did not show any interest in his surroundings, appearing completely withdrawn and living an almost vegetative existence. In the dining room, however, he showed great interest, grabbing food and eating in a ravenous manner. On December 31, 1939, he died of acute intestinal obstruction. At autopsy fourteen spoon handles were found in his colon and two spoon handles and a suspender clasp in his stomach. In the terminal ileum there was a rolled piece of shirt collar, which was the cause of the obstruction. The collection of foreign bodies in the stomach of this patient is very modest in comparison with that found in many other patients. (Arieti, 1974, p. 425)

Finally, schizophrenia in its most advanced stage is characterized by an apparent insensitivity to pain, to temperature change, and by the loss of taste. Arieti notes that it is the combination of increased activity and lack of reaction to painful stimuli that causes many such patients to harm themselves. Indeed, he remarks that he had many times sutured wounds caused by violent behavior without any sign of pain or resistance on the part of the patient. Such anesthesia also accounts for other injuries; for example, patients may sit too close to a source of heat and if they are not moved they will be burned. Taste may also be lost at this point, the individuals failing to show a pleasant or unpleasant reaction when given sugar, salt, pepper, or quinine.

Varieties of Schizophrenic Reactions

We have noted that although cognitive slippage, interpersonal aversiveness, anhedonia, and ambivalence appear to characterize schizophrenics at large, many of those affected show different secondary symptoms. At onset of the disorder (usually from the time of puberty to the early thirties) the individual begins to show conspicuous, unusual behaviors. Important decisions are made with a lack of logic. Affected persons may feel that others are unfair to them, wish to make them seem less than they are, or dislike them for special reasons. Such behavioral anomalities become more and more manifest, sometimes slowly and sometimes acutely. There may then follow a period of confusion and excitement, wherein there is an apparent attempt to reestablish connections with an escaping world. Finally, coherence of speech may be lost and thus the abnormality becomes obvious.

In other instances the individual becomes hypochondriacal, overly concerned with his or her health or physical appearance. These concerns are soon revealed as somatic delusions. In most instances the individual appears less interested in life than previously and seems to concentrate upon specific problems. He may think that certain everyday matters are especially related to him or that coincidentally connected events have special meaning (ideas of reference). Adventitous relationships appear ordained; if he thinks about a particular subject and later sees it mentioned in a newspaper or hears about it on radio or television, his suspicion is aroused. His suspiciousness of other people also increases; he feels that they look at him strangely and make fun of him or talk about him when he is not present. He may begin to feel that he is under the influence of external agents who make him experience peculiar sensations, think foreign thoughts, and behave in alien ways. Finally, he interprets events in a manner inconsonant with the observations of others: his house is electronically monitored, his thoughts are being recorded, his food is being poisoned. Such delusions are initially negative, insofar as they support his notions of persecution. Later, however, the delusions may become pleasant, and even grandiose: he is a king, a millionaire, a great actor; he has discovered the secrets of the universe, the essence of life. Now it is *he* who can control others by his thoughts.

He may also have illusions, in which things are misidentified, and strange resemblances are observed. Frequently there are hallucinations, auditory and visual, in which voices are heard and visions seen. He will also manifest thought disorder, his speech and writing being characterized by extraneous elements, word salad, perseveration, and neologisms. Insight that he is in an abnormal condition is lacking.

From this point on the disorder is manifested by congeries of symptoms that though frequently overlapping, are for analytical purposes put into four main categories: paranoid, hebephrenic, catatonic, and simple.

• *Paranoid.* The number of individuals suffering from the paranoid form of schizophrenia is far higher than for any other, and the disorder often begins later in life. Usually patients feel unfairly accused or victimized, and their thought content is characterized by ideas of reference and delusions, usually persecutory, grandiose, or hypochondriacal. In many cases such delusions become systematized; the patient explains them more or less logically in relation to what has happened in his or her life. A whole delusional system may be constructed around the notion that the patient is persecuted because of his or her beliefs, and he or she may then attempt to give this system a plausible scientific, philosophical, or theological structure. Delusions may have all kinds of content; believing that one is Christ, Moses, Saint Peter, or Saint Paul, or the Virgin Mary is common. (Arieti notes that he has not, however, encountered a patient claiming to be Napoleon.) Hallucinations, particularly auditory ones, are frequent.

Paranoids generally do not regress to the advanced stages of schizophrenia as rapidly as do those with hebephrenic or catatonic forms of the disorder; many in fact remain as they are. They maintain better contact with reality, and their activities may often be channeled into useful work. On the other hand, their suspiciousness and delusions may make them antagonistic and even violent. Escapes from hospitals and homicidal attempts are thus more frequent in this group.

• *Hebephrenic.* Hebephrenia literally means mind of youth, although what is found is more a caricature of youthful behavior. Rather, silly, disorganized actions are common, along with smiling, giggling, and laughter that appears incongruous and empty. Apparent sadness seems shallow, and if there is crying it appears inexplicable. Thought content is characterized by unsystematic, poorly rationalized and in many cases completely disorganized delusions. Hypochondriacal ideas and preoccupations with the body image are more common than in paranoids. Hebephrenics may think, for example, that they have lost their bowels, that their hearts have changed place, or that their brains have melted. Hallucinations are frequent, and are more pleasing in content than in paranoids. Word salad and neologisms are very frequent. Hebephrenics, as noted, disintegrate to preterminal and terminal stages far more rapidly than do paranoids.

· *Catatonic.* After an initial period of excitement in which there is apparently aimless behavior, catatonics move ever more slowly until they eventually reach a state of almost complete immobility. They may become so inactive as to fail to respond to their physical needs and must be confined to bed. A catatonic's body may be put into awkward positions and will so remain for hours (the phenomenon known as waxy flexibility). At other times catatonics put themselves into uncomfortable, sometimes statuesque positions, maintaining them until they are put to bed, and then resuming them the following day. Often they respond negatively to requests by doing the opposite of what is asked. Delusions and hallucinations are present in many cases (although these have been determined only in improved patients, because in the catatonic condition they do not communicate). Echolalia—repeating questions instead of answering them—is prominent.

· *Simple.* Simple schizophrenia rarely occurs acutely; rather, it is manifested slowly, usually from a period before puberty. The individuals gradually become inactive, refusing to go to school or to work, and their lives become restricted. There is generally no looseness of ideas or illogical sequencing in their thinking, but there is *poverty* of thought. The affected person is able to deal only with a few concrete notions, abstract thinking having disappeared. Hallucinations, delusions, ideas of reference, and so forth are absent. However, behavior is still odd and inappropriate, with little emotion expressed. It is as if the world is too much.

The decline may cease at some low level of adaptation, wherein the individual leads an idle, ineffectual, apparently meaningless life. If hospitalization it not required, he or she may sit around the house or wander aimlessly from place to place. Intellectual functions, insofar as they are measured by intelligence tests, are surprisingly unimpaired. Some individuals even hold temporary jobs; others, if unattended, become vagrants or prostitutes, and may be exploited by organized crime (Arieti, 1974; Cameron, 1963).

There are a number of levels of schizophrenic reaction, as well as a number of symptoms characteristic of different forms of the disorder. Some patients have acute attacks and recover, some succumb slowly and deteriorate to the terminal stage, and some are only mildly affected; in fact, almost any combination of symptoms and progress seem possible. Thus, the disorder is hardly open to immediate, polished analysis. But few behaviors are—and, not surprisingly, we must therefore struggle to find a comprehensive view. To this we now turn.

The Etiology of Schizophrenia

It is estimated that 1 to 2 percent of our population has had a schizophrenic episode (Erlenmeyer-Kimling, 1976), and although by no means all those affected descend to the late stages of the disorder, the fact that these victims exist at all gives cause for deep reflection upon the human circumstance. What explains this durable epidemic? Why should it be that so many of us are regularly afflicted with such horror? Alas, there are no answers. We may gain some perspective, however, by examination of what data there are, and we shall in the course of this examination seek the most inclusive explanatory exposition. Such an endeavor will naturally involve substantial speculation.

As we must now expect, there are both organic and environmental theories of the etiology of schizophrenia. At one extreme, the environmental view holds that the disorder is a learned response to untenable, usually stressful, anxiety-producing conditions. At the other extreme, organic theory holds that it is due to an irremediable congenital deficit. Both approaches thus seek single causes, the former settling upon some agent of the environment (such as a schizophrenogenic mother) and the latter upon some (undefined) miscarriage of neural development. Neither is sufficient. There is unquestionable evidence for a genetic component in the etiology of the disease; on the other hand, there is evidence that without appropriate environmental input the genetic potential may not (although in some cases it could) be translated into the clinical syndrome. The most reasonable model therefore is one of organism-environment interaction.

Rosenthal (1970) presents data strongly supporting such a model. He notes that, first, studies of the families of schizophrenics show a clear correlation between the degree of relationship to the affected individual and the incidence of the disorder in the relatives. The more closely related one is to a diagnosed schizophrenic, the higher the probability that he will have the disease. Using Falconer's (1965) method for estimating the inheritance of liability to disease, Rosenthal concludes that when environmental conditions are uncontrolled or unspecified the heritability of schizophrenia is .73. This figure is of course high, and its accuracy is at issue just because the environment does play a role in instances wherein the affected individual comes into long-term contact with his relatives during development of the disorder.

Second, the probability of schizophrenia in each of the pair is much higher (a ratio between 3:1 and 6:1) in monozygotic than dizygotic twins. Each monozygotic twin set has the same genotype, having developed from one fertilized egg, whereas dizygotic twins share no more of their genes (50 percent on average) than do other siblings. This suggests a positive relationship between genotype and schizophrenia. There are, however, a number of complications that prevent us from concluding more. For example, it is possible that the monozygotic pairs were treated more alike than dizygotic pairs. This would increase the probability, relative to the dizygotes, that if one of a monozygotic pair were affected, the other would be. In addition, it is not the case that if one twin of a monozygotic pair is affected the other one inevitably is. The concordance rate for monozygotic twins has always been found to be less than 100 percent, often considerably less. This means that both individuals, who have the same genotype, do not necessarily contract the disorder if one does. It follows then that genotype does not determine the outcome entirely.

As noted, a major problem with many earlier studies is that they failed to separate genetic and environmental influences. Clearly an individual raised in a family where stress and inconsistency are prevalent could acquire some peculiar behaviors. It is difficult under these conditions to uncover what the genetic and environmental contributions are, much less to find how they combine. Another problem is that until fairly recently it has been difficult to be certain whether twins were mono- or dizygotic. Early investigators were forced to rely on similarities among anatomic features to estimate zygosity, and mistakes are likely with such methods. We do not know, therefore, how many monozygotic twins were classified as dizygotic, or vice versa. In addition, samples from the earlier studies were often biased. For instance, if one merely counts the number of monozygotic twins who have been hospitalized for schizophrenia, the concordance rate may be high both because the hospital population is older (allowing time for both twins to have been affected) and because the severity of the illness in such a population is greater. On the other hand, if one merely asks immediately when one twin has been diagnosed as schizophrenic whether the other one has also, the concordance rate among monozygotes may be lower, for the opposite reasons. The means of discovery of the "index case," the time the relatives are examined, and the population of relatives examined have thus presented obstacles to firm conclusions in both twin and family studies.

A final obstacle to such conclusions occurs because different investigators have required different degrees of illness before schizophrenia was diagnosed. One investigator might therefore decide that the disorder did not appear frequently among the biological relatives in a sample, whereas another investigator, using the same sample, might decide that it did. This doubtless accounts in considerable measure for the inconsistency among various reports.

A way of circumventing the confounding of genetic and environmental influences and of defeating some problems of sampling has been found in adoption studies. Here children and their consanguineous relatives who were separated from them early in the children's lives are examined. Environmental factors can thus be methodologically controlled, if not manipulated. Adoption studies, correcting for biases of past research and taking advantage of the mental health records maintained in some countries, also tend to use large and complete samples. The investigators now involved are also aware that schizophrenic symptomology is not discretely distributed; it is not a matter of it either being present or not, but rather it may be present in forms ranging from the very mildest to the most severe. They are thus sensitive to and more likely to detect schizophrenic symptoms.

The first adoption study was reported by Heston (1966). He compared the children of mothers who were actively schizophrenic, having given birth while hospitalized, with children adopted from foundling homes. Both sets of children were thus raised by families who had adopted them, and the children of the schizophrenic mothers were not exposed to a discernible schizophrenogenic environment, having been separated from their mothers immediately. The control children, further, were matched with the children of schizophrenic mothers on such variables as sex, age, type of home placement, and length of time in child care institutions. Both sets of children, who at this point averaged 36 years of age, were then interviewed and tested on several measures.

The results are enlightening. Of the 47 adopted children born of schizophrenic mothers, five had been hospitalized for schizophrenia and three were diagnosed as chronic deteriorated. Of the 50 adopted children born of normal mothers, there were no instances of schizophrenia. Moreover, there were significant differences between the two groups on a number of other dimensions, all to the detriment of those born to schizophrenic mothers. Among the latter there was a much higher incidence of mental deficiency, sociopathic personality, and neurotic personality disorder, as well as felony convictions and military service discharges for psychiatric or behavioral reasons. This suggests an underlying genetic distribution for liability to schizophrenia and possibly related disorders. We shall continue to develop this notion.

In another study, using a slightly different approach, the relatives of schizophrenics who had been adopted were examined for disorders in the "schizophrenic spectrum." Kety and co-workers (1976) first found 5483 Danish adults who had been adopted early in life by persons not biologically related to them. Of these, 33 were identified from hospital records as schizophrenics, following a review of the 507 cases among the adoptees who had ever been admitted to a mental institution. A control group in which there was no instance of mental disorder was then selected from the adoptees and matched to the schizophrenic group for age, sex, socioeconomic status of the rearing family, and time spent with biological relatives, in child care institutions, or in foster homes, before transfer to the adopting family. More than 300 relatives of the schizophrenic and control groups were then interviewed. None knew the mental status of the index case.

This procedure produced four samples: the biological relatives of the schizophrenics, the biological relatives of the normals, the adoptive relatives of the schizophrenics, and the adoptive relatives of the normals. The incidence of schizophrenic spectrum disorders (from the most severe to the mildest detectable cases) among the *biological* relatives of the schizophrenics was significantly higher than in any other group. For any form of schizophrenic illness, the prevalence among those genetically related to the schizophrenic index cases was 13.9 percent, compared to 2 percent in their adoptive relatives and 3.8 percent in all relatives not related to the schizophrenic index cases. This again indicates a genetic component.

Such a proposition is made even more likely by further analysis. It is possible, that is, that environmental influences in utero, birth trauma, or early mothering experiences may have contributed to the foregoing outcome. However, among the relatives of the schizophrenic and normal adoptees were 127 paternal half-siblings, who did not have the same mother or postnatal experiences as the index cases, but shared only the same father. Should there be a higher incidence of schizophrenic spectrum disorders among the half-siblings of the schizophrenic adoptees, it would be powerful evidence for a genetic influence. This indeed occurred: There were 14 instances of schizophrenia among the paternal half-siblings of schizophrenics, as opposed to two in the paternal half-siblings of controls.

Granted that the potential for schizophrenia is influenced by genes, we must now examine more specifically its distribution. What, for instance, is the relationship between the distribution of genetic potential and subtypes of the disorder, and between the distribution of genetic potential and degree of severity? Is schizophrenia biologically the same "thing" in paranoids and hebephrenics, and in deteriorated and ambulatory or borderline cases? May we conclude that the most severely affected individuals are likely to have more genes that enter into the etiology of the syndrome, and that the less severely affected individuals are likely to have fewer such genes? Rosenthal argues that we may.

We have noted previously (page 403) that those diagnosed as

having simple schizophrenia are generally considered less ill than those diagnosed as paranoid, catatonic, or hebephrenic, and further that paranoids usually do not degenerate as far or as fast as catatonics and hebephrenics. The subtypes may thus represent different degrees of the same disorder in which different symptoms are prevalent. This notion is supported by the fact that monozygotic twins who are affected typically manifest the same subtype. Since they have the same relevant genes, this would be expected. We may thus envisage a genetic distribution that is quantitatively related to symptomology, those individuals having more of the relevant genes showing the more severe behavioral characteristics, being more ill, and being most likely to deteriorate to terminal stages. Rosenthal (1970), indeed, suggests just this: "We may therefore assume that the pathological genes tend to cluster in some families and that those who have most of these genes will show hebephrenia and/or catatonia; whereas those who have fewer of these genes will show paranoid, simple, atypical, or possibly borderline forms of the disorder" (p. 144).

Important to this argument, and highly meaningful to our assessment of the conduct of human affairs, is the question of schizophrenic behaviors among individuals in the population at large. That is, if we stipulate a distribution of genes that underlies the disorder, we must concede their influence not only upon those who are clearly ill but also upon those with similar but less severe symptoms. We would then expect in the general population a fair proportion of individuals who manifest mild schizophreniform behavior.

With what frequency are such individuals found? Unfortunately, we do not know. Most research has concentrated upon those with diagnosable schizophrenia, and only recently, particularly in the adoption studies, has the incidence of lesser symptomology been systematically explored. However, many investigators have been impressed with the number of people not hospitalized because their symptoms did not warrant it but who nevertheless showed behaviors attributable to "an insidious schizophrenic process" (Arieti, 1974). Bleuler was one of the first to note the existence of such "latent" cases; indeed he believed their incidence to be far larger than that of those ill enough to require institutional care. Other researchers have since found much the same thing, through the years having described the presence of relatively benign but identifiably schizophrenic symptomology as schizophreniform psychosis, and symptomatic, atypical, peripheral, or reactive schizophrenia. The description in use now is reactive schizophrenia, as opposed to that of the traditional subtypes, collectively called process schizophrenia.

"Schizophrenia" is thus not a discrete entity; rather it is a conceptual congeries of symptoms that out of diagnostic necessity and for purposes of treatment has come to be regarded as such. We must classify to deal with life; we know this. Categories simplify decisions. Indeed, that is how we progress. Much of scientific and social development is attributable to new and better categories, to better classification. Often, however, these divisions get in our way and must be disposed of or reworked. That is the case here. Schizophrenia is not an all-or-nothing affair. The operative notion instead appears to be one of a continuum of behavioral symptoms that to be sure is separable from other continua, but within which is marked more by difference in degree than kind.

This is exactly what would be predicted if an underlying distribution of genes were involved. A sobering fact, moreover, is that, as Bleuler observed, the number of latent cases, with relatively mild symptoms, would be far in excess of the number of cases with symptoms intense enough to gain our institutional attention. That means that there are far more people with schizophrenic characteristics in the population than one would imagine from knowledge of the incidence of those hospitalized. We see them daily on the streets, at our jobs, and, inevitably, in our homes.

Bleuler recognized this also, noting that schizophrenic symptoms are evident even in "normal" persons when they are preoccupied or distracted. We are all capable at times of peculiar associations, incomplete concepts and ideas, logical inaccuracies, and stereotypy. Schizophrenic traits thus blend imperceptibly with behavior regarded as normal, the individual symptoms themselves being less important than their extensiveness, their intensity, and their relation to the psychological setting (Arieti, 1974). The latter is particularly germane. For instance, drawing doodles in the presence of an unentertaining speaker may be indicative of nothing but boredom, but including them as part of a serious communication would be indicative of serious pathology (Rosenthal, 1970).

How then may we conceptualize the basis of schizophrenia? First we postulate an underlying distribution including polygenes through which it is potentiated. The more such polygenes possessed by an individual, the more liable will he or she be to manifest schizophrenic symptoms. Individuals possessing such genes in a number that places them at one extreme of the distribution will be those most likely to be severely affected. Second, we postulate a distribution of environmentally derived stress. It is therefore the interaction of genotype and environment that finally determines the behavioral outcome. Some persons may be genetically vulnerable but never experience a sufficient amount of stress to force them across the diagnostic threshold (although they may show lesser schizophrenic characteristics), whereas other persons may be genetically less vulnerable but have life experiences that in combination with genotype lead them to more ob-vious pathological behavior. For still others, who are genetically quite vulnerable, little out of the ordinary exigencies of living may be re-quired in order that they become manifestly affected. For such people *not* to become overtly schizophrenic, they would need an especially anxiety-reducing environment. Finally, at the limit of genetic vulner-ability a substantial degree of symptomology is probably inevitable under any environmental regimen. The perception of stress and the provocation of anxiety lie so much within the individual that the crea-tion of an environment free of stress is virtually unimaginable. We can be anxious about anything, including whether or not we are anxious.

Individuals at the other end of the genetic distribution, however, may be able to contend with the grimmest environmental offerings with relative equanimity. Indeed, from the viewpoint of environmental theory, it is a wonder that such a small percentage of people under uniquely powerful pressure—one thinks, for example, of the inmates of concentration camps—succumb to damaging psychoses. Similarly, the huge proportion of children raised under more common but seem-ingly unbearable circumstances who do not become schizophrenic is potent testimony to their resistance to the disorder. This is not to sug-gest that they are completely unaffected by the conditions, but that the break with reality that characterizes schizophrenia is not apparent. It is thus clear that at least some degree of genetically mediated potential is necessary if schizophrenic symptoms are to appear in a clinical syndrome.

The picture that then emerges, the one most in accord with what psychiatric facts there are and the one with the greatest explanatory power from all viewpoints, is that of a distribution of gene-mediated vulnerability overlaid by a distribution of environmental stresses, the two combining in a way such that the realization of schizophrenic symptomology is a product of their interaction. Rosenthal (1970) de-scribes this model as one of "diathesis-stress," where "diathesis" stands for the organismic variables that predispose individuals to the disease. What actually constitutes the diathesis, however, is as unknown as what constitutes the stress. The model therefore is, as it must be, a conceptual shorthand for neural-environmental interactive processes—which, bluntly, we do not understand. Schizophrenics have of course been examined for neurophysiological and neurochemical abnormal-ities, and a number of proposals have been made (Axlerod, 1972; Fried-hoff & Van Winkle, 1967; Kety, 1975; Osmond & Smythies, 1952; Stein & Wise, 1971; Woolley & Shaw, 1954); however, a specific etiological factor has yet to be identified with certainty. There may in fact be a number of different chemicals involved, under the influence of a num-ber of different genes. Their actions, separately, in concert, or in combination with environmental input may account both for the apparent threshold effects evident in acute schizophrenic episodes and for the variety and severity of the symptomology.

The importance of such a model, especially as compared with the older, single-cause approaches, is that it opens avenues for superior research. We expect good theory to propose conceptual solutions and to suggest the means of their experimental test, and the diathesis-stress model succeeds in these respects. We need now to examine schizophrenia in all aspects from this point of view. It is more than an understatement, however, to note that this is a difficult task. For all that we know about the disorder, it is not nearly enough to draw firm conclusions about the mechanisms involved. It must be studied from a perspective not appreciated hitherto.

Such study, though lacking sufficient time to deliver specific answers, is now well under way. Neurochemists, geneticists, psychiatrists, psychologists, and pharmacologists have begun in earnest to combine their talents to search for the processes that lead to schizophrenic behavior. The search will be enhanced especially as more and more psychiatrists and psychologists overcome the biophobia so evident in their fields. Rosenthal (1970) remarked not many years ago that he had been "continually surprised to learn how little most mental health devotees know about the possible hereditary contributions to the phenomena they are studying and teaching. Moreover, many do not want to know" (p. ix). But in science, as elsewhere, times are changing. Perhaps we are prepared now seriously to confront the enigmatic, often confusing, sometimes horrible disorder of schizophrenia in a more productive way.

In the meantime the modern notions of a spectrum of schizophrenic symptomology, of an underlying genetic distribution, and of interaction with environmental stressors provides us with insight into the larger reaches of human behavior. We must think, that is, not in terms of a dichotomy between the healthy and the ill but in terms of a continuum of behaviors that influence the ordinary and extraordinary events of our lives, and must include actions schizophrenoform in nature. Our image of humanity, our thoughts about the acts of our neighbors, our friends, our leaders, our enemies, and of course, insofar as we are capable, our thoughts about ouselves must then be reappraised. This is hardly to say that all we find to be disagreeable or peculiar (usually in others) is schizophrenic; rather, only that there genuinely is an insidious element within the "normal" range of human behavior that historically has not been and is not the result merely of distraction or ambition but owes its origin to those same processes that in greater measure produce a debilitating disease.

An intriguing question, the answer to which bears equally upon our image of ourselves, is why schizophrenia exists at the level it does.

One can see, that is, that there will normally be a certain number of genetically related developmental errors that recur in each generation and that natural selection will keep at a minimum, but it is difficult to understand why something like schizophrenia should be so prevalent, considering that as such it is markedly maladaptive. Schizophrenics do not survive well without help and protection, and even so they (and their biological relatives) fail to reproduce at the same rate as do others in the population. That means constant strong selection against the genes that potentiate the disorder, which in turn should make them relatively uncommon. However, the genes cannot be that rare, because the disorder itself is so common. How can this be?

No one knows. But the fact that a substantial number of genes are lost through differential reproduction and yet the incidence of schizophrenia remains essentially unchanged in each generation can only be accounted for by mutation. Enough genes must mutate from some other state to one that enters into the potentiation of schizophrenia to match the loss of schizophrenopotential genes that occurs. This requires either an excessively high mutation rate for one or a few genes, or, in a polygenic system, which we have assumed to hold, that a large number of revelant genes at different loci are constantly mutating. Considering once again that schizophrenia is maladaptive, the question is why?

An appropriately paradoxical and evolutionarily sound (if highly speculative) proposition is that the genes that potentiate schizophrenia maintain their frequencies because in combination with other genes they produce characteristics that are beneficial to the species. That is, we may be seeing another instance in which individuals at one end of the distribution are sacrificed for the benefit of the remainder. What might that benefit be? Loosely, creativity.

Genius and madness have in popular thought long been considered to run together in some way, to be manifestations of the same process. This may not be too unreal. What characterizes a creative act, as opposed to the ordinary application of intelligence is, in a truly meaningful sense, a break with reality, a break, that is, with the standard conceptual form of the world of the day. "Seeing things differently" is thus the sine qua non of creativity. It is also the sine qua non of schizophrenia. Arieti (1974) notes this, commenting succinctly upon "the bond between those human beings who achieve the pinnacles of creativity and those human beings who are seriously ill and in some cases locked in the back wards of psychiatric hospitals. Great artists and the mentally ill are shaken by what is terribly absent in our daily reality, and they send us messages of their own search and samples of their findings" (p. 368). Further, he observes that "even the schizophrenic who draws or paints for the first time in his life may immediately find at his disposal an emerging personal style and a new technique" (p. 362). Outside the realm of art the relationship is less clear, partly because of the limitations imposed by means of expression, and also because the capacity for abstract thought disappears as the disease progresses. A hospitalized schizophrenic is far less likely to engage in scientific pursuits than to paint or write poetry.

The freedom to see things differently is nevertheless a part of creativity at large. But such freedom may be an elixir or a poison. The difference lies in the individual's capacity to retain control, to maintain contact with the standard world, and, indeed, to relate his or her visions to it. This capacity must also be gene-related. There will be individuals, that is, with genetic constitutions such that in effect they can make good use of schizophrenopotential genes. Because the behavior of these individuals is largely beneficial to our species, it makes sense that schizophrenopotential genes remain with us.

For such a proposition to be at all viable, it must conform to evolutionary principle. That means, first, that we must think of creative capacity not as restricted to a few but as distributed among a large proportion of the population. The little insights that occur to so many of us must thus be seen as species-beneficial, the products of thundering geniuses representing only creativity at the limit. In such a case we may expect that the genes that combine with those of schizophrenic potential to produce creative individuals will be many and will be spread widely throughout the population. Second, there must be differential reproduction in favor of creative individuals. Were this not so, the genes they carry would not be in the population to the extent necessary to offset, much less take advantage of, those that potentiate schizophrenia. Those most inventive and adept in their relationships with the physical and social world therefore had to have left a relatively larger percentage of surviving offspring throughout much of our evolutionary history.

Is this likely? It would seem to be. We have seen that among other animals there is a positive relationship between dominance and reproduction, and there is every reason to believe this to have been true throughout most of human evolution. We know indeed that the vast majority of human cultures even of recent times (being polygynous) have rewarded those in charge with greater reproductive opportunity. In earlier periods much the same undoubtedly occurred, and occurred under conditions wherein the relationship between creativity and positions of leadership was more direct than we find it today. The more resourceful hunter, the more inventive gatherer, the more originative processor of food, constructor of shelter, and raiser and protector of offspring would not only be respected and have his or her methods copied, to the benefit of all, but would leave more of his or her genes in the next generation as well.

In this view then it is suggested that the genes that in too great concentration lead to schizophrenia will in more moderate numbers and in combination with other specific genes potentiate human creativity. We should therefore expect to find all manner of behavioral combinations, depending upon the overlay of the two genetic distributions and their interaction with the environment, from the stolid and unshakable through the solidly original, the bizarrely inventive, the brilliantly innovative, and the dully but mildly schizophrenic to the clearly insane. Without question such combinations exist.

Whether they exist for the reasons advanced here, however, is open to debate. Closure is obtained upon the problem of the high incidence of an otherwise maladaptive condition with this view, and some loose-end notions about creativity and madness are conveniently incorporated, but reliable data are scanty and difficult to interpret. Heston (1966) did report that the unaffected children of schizophrenic mothers were more artistic, spontaneous, and interesting than ordinary, and Karlsson (1966) found a higher than expected number of creative individuals among those biologically related to schizophrenics. Are these people protected in some way by their other genes, such that they can, metaphorically, make use of their schizophrenopotential ones? We can hardly say. Hopefully we may someday know. In the meantime we are left with at least the outlines of another fine, bitter piece of the ironic evolutionary way, another fragment of our mirror to the past.

RACE AND INTELLIGENCE

Mammals are creatures of status, and with good reason. Status is correlated with priorities of breeding and with allocation of the resources required for maintenance of offspring. It is obtained largely through dominance, which, we have seen, involves a fair measure of aggression. Throughout the mammalian order, then, there has been and continues to be selection for the dominant and aggressive, for they are most likely to survive and successfully reproduce. The competition among organisms of the same species upon which natural selection depends thus derives in considerable degree from the behavior of those organisms and the outcomes of their aggressive initiatives against one another.

It is within this context that the human racial problem must be viewed. Discriminatory acts are generically acts of dominance. They are thus not inventions of the "ecologically displaced" human animal nor aberrations peculiar to culture; rather, they result from and are a profound part of the evolution of all higher organisms. They serve behaviorally to differentiate those who will have more from those who will have less, with the final effect of differentiating biologically those who will leave more from those who will leave less. As the evolutionary process has itself evolved, they are necessary.

The importance of this observation with regard to race lies in the urgency for rectitude with which it permanently imbues any proposal that meaningful racial differences exist. Discrimination in general being intrinsic to the status distribution and the allied dominance-derived apportionment of perogatives and resources, we are required to be particularly cautious in our advocacy. We must realize that compensatory ideals such as equality under law or rights of individuals, though ethically indispensable, are tenuous at best in the face of our propensity to discriminate. Given that propensity, we easily find warrant to circumvent any right or law. (History is not sanguine on this: The events in Europe between 1939 and 1945 alone are surely sufficient to remind us that the crust of custom is very thin.)

The sad fact is that we are prone to seize upon differences of any kind in order to promote our own above others. We have always done it and we have always known that we do it. Indeed, there are few peoples who at one time have not thought themselves superior to some other, and this more often than not has proved adequate cause for war, the institution of slavery, and even genocide. With that at our backs we must look ahead with considerable care, for in making us what we are the evolutionary process has simultaneously provided us with a behavioral apparatus that can unmake us altogether.

This is not unusual. The conservative elements of evolution are always at odds with those generating change, with the result that species maintain their capacity to adapt to long-term alterations without losing their genetic integrity. But again, as usual there is a price, and it is assessed against individuals and groups within for the benefit of species as a whole. Our species has benefited from internal differences and the competition (which includes discrimination) that has been essential to our evolution. However, some individuals and groups have found the price to be high.

Racial discrimination is in this view no different from any other and doubtless has been practiced when the opportunity occurred from the time that there were races. It is therefore not special; it is normal. But as has also been noted, it is just this which must give us pause in our examination of race differences, for, no matter how it is phrased, a scientific indication that they have meaning for us will in the minds of many tend to legitimize the practices that have gained for them social-economic advantages.

We thus find ourselves in an unenviable position. We must first face the fact that discrimination in general was an evolutionary necessity and therefore undoubtedly has a biological basis. However, we find now that racial discrimination is not only ethically undesirable but in modern society has no evolutionary value, because it no longer leads to the differential production of offspring by the favored. (In that sense it is vestigial, although the matter is complex, for we must remain competitive unless we intend to stop evolving.) We must next face the fact, as mentioned, that given our penchant, a scientific statement to the effect that racial differences may be relevant in any way to our current existence must be grist to the mill of those who gain in other ways. Indeed such a statement offends our new morality on human differences, except—finally—that we may accede to a presumably higher morality, which is to tell the truth.

We see good reasons all around that, juxtaposed, leave us encircled by a hard choice. The question is: Given our tendencies and our history and the justifications some would seek therefrom, and given our present social commitments and the faint, costly, but genuine progress that emanates from that commitment, it is wiser simply to avoid raising the issue of racial differences at all?

Many, evidently, would say yes. It is, for one thing, comforting to assume that racial differences in behavior are environmentally determined and, since they would then disappear upon proper social adjustment, are already adequately understood. For another, it is frightening to contemplate the social loss that could occur should the assumption that racial differences in behavior are environmentally determined be seriously challenged. From this position there appears to be great risk in the study of race, an unnecessary risk at that. The other opinion holds that there should be no limits to scientific inquiry, even with regard to unpleasant and unhappy subjects, the risk always being greater in the long run when supposition, however gratifying, is substituted for knowledge. Supporters of the second view argue that the answer to the question is no. It is not wiser.

Which is it to be? We have taken pains to suggest that the fears of those who would, essentially, close off debate are justified. On the other hand, the excommunication of topics from research because of their social sensitivity presents an even more fearsome prospect. *Lasting* social justice indeed depends upon the very intellectual freedom that some, well-intentioned and working in the interests of amelioration and change, would deny. This is inevitably true. Theirs is a noble myopia, but a myopia nevertheless.

We must therefore remain committed to the goal of complete freedom of inquiry, for only in our constant struggle for its attainment do we find the intellectual and moral tension—and consequently the honesty—required to deal with great matters. Repression is never to be condoned; it leaves a dark hole in the fabric with which we cloak ourselves against the cold of ignorance. If we can say one thing safely, it is that knowledge is forever our ally. We must strive for it as surely as we breathe.

This, however, can be accomplished properly only in an open climate. Scholars especially must behave as scholars, not as executioners, direct their attention to the substance of the debate, and above all refrain from calling upon the emotions of the larger community at every turn. Legislators must behave similarly, principally refraining from the selection for use only of the bits of scientific evidence that support their preferences. Finally, the public must understand and be tolerant of the weaknesses of the scientific enterprise; it can respond only as quickly as the tractability of its subject matter and the excellence of its practitioners allow.

We are compelled therefore to examine the subject of race differences, not because there is so much to learn, for there is not, but because our options are so few.

The Concept of Race

Evolutionary geneticist Theodosius Dobzhansky (1962) remarked some time ago that

> The scientific study of human races is at least two hundred years old. There are nevertheless few natural phenomena, and probably no other aspect of human nature, the investigation of which has so often floundered in confusion and misunderstanding. And this is only partly due to the biases and passions engendered by race prejudices and consequent defensive reactions. (p. 253)

Concern with who breeds with whom and what happened to their progeny is obviously much more than 200 years old, as evidenced by some of our earliest written records. We have until recently been careful to state our familial, tribe, clan, and other consanguineous identifications when possible and to describe our lineages. For all this investment, scientific or otherwise, however, we have purchased little in the way of knowledge. We still do not know whether there is much significance in the fact that as a species we are divided into races or as individuals we are descended from this or that large breeding group, no matter what its past influence upon pride and profit. It is obvious that, simple formulations to the contrary, easy answers to the meaning of race have not been and will not be forthcoming.

This is well illustrated in the difficulty we have even in defining the term. Surely there are physical differences among humans—we see them all the time—and certain charactristics are shared more in some groups than others. Blond, pale, blue-eyed people are not frequently native to Mongolia, for instance, nor are people with black kinky hair, very dark skin, and brown eyes frequently native to Sweden. But what beyond these sometimes loose correlations of traits in different populations may we associate with the concept of race? Not much. To be sure there are other distinctions (in blood groups, height, body build, and so on), but variation in all of these is also to be found *within* "races." Further, there are always intermediates *between* them. A race is therefore not a discrete entity, but rather the members composing it vary in the same features that are proposed to distinguish it from others. Further, the more features one uses to categorize, the fewer the number of individuals who will qualify as members of the race. When a large inventory of such features is used, almost no one may.

The lesson in this is that "race" is an abstraction that lends to real phenomena a precision they no not themselves possess. This is not to say that differences do not exist—clearly they do—rather it is to say that the differences are not found among five or six (or 50 or 60) more or less homogeneous groups, with little in between, but are found among individuals and small groups who form a graded series. There is of course a certain amount of clustering of characteristics, but the basic observation is that the variation is continuous, not discrete.

This leads us immediately away from an old error that, as noted earlier, has produced its share of perniciousness. I refer to the notion of racial purity. Purity is something that for better or worse we tend to treasure, and we treat its loss gravely. This has been true in many affairs, but particularly with regard to breeding. Little wonder then at the appeal of the proposition that the fall of great civilizations was the result of vitiation of the founders' stock because they mated with lesser peoples. However, according to Dobzhansky (1962), there is no evidence that races were ever pure, and certainly they are not now. We need not fear for racial chastity. As far as we can tell, there is nothing to lose.

What, then, are races? If they are not genetically homogeneous groups, and if there are intermediates between them, what could they be? They are breeding populations which differ in gene frequencies. We have noted previously that among south African Bushmen, for instance, the relative frequency of the cDe allele of the Rh blood group gene is 89.0, whereas among Japanese it is 0.0. On the other hand, among Japanese the frequency of the CDe allele is 60.2, whereas among Bushmen it is 9.0. All Bushmen are not cDe and all Japanese are not CDe, although the differences are considerable. But the point is that Bushmen do not have all the human cDe alleles, nor do the Japaneses have all the CDes. CDe is in fact quite frequent outside of sub-Saharan Africa, whereas cDe is found in high frequency among a number of populations there, and occurs in lesser frequency in all parts of the world. These two characteristics are thus shared by most breeding groups; they differ merely in prevalence.

Differing gene frequencies across populations clearly reflect the tendency of the members of those populations to breed with one another rather than with members of other populations, and this results in the clustering of gene-determined traits that gives the race concept what coherence it has. Nevertheless, it cannot be emphasized too strongly that it is the uniqueness of the combination rather than of the genes themselves that is the foundation of the racial differences we see. (Some genes, such as the sickle-cell allele, are limited to a few populations, but their presence is often ascribable to peculiar external conditions, such as a severe malarial environment.) Thus, all human populations partake of a gene pool common to the species. This is reflected in the fact that all human populations can successfully interbreed.

Two Mendelian (breeding) populations then may be said to be racially distinct if they differ in gene frequencies. But to what extent must they differ? How many genes must be involved and how different must the frequencies be? It is entirely arbitrary. One investigator might on the basis of several broad traits conclude that there are nine races, while another might, on considering more traits, conclude that there are 90. This, and the fact that very few frequencies are actually known (genes for such popular items as cranial and postcranial bone structure, hair texture, and even skin color have not been analyzed individually) lend the race concept a flexibility that leaves many investigators uneasy. Therefore, while no one disputes the observation that individual humans are genetically different, neither has anyone produced a wholly satisfactory means of treating the differences in racial terms.

The gene frequency concept of race derives from population genetics, clearly reflecting the "population thinking" (page 272) of Mayr (1942) and other modern systematists. Older concepts derived from the "typological thinking" common in biology before the implications of Mendel's work were understood. Thus there were thought to be racial types (just as there were thought to be species types), which more or less personified the investigator's intuitions about the distinctive characteristics of the races, and from which others in a race varied. (Needless to say, notions of racial purity are not inconsonant with, and doubtless received support from, the typological view.)

However, the difficulties of typological thinking with regard to race are no different from those with regard to species: It has not proved productive to conceive of a representative type for each race from which individuals are variants, for one reason because individuals completely corresponding to each racial type are conspicuously rare, the number of those having some but not all of the features of the presumed type constituting the majority of the population by far (Dobzhansky, 1962). Another difficulty arises because the typological approach does not distinguish between differences within and between populations, making the categories that derive therein "arbitrary slices of essentially continuous phenotypic variability" (p. 257).

The typological approach is still with us, nevertheless. Baker (1974), for instance, argues against the population-genetical views of Mayr and Dobzhansky and for the older analysis by types. Baker actually substitutes "typical" for "type," but the result is exactly the same. Recall that Mayr (1942) defined typological thinking as that which treated the individuals of a population as deviants from a particular representative type. Baker clearly does this, asserting that "The grouping into taxa, large or small, can only be done by persons gifted with the capacity for induction, who are capable of recognizing such resemblances as exist. They will form in their minds an idea of a 'typical' member of the group, and will note the departures, in various directions, from this form" (p. 118). He also rejects as a problem the fact that very few "typical" members of a group may exist, suggesting that "The idea of the typical is valuable in taxonomy; and it is inescapable, because anyone who looks will see it, even though the perfectly typical form may not be found" (p. 123).

Finally, Baker rejects the kind of analysis upon which the gene frequency definition of race ultimately depends; that is, upon measurement of many specific gene-related characters, such as blood groups, claiming instead that "It is important to realize the limitations of statistical methods in taxonomy. One cannot idealize a 'typical' member of a taxon by simply counting or measuring and then subjecting the figures obtained to statistical analysis" (p. 119).

Baker's notion then is that people may be sorted into races by gifted inductors working on a rather global scale with abstract, typical forms that may not actually exist. If there is any difference between this "typicalogical" thinking and the typological thinking described by Mayr, it is difficult to see. This is in no way intended to denigrate Baker's efforts, or those of any classifier, for, as has been noted repeatedly, we live by categories whether we wish to or not, and certainly classification has played and continues to play a most important role in science. Some methods have proved less helpful than others, however; indeed, some have distracted us from further analysis at best and restricted whole scientific and social endeavors at worst. The typological approach to race appears in the judgment of most modern students of the subject to be an unsound method for the reasons given.

We can say then with confidence that gene-related differences occur among human individuals and among human populations. But exactly how these differences are to be cast meaningfully into the framework of differences due to race we are unsure. A technically valid scientific definition of race can be made, but how this clarifies the social issues centering upon the subject, with which we seem interminably preoccupied, remains obscure. This must be kept foremost in mind throughout what follows.

The Origins of Racial Differences

Earlier we noted Dobzhansky's (1962) assessment of the success of racial classifications. His conclusions with regard to our understanding of how racial differences arose is no less direct: "If the classification of human races is in an unsatisfactory state, the understanding of their origins and biological significance is still more so" (p. 269). It is obvious, that is, that populations differ in color, in skeletal and other physical features, in frequencies of the alleles that determine blood groups, and so on, but just what this means to us, or should mean to us, is unclear. Are the differences the products of natural selection, and should we then conclude that they represent adaptations to different environments? Are they the results of genetic drift, changes in gene frequency because some alleles are lost in each generation due to small population size? Why *are* there racial differences, and what do they matter?

These problems did not trouble earlier researchers. It was assumed that racial differences derived through adaptations to local conditions, chiefly by the inheritance of acquired characteristics. Black people were black, for example, because their ancestors had acquired deep suntans and passed them on. This proposition we know today as Lamarckism, and we know it to be false. But surely there must be some correlation between the traits that populations exhibit and the conditions under which they live. What interpretation *is* correct?

We do not know. Again, as Dobzhansky (1962) notes: "Shocking though this may be, solid and conclusive evidence concerning the adaptive significance of racial traits in man is scant in the extreme, and the best that can be offered are plausible speculations and surmises" (p. 271). If racial differences are to be ascribed to natural selection, so that different races are seen as adapted to the conditions that surround them, then the advantages of the racial characteristics must be demonstrable. In some cases there may such advantages. In others, however, they are not obvious. What is the advantage, for instance, in having straight or very curly hair, or thin or very full lips?

What of the blood group differences? Are they fortuitous, indicating merely who has been mating whom, or is there more to it? Once again we are unsure. What is clear, however, is that it is not by any means safe to assume that every interpopulational difference has evolutionary significance. Indeed, the truth may be quite the opposite.

Perhaps the best case for a racially adaptive feature can be made for color of the skin. It was long thought that the dark skin of those living near the equator evolved as a protection against harmful effects of the sun. Recent evidence does support this notion.

It is sunlight in the ultraviolet portion of the spectrum that damages the cells beneath the outer layers of the skin if too much passes through. Protection from ultraviolet is gained by an increase in the external (horny) layer, and by the deposit of pigments, formed by cells beneath, that collect and defract the ultraviolet rays. The pigment granules are of two sorts, melanin (black) and keratin (yellow). Brown results from the combination of the two. It is hypothesized that skin color is correlated with the amount of direct sunlight received. In northern latitudes there is little direct sun, more ultraviolet in particular being absorbed by the ozone layer through more of which the rays must travel. Skin in these latitudes therefore lacks pigment, either black or yellow. In more southern latitudes yellow and/or brown skin begins to appear, while close to the equator, black skin predominates. This hypothesis is supported not only by the fact that the skin color gradient is fairly close to what it is supposed to be, but by the fact that the number of pigment-forming cells is about equal in all human skin. It is thought to be the inhibition of the enzymes that catalyze the formation of pigment that results in paleness in some populations.

There is more to this proposal, however. It has been suggested that more important than the protection against cellular damage by ultraviolet rays, which pigment is known to provide, there is also in pigmented skin protection against oversynthesis of vitamin D. Radiation of the skin by ultraviolet light is apparently the sole mechanism for production of vitamin D in humans. Too little of this substance leads to rickets, skeletal deformities due to softening of the bones, while too much of it leads to multiple calcifications of the soft tissues, kidney stones, and ultimately death. It is Loomis's (1967) thesis that skin color is an adaptation that allows maximal penetration of ultraviolet in those living in northern latitudes and prevents undue penetration in those living in southern latitudes, so as to maintain the rate of vitamin D synthesis within acceptable physiological limits.

The rosy winter cheeks of the children of northern Europe act in this formulation as processors of ultraviolet, daily exposure of only 20 square centimeters of such thin, unpigmented skin being enough to generate an antirachitic dose of vitamin D. (This explains perhaps why in northern habitats even infants are put out of doors during the winter when there is sunshine.) The cheek skin of a heavily pigmented child living in Scandinavia under the same conditions would synthesize too little vitamin D to meet his or her body's requirements. On the other hand, untanned northern Europeans living near the equator and exposing most of their body surface to ultraviolet would synthesize up to 800,000 International Units per day. Doses of more than 100,000 IU per day lead to hypervitaminosis D. Deeply pigmented Africans synthesize only 5 to 10 percent as much, and their daily production thus falls in the acceptable range.

Pigmentation therefore protects against both sunburn, as originally thought, and against hypervitaminosis D. The reversible pigmentation response of pale skin (tanning), in addition to limiting the harmful effects of sun upon the skin itself, obviously serves to reduce its output of vitamin D in the summer and to allow for an increase during the winter. Tanning is then presumably an adaptation that has evolved as humans began to inhibit regions having direct sunlight for only a few months during the year.

If the loss of permanent pigmentation is an adaptive characteristic for life in northern latitudes, and if, as the distribution of fossils and stone implements suggests, Europe was inhabited relatively late, then it follows that we evolved with pigmented skin, some populations losing it as they moved north and conditions changed. This does not mean, however, that pigmentation is in any sense more "primitive." All populations continued to evolve, with gene exchange among them, some doubtless becoming more pigmented and some becoming less pigmented as they inhabited regions new for them.

In both Europe and China, skin is lighter in northern latitudes. (It is also lighter in young children, darkening as the child grows older. This reflects the declining need for vitamin D with maturity.) The one exception to the correlation between latitude and skin color in the Old World is the Eskimo. Eskimos have skin that is medium dark, but although they live in northern latitudes they are free from rickets. In this case, however, food plays an important role, for the Eskimo diet, which includes fish oil, contains several times the minimum preventive dose of vitamin D. The same applies to any coastal people; selection for nonpigmentation would be unnecessary as they moved north so long as they continued to subsist substantially on products from the sea.

It has been noted that pigmentation results from the production of one or both kinds of pigment granule: melanin and keratin. In black people it is melanin that filters out excessive ultraviolet radiation. Among Mongoloids, however, keratin (which gives a yellowish tint to the skin), serves this purpose. Mongoloids thus may live within 20 degrees of the equator even though their skin at this latitude contains little melanin. But nearer the equator Mongoloids and their derivatives have been selected for further pigmentation: hence the brownish skin of persons of Mongol extraction who entered the Americas over the Bering Straits 10,000 to 20,000 years ago.

Loomis thus suggests that differences in skin color are primarily accounted for by selection for the capacity to maintain the synthesis of vitamin D within an acceptable range. (Since the colors of skin, eyes, and hair all result from the same pigments, we would expect and do find a substantial correlation among them.) An interesting and otherwise inexplicable problem is accounted for by this hypothesis, and makes it even more feasible. Loomis notes that black skin absorbs more heat than white skin, because white skin is more reflective of incident light. On this basis alone, we should expect the reverse of what we find; that is, heat-absorbing black skin should occur in the north and reflective white skin should be found in equatorial climates. Similarly, we should expect to see tanning in the winter and depigmentation during the summer in northern peoples. The fact that we find the opposite in both cases suggests that there must be a counteradvantage that is more important evolutionarily than heat loss or retention due to skin color. Regulation of the synthesis of vitamin D is evidently that advantage.

A fair case can thus be made that an obvious physical difference among human populations has its origins in natural selection. Are there others? Probably. The bulk and shape of the body is one. Study of infrahumans has disclosed a general proposition that may be applicable in the analysis of racial differences: Those who inhabit the colder parts of the species range tend to be larger in size but have smaller protruding parts than those who inhabit the warmer parts of the range. This is seen as an adaptation for the conservation or dissipation of heat. A comparatively small skin surface relative to mass conserves heat, whereas a relatively large skin surface dissipates it. Applied to humans, the proposition leads us to expect that those native to hot, open countries would be lanky or slight, whereas those native to cold climates would be bulkier, and sometimes shorter. This is pretty much what is found.

Beyond these and perhaps some other similar characteristics, however, there is little firm evidence relating physical differences in race to natural selection. Many traits may indeed be unimportant in this respect, being due in part to genetic drift or only being correlated with the actual traits selected for. These in any case are of less concern to us now, for it is the possibility of racial differences in behavior that excites, and has always excited, our interest.

Racial Differences in Behavior

The first question here is why we consider that there might be racial differences in behavior at all. How, that is, did the issue ever come up? The answer derives from several observations. First, populations do behave differently, and there have been apparent correspondences between those behavioral differences and the physical differences by which races are denoted. Geographically, not only the appearance but the customs of people varied, and it would be remarkable if the two had not at some level been associated. Since animals differ in both appearance and behavior, why not humans? Too, we bred domestic animals and were able to observe the effects of lineage upon appearance and temperament. The extension of the conclusion drawn from these efforts to humans thus seemed straightforward and proper. We still do it.

A more insidious, and, one must imagine, irremediable reaction derives, as noted earlier, from our own evolutionary past. We do have a penchant for discrimination, both within our own populations and against other populations, and this penchant finds a most suitable avenue for expression when physical human characteristics are linked to behavioral ones. It is far easier to know what to do, that is, when one can see immediately the group to which another belongs. Physical appearance in this case serves as a behavioral stigma: we really discriminate not on the basis of color, but rather read color as a sign of racially controlled behavior. Barbarians, heathens, infidels, foreigners, inferiors—those not of the chosen—reveal themselves to us most quickly and conveniently through their physiognomies.

We thus find entries in our ledger referring to warrior races, races of scribes, races of merchants, and so forth, alongside entries listing the relative behavioral debits and credits of the known peoples of the world. Peoples then have always been rated as to their qualities (primacy on balance going, naturally enough, to the one with whom the particular accountant was affiliated). All manner of traits were considered, certain "races" being for instance notably cunning, industrious, lazy, scholarly, tractable (which usually meant that they made desirable slaves), or wild (which usually meant that they did not).

Most of the time, therefore, intelligence was not the preeminent concern; it assumed that role more by default than by contrivance. Its preeminence is, however, no less real for that. And it is no less important. Having over the decades generally rejected the notion that races differ in other behavioral attributes our forebears believed them to, we are left with intelligence as the fulcrum for judgment. And that a judgment will be rendered we must have no doubt. It will be no good then saying that we should not let it influence our treatment of others, for inevitably it will. It will be no good either saying that equality under law will prevent undue transgressions, for inevitably it will not. Discrimination, given the smallest writ, will prevail.

If this is horrifying, it is not unduly so. We live with these things all the time. There is a vast distinction, however, between acting upon fact and selecting facts to suit actions. The proposition advanced here is that in matters concerning the allocation of perogatives and resources we rarely do the former because, frankly, we have been successful doing the latter. If we can but keep this knowledge before us at crucial junctions we will perhaps be less likely, if not unlikely, to make precipitous, and, in the present world, self-defeating decisions.

Black-White Differences in IQ

Given all this, what can we say of racial differences in intelligence? The answer is made difficult by a number of considerations. First, we must decide what we mean by race. We have noted the problem in this. Let us suppose, however, that we proceed on the basis of color, and choose to compare blacks and whites. Which blacks should we use then, and which whites? Is it reasonable, following this line, to compare the blackest with the whitest? Superficially it would seem so. We use populations from Scandinavia and central Africa. But another difficulty ensues because there are cultural as well as genetic differences between such groups. Can we be certain that in attempting to measure intelligence with a standard instrument we are measuring the same thing in both? No. Mental testing is a North American and western European movement and thus reflects the biases of the cultures that produced it. It is unreasonable to compare those who are raised in the tradition of the test makers with those who were not. This has been recognized. We might then attempt to produce less biased tests, tests that are not culture bound. This has been attempted. In the judgment of most investigators, however, the outcome is quite unsatisfactory (Spuhler & Lindzey, 1967). Culture-free or culture-fair tests indeed appear to be impossible to construct.

We thus require at least two large populations, having the same cultural background, that are genetically different enough overtly to be considered races. They are not easy to come by. Even when they are available there is a third difficulty, for environmental differences between the races, especially with respect to variables that might affect scores on the tests, must be minimal. In almost no society does this occur. Finally, we do not actually know what "intelligence" is. We could be measuring something on the tests that for example is relevant to success in the formalities of school or at certain occupations but does not cover many of the other important factors that contribute to success in life.

How then do we measure racial differences in intelligence? Not very well. The hundreds of studies that have been undertaken on possible black-white differences in IQ in the United States, for example, have, in the opinion of competent researchers, proved little. Loehlin, Lindzey, and Spuhler, in their recent (1975) exhaustive book, can thus conclude only that it is a matter of personal preference what weight one assigns to deficiencies in tests, differences in environment, and differences in genes as causes of the difference in IQ scores between blacks and whites. They do note that they consider it likely that some genes affecting some aspects of intellectual performance differ in frequency between U.S. racial-ethnic groups. This is still a long distance from the definitive statement we might at this point expect; and we must remember that IQ may not be an adequate, and certainly not a complete, measure of intelligence. Spuhler and Lindzey (1967), in a previous publication, remarked that

> For the areas of human behavior that are vital in everyday life, for the varieties of behavior that allow individuals to participate satisfactorily in their society, there is no evidence for genetically determined racial differences. Indeed there is at least the possibility that selection acting over the past two or more million years has made genes adaptive for symbolic behavior, for behavior associated with language, and consequently has made it very unlikely that such racial differences exist. (p. 414)

It appears therefore that little can be unequivocally said. That does not mean that there is nothing about which to talk. In recent years the issue of racial differences in intelligence has generated a considerable volume of scholarly discussion, as well as a considerable amount of unscholarly heat. No person has in all this been more prominent or controversial than Arthur Jensen. Indeed, the word "Jensenism" has been used by some critics to describe his views. What exactly did he do?

In 1969 he wrote, at the invitation of the student editors of the *Harvard Educational Review,* an article of 120 pages, less than one-tenth of which dealt with race. His interest was and is in finding an educational program that will increase the performance of all children. He felt and feels that if children do differ in intellectual capacity, then differential methods of training might be considered. If races differ in intelligence, then it might be useful to look for differential methods of training for them as well. He believed that there was evidence to support that view.

To say that Jensen's remarks were not received kindly in some circles is manifestly to understate the case. The editors of the magazine had indeed arranged for comments from a number of other academicians, the effect being to serve a quasi-philosophical feast, with Jensen as the entrée. A short time later on his own campus sound trucks appeared, exhorting all to "Stop racism. Fire Jensen!" (Cronback, 1975). It was as if he had thought the unthinkable. The reaction produced in some certainly reflected this—and also, sadly, exposed the all-too-fragile substructure upon which antidiscrimination measures lay. Nothing is unthinkable to those who are secure.

No one wants to hear that the group with which he is most obviously identifiable is in any way less than some other obviously identifiable group. The feeling is understandably magnified when the first group has through moral argument just begun to acquire the means to emerge from social and economic domination by the second, for the proposal will suggest to some that there may have been good reason for that domination originally and that it might in fact be something to think about again. Little wonder that black people and those who actively espouse the minority cause took umbrage at Jensen's claim. They had no choice.

Neither did Jensen. The central observation, which few deny (Loehlin et al., 1975), is that U.S. blacks score, on the average, 15 points less on standard IQ tests than do whites. The mean IQ for whites is about 100, and for blacks it is about 85. Even though there is substantial overlap of the two distributions and a considerable percentage of blacks fall above the white mean, most blacks clearly fall below it. This, plus the failure of large-scale compensatory education programs to improve the performance of black children in school, led Jensen to consider that there might be a genetic involvement.

The debate from this point on focused on whether the IQ difference could be explained by the environmental disadvantages long suffered by blacks. Jensen's opponents argued that it could; Jensen of course argued that it could not. We shall not discuss the matter in detail. Jensen has presented his case cogently (e.g., 1973), and the numerous counterclaims may be found by consulting the references therein and elsewhere. There are, however, two overridingly important points.

First, the hypothesis that genetic differences underlie racial differences in intelligence cannot be demonstrated in any strict sense without the proper breeding experiments. We would have to do what we do with plants and other animals when two populations differ on some attribute and we wish to estimate proportions of genetic and environmental variance: we crossbreed males and females of the two populations and see where their progeny lie relative to the parents and to one another. Further tests are then possible. The genetic hypothesis would be supported if the derivative generations score roughly what would be expected of them on the basis of the percentage of genes they share with one or the other parent. The environmental hypothesis would be supported, obviously, if rearing conditions rather than genes appeared to be correlated with the scores obtained by the offspring. Such tests have not been made. Jensen has therefore had to argue for the plausibility that differences in IQ scores between blacks and whites are largely genetically determined, given other evidence.

What other evidence? Chiefly that IQ is a highly heritable trait in North American and European white populations. The argument is that if IQ is a highly heritable trait within populations, it is also highly heritable between them. This, again, is not necessarily so. DeFries (1972) demonstrated that one could estimate between-populations heritability ($h^2\ b$) from within-population heritability ($h^2\ w$). To do it, however, the correlation of the additive genetic values of members of the groups (r) must be known. It is not. Nevertheless, using different hypothetical values for r, DeFries showed that only when it is relatively large is a high value for $h^2\ w$ associated with a high value for $h^2\ b$. Thus a trait that is highly heritable within populations is not necessarily highly heritable between them. We therefore do not strictly know how much, if any, of the variance between races in IQ scores is due to differences in genes.

Jensen is now aware of this. But he maintains that the circumstantial evidence for genetically based racial differences in IQ nevertheless lessens support for the environmental hypothesis, and should at the very least caution its advocates. In this he is undoubtedly correct. We are accustomed to accept the smallest environmental difference as causative, whereas the possible effects of genetic difference are rejected in principle. But the point is that the research that could convincingly demonstrate either the genetic or the environmental hypothesis has not been done.

It is extraordinarily difficult to demonstrate anything about behavior, much less human behavior. More often than not we assume what we set out to prove, only to have the logical and methodological flaws in our work exposed by the next generation. And theirs is exposed in turn, until it seems that we have explored all approaches and none, old ideas now sounding new and new ideas old, and we end by doing the usual things but speaking of them in unusual ways. The more things change, says a bitter but honest little aphorism, the more they stay the same.

Scientists are immune to none of the normal human frailties. This is as it otherwise should be, but it can result in abnormal amounts of passionate confusion. Scientists are, after all, in the business of offering proof, which is frustrating enough: To try it with their own species, in which their other investments are so enormous, is to court madness. This they do, however, some with more grace and some with less, leaving the corpus of humanity as both beneficiary and victim.

A current instance—which, if it does not fundamentally harm the genetic hypothesis, certainly provides no help—concerns the late Cyril Burt. Burt was a prominent British psychologist who in many years of research (he died at 88) published an impressive set of data supporting the relevance of genes to IQ scores. He was also influential in implementing a system of education that assigned children at the age of 11 to one of three levels based on such scores, so his views had more than academic impact. Burt's data indicated that the heritability of IQ within the British population is on the order of .80, which means the genetic differences account for 80 percent of the variance.

Burt died in 1971. Shortly thereafter Leon Kamin, an experimental psychologist who had not previously concerned himself with the issue, noticed highly improbable correlational consistencies and methodological oversights in Burt's reports and began to discuss these in lectures. The first to publish an account of them, however, was a person who both knew and admired Burt and who had been a student of one of Burt's most influential pupils, Hans Eysenck. It was Arthur Jensen.

Jensen had set out to make a compendium of all the data on the correlations between relatives that Burt had published (and which were scattered over numerous articles in several journals). In tabulating these data he noticed that the size of the samples changed from one publication to the next as Burt's research progressed, yet in no fewer than 20 instances the values for the correlations remained the same. This is most unlikely. As the data base changes (i.e., more and more people are tested and their scores are added to the original set), figures derived from those data are expected to change. Rarely, and for a single such figure, this may not occur. But for the correlations reported by Burt to have been invariant in 20 instances exceeds imaginable likelihood. It meant error or, as some would have it, fraud.

To worsen the situation, Burt's original data are unavailable. Jensen had attempted to obtain them, but

> . . . nothing remained of Burt's possessions save various notes, letters, manuscripts, reprints, and books. I was told that shortly after Burt's death many of the books and journals had to be sold and donated to libraries, and that many boxes of old data, which Burt had kept for many years, were disposed of in the course of vacating his flat in Hamstead. These boxes, etc., I was informed, were either poorly labeled or not labeled at all, so that their exact contents were not apparent to casual inspection. (Jensen, 1974)

There was thus no way to correct Burt's mistakes or even to verify that any mistakes had been made. Jensen was forced to conclude that Burt's correlations were "useless for hypothesis testing."

It is debatable whether the inadmissibility of Burt's data alters the scientific assessment of the genetic hypothesis, for much of what Burt reported has been substantiated independently by others. It may be that the heritability of IQ in white populations will be considered now to be less (possibly of the order of .60 rather than the .80 calculated by Burt), but it remains large enough certainly to warrant investigation. What is worrisome is the extrascientific effect centering upon the revelation of Burt's inaccuracies. Whether due to innocent error or fradulent calculation, the fact that they did occur and were so long unquestioned will unfetter the nemesis of all men in affairs of this kind: guilt by association. This is always a danger. With an issue as sensitive as racial differences in intelligence it is a certainty. We must expect its presence in future discussions and resist its undeniable appeal.

Jensen's behavior has been on the whole most laudable—and if occasionally testy, not without reason. His name has been subverted to the status of an "ism," his university classes have been disrupted, his dismissal has been sought, and he has been publicly accused of undergoing a metamorphosis similar to (and with the moral ramifications of) that of Dr. Jekyll and Mr. Hyde (Dworkin, 1974). It likewise has been said in private that his research is a ruse behind which lurk despicable motives. For example, as noted by black economist Thomas Sowell:

> Too many of Jensen's critics have tried to reduce the argument to [the level of Archie Bunker], claiming that Jensen "really" set out to promote racism disguised as research, as a means of halting recent black advances and/or justifying the failure of the public school system to educate black children. Crude as it may seem, I recently heard it stated just this baldly by a psychologist whose name would be familiar to any reader of this magazine. (1973)

It is hoped then that the agents of this particular inquisition miss their mark.

Otherwise, what may be said of the substance of the dispute? Can the disadvantages to which blacks have been subjected explain the difference in black-white IQ obtained by many investigators? We do not know. It does seem that when socioeconomic variables are manipulated (Scarr & Weinberg, 1976), rather than being statistically negated through matching, the difference narrows considerably or disappears. Even statistical negation of socioeconomic factors reduces it. The decisive consideration, however, as noted by geneticists such as Bodmer and Cavalli-Sforza (1970) and Dobzhansky (1967, 1973) is that we can never know the capacities of different individuals, populations, or races until they have had something close to an equal opportunity to demonstrate those capacities. Statistical manipulations, as valuable as they can sometimes be, are only as good as the observations upon which they are based, and we do not know with regard to our tests and measurements how the performance of one race is affected when that race has only recently emerged from psychological subjugation by another. As Sowell remarks, "being black in America is something more than making a few thousand dollars less or averaging fewer years in school."

This means that, in spite of the furor surrounding them, the present data are quite limited. We will not even have satisfactory preliminary data until we do what we do with plants and other animals; that is until we first provide a reasonably satisfactory and similar environment for both races. We cannot otherwise validly estimate the effects of genetic differences between them. This has always been so (and, as often remarked by Dobzhansky, is generally overlooked by political theorists of all persuasions). The truth is that genetic differences are maximally expressed when the organisms live in the same environment. With regard to race, such a condition does not exist and has never existed in the United States, and probably has never existed anywhere. We shall have to wait a few generations until, with good fortune, it does.

In the meantime we are presented with a set of data and an analysis that on almost any other topic would have drawn scholarly approval, if not agreement. Jensen has in no way violated scientific propriety—indeed in many respects he has been a model investigator. But his conclusions are not *liked*, and for that he could accumulate ten times the data and still meet the same opposition. This is quite normal and derives in part, as mentioned previously, from the unreliable foundation that necessarily underlies the attempted correction of differences in opportunity between formerly dominant and dominated groups, especially when the groups are physically distinct and when the latter is a numerical minority.

As Sowell (1973) again notes, "The attempt by liberal intellectuals to make racial equality a fact by fiat only set the stage for Jensen's demolition of their position" (which, he continues, is not to be confused with establishing Jensen's own). These persons were thus cognizant of the fragility of what they promoted and reacted in the usual defensively aggressive way when it was challenged. We are thereby made aware of the origin of the worst accusations against Jensen, and of at least one reason why his work attracts such critical attention: For important, emotional issues we raise our theshold of conviction; the standards for acceptance of unwelcome ideas become much higher than for others. Although this is to be expected, and is certainly understandable, it should also be borne in mind when Jensen is discussed.

The second overriding point has been mentioned briefly before, but it very much requires mentioning again. We have noted (page 169) that theoretically the genetic component of an attribute under strong selection over a long period will consist chiefly of dominance and/or epistatic variance, most of the additive variance, which responds to selection and accounts for the resemblance among relatives, having been "used." It is thus true of fitness characters that the proportion of additive genetic variance is small. It is therefore noteworthy that not only the total genetic component of variance (heritability in the broad sense or the degree of genetic determination) of IQ has been found to be so large, but that the proportion of additive variance within that component has been found to contribute the most to it. Jinks and Fulker (1970) have, for instance, estimated the proportion of additive genetic variance of IQ to be .71. Although this was based upon Burt's data and will have to be reconsidered, it nevertheless is not out of the range provided by other estimates (Jensen, 1973). However, the question is: If IQ is a fitness character, why should the additive variance be anywhere near .71?

The answer of course is that it should *not*, if indeed IQ is closely related to fitness. If it is not so related, then presumably it has not been selected for throughout human evolution. If it has not been selected for, then it evidently has not played a very important role in that evolution. But human evolution from its inception has been shaped by selection for behavioral characteristics, not the least of which is the capacity to comprehend, symbolize, and manipulate the psychological, social, and physical environment. We call this capacity intelligence. But then the question becomes: What does IQ have to do with intelligence?

The fact is, we do not know. IQ is a convenient measure that correlates well with achievement in school (which is not surprising, since IQ tests were constructed originally to do just that) and moderately well with occupational status. Surely it reflects some portion of intelligence as evolutionarily conceived. But what portion?

Again we cannot say. As Sarason (1973) remarks, "It has not even demonstrated that the level of problem-solving behavior in non-test situations is highly correlated with the level of similar types of problem-solving processes in the standardized test situation." It could be, however, that much of what we were selected for, much of what separates us from other species and has contributed to differences in reproductive capacity among individuals and populations within our own is not measured on IQ tests. If it is not, then it is difficult to find evidence in

IQ differences for natural selection having produced racial differences in intelligence. IQ measures something, but what it measures seems *too* heritable to have responded to natural selection for millions of years.

This is a nice irony, upon which proponents of the genetic hypothesis do not dwell. It is there nevertheless. Thiessen (1972) has noted that throughout behavioral research traits are investigated that superficially seem to be fitness characters and therefore important to survival but, like IQ, show high proportions of additive genetic variance. He suggests that these traits may actually not be very relevant to the functioning of the organisms concerned and that what is underlying them is really "genetic junk." Convenience of measurement may thus be leading us to deal with characters that in an evolutionarily meaningful sense are not "there."

This happens all the time in science. Organisms do not hand us their behavior to be disjoined, analyzed, rejoined, and returned. It is we, who in our attempts to understand, conceive of variables that limit and define and can be measured. Sometimes these are valid—we have guessed right—and sometimes they are not. In the often stupefying heat of research pursuit it is easy to confuse measurement with meaning. Such confusion was indeed formalized in psychology by its acceptance of the operational definition. Whatever could not be measured with the tools at hand was treated as if it did not exist. This leads to a restrictive view of the world, exemplified by the rather silly but once popular notion that intelligence, by definition, is what is measured by intelligence tests. Although increasingly less so, we still find ourselves confined by this notion today.

Jensen (1973) curiously rejects the larger, evolutionary concept of intelligence as a fitness character because it is "a broadening of the concept of intelligence far beyond its meaning in psychology" (p. 24), while at the same time he concedes that "the correlation between IQ and our various criteria of success in life is far from perfect" (p. 34). It is true that other factors influence success, but it is equally true that "school-smartness" does not guarantee it. Yet it is school-smartness that IQ scores most significantly predict. And even in endeavors that would seem to draw heavily on that variable there are too many instances where prediction has failed for us to have great confidence in it. Einstein, for example, was once considered backward in school—and Darwin barely adequate.

We can thus conclude that there is more to intelligence than the intelligence quotient, and indeed that theoretically what is measured by IQ tests resides in genetic limbo. Therefore, far from being convinced by the high proportion of additive genetic variance reported for IQ scores, we find in it only cause for concern. This does not mean that on these grounds the work of Jensen and others is refuted; it means that there is one more reason to see complexities in the matter.

Race and Civilization

Cognizance of racial differences of course antedate the development of IQ tests by some millenia—we have never needed numbers. The foremost consideration has for far longer been the apparent capacity or incapacity of certain populations to produce the cultural phenomenon we have come to call civilization. To be sure we do not know exactly what civilization is, and, as with intelligence, we tend to acknowledge it when we see it rather than depend upon formal definitions. This is not to say that formal definitions have not been proposed, but that, as the concept is actually used, "civilization" seems to be more a loose state of mind in the proposer than anything else.

Nevertheless, we must agree that there are criteria that if met more or less by one group and not by another do differentiate the two with respect to the cultural level at which they live. Thus a population using crude stone implements, cultivating neither plants nor animals, having no form of written communication and practicing magic could legitimately be contrasted with one having finer implements (possibly of metal), cultivating food crops, having a written language, and having a religious system with at least some elements of ethical content. There is no point in denying that we can recognize differences in cultural advance and that they reflect differences in accomplishment.

The question is why the cultural differences exist? If humans are everywhere the same, why is the level of culture not everywhere the same? The truth once again is that we do not know. However, as usual we can find answers that emphasize either a genetic or an environmental origin. Simply, the genetic argument would have it that some peoples are endowed with whatever capacities are necessary to create and sustain cultural innovation (and they must be many and varied— intelligence, creativity, persistence, and so forth) and some peoples are not, or are less so. Thus the great civilizations, which have changed the lives of all who followed, have tended to concentrate in relatively few lands occupied by a few (often related) peoples. The environmental argument would have it that some set of external deterrents— geographical isolation, disease, and so on—prevented the exchange of cultural novelties that normally occurs between the peoples of one area and the next. Thus the advance of certain populations was precluded not for lack of capacity but for lack of interchange of ideas.

Obviously neither proposition has been demonstrated, and we shall not consider them at length here. It would be well, however, to spend a little time with two versions of the genetic view in order better to comprehend certain of its representations. In the process we shall take the opportunity briefly to examine arguments on the environmental side.

Baker (1974), in his recent book *Race,* presents an erudite but standard typological (or as he would have it, typicalological) approach. We have noted previously that he rejects the populational concept of modern evolutionary systematics for the older notion of typical forms. He distinguishes six human races. These are Australasid (to be found as the original inhabitants of Australia, Melanesia, and Tasmania), Europoid (in Europe, Arabia, northern Russia, Ethiopia, and so on), Negrid (in central and southern Africa, e.g., Zaire, Senegal, southern Sudan, and Rhodesia), Khoisanid (the Hottentots and Bushmen in southwest Africa and the Kalahari Desert), Mongoloid (in Mongolia, central China, and Vietnam), and Indianid (the original inhabitants of Mexico, Peru, and Brazil). Other peoples are derived from these.

Baker's basic contention is that forms that differ in physical characteristics are very likely to differ in mental characteristics and that some will be superior to others. After reviewing the evidence for morphological differences, he thus states that

> It is not to be supposed that genes conferring genuine "superiority" of any sort, if such exist, would be easily susceptible to genetic analysis. One would anticipate the cumulative effect of many genes, each having a small effect. For the solution of the ethnic problem, however, one is not immediately concerned with the question whether the analysis of human polygenes is possible, or likely to become so. One wants to know whether it is conceivable that members of two taxa may differ in large numbers of genes affecting many parts of the body, but not at all in those that affect the nervous and sensory systems and therefore play a part in determining mental qualities. (p. 426)

In addition,

> . . . one must ask himself whether it is conceivable that the mental qualities of each human taxon, though differing, must somehow add themselves together in such a way that all taxa are necessarily to be regarded as "equal" mentally, in the special sense that no taxon is superior to any other. What known cause of evolution could have produced this result? Is it not more probable that natural selection has adapted taxa to different environments, and that as a result some of them have a greater tendency than others to produce persons possessing special agility and versatility of mind? And if one introduces a value-judgment by predicating that agility and versatility of mind are superior to mental sloth, is it not at least *likely* that superiority and inferiority are realities . . . ?" (p. 427)

What prompted selection for mental agility and versatility? Habitation of the temperate zones. Thus,

> The retreat of the ice after the last glacial epoch opened up vast areas of the northern hemisphere and offered great opportunities to people who could adapt themselves to a new mode of life and develop the foresight needed for the maintenance of large populations under strongly seasonal climatic conditions, very different from those in many parts of the world, where a more equable environment made it less necessary to look far into the future. Certain potentially favorable environments are more demanding than others, and thus favor the natural selection of an enterprising type of mind. (p. 428)

We should therefore expect the major civilizations to have been founded by north temperate peoples. According to Baker, this is so. He distinguishes five civilizations that he believes to have originated independently and to have given rise to others. These are the Sumerian, ancient Egyptian, Helladic-Minoan, Indus Valley, and Sinic. All were products of Europoid and Mongoloid races.

Further, Baker considers it likely that the provenance of civilizations rested not upon all the persons of particular taxons but upon "a small proportion of very talented people" (p. 525). As an example, he concludes that the greatest intellectual achievement of humans is probably the invention of parts of speech, and especially of the sentence.

> This invention appears to have been made independently in different ethnic taxa, and there is reason to believe that it must have been a product not of any society as a whole, but of its most intelligent members only, for when a language comes to be dominated by mass media or communication . . . the parts of speech tend to be blurred, the sentence corrupted, and ideas to be vaguely conveyed by the mere apposition of words without logical connections. (p. 501)

The "ethnic problem" thus finally reduces to how large a proportion of very talented people various taxa regularly produce. Indeed, the capacity of a taxon to produce a significant number of such individuals becomes in this view a secondary characteristic of the taxon.

It follows that only certain taxa have the capacity: Europoids and Mongoloids. Australasids, Negrids, Khoisanids, and Indianids do not. Australasids in fact are found to be primitive and Khoisanids paedomorphic (the adults resembling the infantile or juvenile states of their ancestors), which accounts for their inferiority, whereas Negrids and Indianids apparently were enervated by their too hospitable climes.

The physical characteristics most clearly correlated with the lack of mental powers necessary for civiliation according to Baker is small brain size, found in both Australasids and Khoisanids. Thus, "It must be kept in mind that certain taxa have remained primitive or become paedomorphous in their general morphological characters, and none of these has succeeded in developing a civilization. It is among these taxa in particular that one finds some direct indication of a possible cause of mental inferiority in the small size of the brain" (p. 428). However, he later notes that "There is no actual proof that all the pongid and paedomorphous characters in the brains of Australasids and Sandids [Khoisanids] respectively are directly concerned in determining the intellectual faculties . . ." (p. 432). This hypothesis then remains unsupported.

Of great relevance to thought on social problems in many other parts of the world is the "cold" theory—the proposition that adaptation to life in the temperate zones led to the "enterprising type of mind" that may later have entered into the foundation of civilizations. There does seem to be agreement among anthropologists that hominid evolution was significantly affected. Cold winters required the mastery and use of fire, the development of warm gear, and, most important, affected social and intellectual behavior. We have discussed this before (page 252). Campbell (1974) thus surmises that

> Cold winters . . . necessitated considerable development of social behavior. It seems inescapable that there would have been a fairly complete division of labor by this time: the men hunting and the women minding the babies and gathering vegetable foods, water (in skins), and fuel. The division of labor and separation of the sexes must have increased the need to communicate abstract ideas by the development of language, and the vocabulary expanded. Perhaps the expression of the emotions (which language could replace) was first inhibited in a closely knit cave-dwelling band, and this emotional inhibition was to become increasingly important. It may prove to have been one of the most fundamental social developments which has shaped the psychology of modern man. (pp. 387–388)

He notes also the "The cold made demands on man's ingenuity to devise protective facilities such as clothing and tents. It was surely an important factor in the evolution of human intelligence" (p. 388). It therefore seems probable that hominid expansion into the temperate regions significantly affected human evolution.

However, Baker's conclusion does not follow. That is, it appears *not* to have been the case, as he assumes (in common with others who hold to this racial theory) that those who left the tropics and those who remained behind evolved separately. Our paleospecies, *Homo sapiens sapiens* is considered to be approximately 50,000 years old. But is was not *H. s. sapiens* that first moved north to exploit that environment in adaptation to which our intellectual capacities increased. It was our ancestor, *Homo erectus*. This form of human indeed appears to have been well established in the temperate zones between 700,000 and 400,000 years ago (Campbell, 1974). One part of *H. s. sapiens* therefore did not break off to derive the evolutionary benefits of temperate life, leaving the other to stultify in the ease of the tropics. Rather, the species to which we belong today evolved long after the migration began.

It may seem injudicious to note the above in the present context, for alone it appears to support Baker's proposition. The different populations would simply have been in place longer. However, another observation, considered in conjunction with the amount of time involved, makes this not so. Over the 700,000 to 400,000 years in question there were numerous glacial and interglacial periods in the temperate zones, forcing populations back to tropical regions and then allowing expansion to the temperature areas once more. Indeed, Washburn (1963) estimates that even as late as 15,000 years ago there were three to five times as many Bushmen as Europeans because the ice sheets of the last glaciation reduced the habitable area of Europe to half that available in eastern and southern Africa. (By recorded times the Bushmen were themselves restricted, because of the southeastern expansion of black populations, to the Kalahari Desert.) Therefore, in evolving from *H. erectus*, *H. s. sapiens* did so moving back and forth between more northerly and southerly latitudes, with various populations exchanging genes all the while. The fossil evidence reflects this. The bones of humans of a species indistinguishable from our own, dating from 30,000 to 60,000 years ago have been found in such diverse places as France, Kenya, South Africa, and Southeast Asia (Campbell, 1974). We were thus by all accounts a traveling species through all but the smallest part of our last evolutionary sojourn, the differences we see now (except possibly in the case of some relict populations) resulting from minor genetic drift and a few very recent physical adaptations.

The point of this is that we seem to have been in the places we found one another when we first began to write upon the subject for far too short a time to account for the severe differences between races that some believe exist. And the differences are severe. Jensen (1973), for example, has noted the importance of the average difference of 15 IQ points between U.S. blacks and whites. The black distribution in this case produces seven times as many persons with IQs below 70 than does the white. As many as one-sixth to one-fourth of the persons in the former thus fall below IQ 70. In present society such individuals are considered to be mentally retarded. However, it is very unlikely that so large and so complex a biological difference could have developed between black and white populations in the limited time they have been exclusively in their current habitats. The skin color difference may have evolved as a result of this, as we have seen, but that requires the selection of at most a few genes that regulate the production of pigment, a simple process when compared to that influencing cognitive ability.

What evidence there is with regard to the proposition that failure to generate a civilization may be ascribed to intellectual deficiency is thus hardly conclusive. Baker as much as admits this in the final paragraph of his second chapter on racial differences in achievement, remarking that

> It is sometimes argued that a few thousand years are a short time to the whole period of man's existence, and that consequently one should not lay too much stress on the time at which the civilized state has been attained. It will never be possible to find out whether any of the races in which civilization did not appear independently would ever have attained it independently if given sufficient time . . . (p. 528)

Considering the exponential rate at which successful cultural innovations develop, considering the dependence of civilization upon cultural innovations, and considering the differential access that a number of populations have had to the exchange of such innovations, we are at the very least enjoined from judgment. We shall discuss this further shortly.

Baker represents the typological (typicalological) approach, and, although his conclusions are not a necessary product, in racial matters they are often correlated with such a view. Typological thinking emphasizes differences between idealized types, which are held to be representative of particular groups, at the expense of differences among individuals within those groups of populations that are intermediate between them. One thus tends to see a discrete distribution rather than a continuous one, to stress separability rather than similarity. To be sure, categories are needed (we have said this all along), but the problem in this case is that the categories, which do not exist in nature but are our own fictions, may to us become reified, and in our efforts to place organisms in one or another we may find differences that are not there. It was for this reason (among others) that modern systematics abandoned the typological view. It serves us poorly here.

Another version of the genetic hypothesis is given by C. D. Darlington in *The Evolution of Man and Society* (1969). Darlington presents what in fact is among the most exclusively genetic interpretations for differences among classes, castes, and races to have been offered in recent times. For him, however, it is not the evolution of fairly homogeneous groups having typical characteristics that originally led to the creation of civilizations but the hybridization of peoples and the subsequent increase in genetic variance. Thus, he suggests that there is

> . . . one principle at the root of man's social and mental evolution. It is that every improvement in the brain by its nature leads to new

ideas and new inventions. Each invention propagates its own inventors and leads to their multiplication, and expansion. And these in turn lead to their hybridization with other stocks and the production of yet more genetic combinations with similarly enhanced capacities. (p. 25)

Darlington, as do other geneticists and anthropologists, therefore sees the brain-behavior relationship as crucial in our evolution. However, for him there exists a unique condition, superimposed upon the normal operation of natural selection, which has effected our history. He argues that we have undergone a directed evolution, directed in the sense that once we began to change, further change along the same lines was enhanced beyond what it usually is. The result was the evolution of certain capabilities at an unprecedented speed. The principles of directed evolution is known as orthogenesis, and it has in Darlington's view "dominated human evolution" (p. 22).

The effect of the application of the principles of orthogenesis to human evolution is to alter the time frame in such a way as to make recent events more important than they could otherwise have been. Thus, for Darlington the last 100,000 years have been the most important. The glacial periods during this era not only caused migrations to unoccupied lands (e.g., the Americas), but more significantly led to the mixing of peoples, which resulted in hybridization. From these movements there also arose changes in the distribution of races.

Significantly, once directed selection had begun it continued along the same lines in all humans, regardless of habitat. Thus,

> Given this situation of directed evolution an important corollary follows. The brain had acquired a dominant position in man's relationship with his environment. Geographical adaptation, of course, was always bound to be important; but the use of the brain and the improvement of its use was bound to become relatively and in the long term more important. Which means that, on the one hand man's external habitat was the agent of the diversified change which controlled the evolution of his races. But on the other hand, man's internal character was the agent of the directed changes which controlled the evolution of his species. The races and the species were evolving by different methods. (p. 27)

As a species, our evolution was therefore directed by the same "internal character." As races we evolved through natural selection in the usual way. This suggests that whereas some of our physical traits responded to the environmental differences found in various geographical areas, our mental traits, dependent upon a basic brain-behavior relationship, did not.

This is not to say, however, that at times certain populations were not gifted with propensities for certain behaviors. This proposition is in fact central to Darlington's reconstruction of history. Thus, there have been hunters, herdsmen, farmers, and so on, who were genetically predisposed to their livelihoods, to the extent that they could learn no other (p. 30); to the extent, for instance, that "just as the crops became adapted to the men (their men), so the men became adapted to the crops (their crops). The whole concern became one system. It became, like the genetic system of a species, fused into one evolutionary unit or entity whose parts are mutually dependent and come to be mutually adapted" (p. 79).

That particular groups of men and particular plants or animals should have evolved together to generate a human such that the character of the men came (rather metaphorically we must assume) to reflect that of their associated flora or fauna is a dizzying prospect. However, there is no escaping such statements as: "The extreme attachment of the herdmen to his animals has often led to extreme specialization in his choice of stock. The Lapps with their reindeer, the Bedouin with their camels, the Masai with their cattle . . . the Scythians, Huns and Mongols, often with a sole attachment to the horse, all these have a character related to their animals" (p. 81). The same was true for other systems, so that eventually there evolved peoples of all sorts with genetic propensities linked to their livelihoods. Iron workers, grain farmers, scribes, priests, stonemasons, governors, warriors—these and others differentiated in the quickstep of orthogenesis from the ubiquitous neolithic hunting bands into occupational genetic castes in less than 10,000 years.

Given that, history becomes the evolution, migrations, and dissolution of such castes, and civilizations are begun and ended as the genetically disposed craftsmen, teachers, priests, and governors move, or rise and fall. Ideas therefore do not spread across geographical space so much as people do—the people who evolved with, invented, and are most capable of using the ideas. Whatever barriers inhibit the movements of such people thus inhibit the transmission of ideas.

Altogether this is an awesome thought. It is not one upon which we need dwell, however, for orthogenesis as a principle in human evolution is almost universally rejected. Dobzhansky (1973), for instance, dismisses Darlington's argument rather summarily by noting that "Even some eminent geneticists who should have known better have been led astray by such reasoning" (p. 93). Doubtless this is true. We know little of the origins of civilization, but there is no evidence supportive of Darlington's view. Selection for behaviors so complex and subtle as stonemasonry and governorship have never been demonstrated, and human evolution according to all other accounts proceeded much more slowly that he proposes.

There is still a point to studying Darlington, for he, like Baker, represents in many respects the limit of the genetic approach, although his and Baker's are quite different perspectives and actually tend to cancel one another out. Darlington, for instance, considers that black peoples "arrived or arose in Africa last of all for it was only they who could successfully occupy the tropical steppes and complete the southward colonization of Africa which they were still in the process of doing when the white man arrived" (p. 44). Rather than being the original Africans, black Africans would thus have come late to that habitat and not have been left behind much earlier by enterprising migrants. Light brown would have been the first human skin color, black being a recent mutation. (In fact, the greater part of Africa was occupied by brown Bushmen until the expansion of black peoples a relatively short time ago.)

Darlington also finds environmental reasons, which Baker does not, for the lack of high civilization in black Africa. He notes, as we have mentioned, that the introduction of Indonesian agriculture in particular affected the development of African peoples because the slash-and-burn techniques by which it is practiced encouraged the spread of malaria and sleeping sickness. Besides the effect on the indigenous human populations, there was an effect upon the animals which civilization has long been associated: cattle and horses. Cattle did get through along the high and dry corridor which connects Ethiopia with southern Africa. The Masai tend them to this day. But cattle could not survive the disease conditions in the African interior. Horses, coming later, were unable to survive at all. And this is more serious than one might superficially think, for it

> . . . had a grave effect upon the development of African societies. For, without the horse, the wheeled chariot and all wheeled vehicles failed to appear and a warrior governing class never took shape. Another great step, the replacement of the hoe by the plough, was also hindered. For ploughing, cattle would have sufficed, but the ease of yam-growing and the diseases of cattle were together discouraging for the development of heavy cultivation. (Darlington, 1969, p. 652)

If disease prevented the introduction of cattle and horses and what they contribute to the burgeoning of civilization, it had the same effect upon the introduction of new peoples. The African centers of Zimbabwe and Timbuktu fell, and scholars and technicians from other areas penetrated no further. The ideas, as well as the animals and artifacts, of other peoples were thus unavailable to sub-Saharan Africans. The messengers could not survive.

Finally, disease affected the Africans themselves. We have mentioned the presence of the sickle-cell allele in populations living in the vast part of the continent known now as the yam belt (page 304). It spread as did root farming, and in the heterozygous state it does protect individuals against the effects of malaria. It is a genetic adaptation to the disease. But the price of protection for the population is paid not only by those homozygous for the sickle-cell allele, who have severe anemia, but by those heterozygous for it as well, for they have sickle-cell disease. Although it is milder, it is still anemia. Development over a huge portion of Africa was thus on this basis alone something of a no-win situation, for three-quarters of the population was anemic to some degree (one-quarter to the extent that the individuals died before the age of reproduction), and the other quarter risked or had malaria. There were in addition other tropical diseases, chiefly due to protozoans and helminth worms, carried by parasites, to which no resistance of a harmless kind has evolved in man. Nor could hygienic instructions, such as those issued by Islam or Hinduism, have been useful in dealing with these parasites. "It is these tropical diseases," Darlington concludes, "which began to cripple the development of African societies just at the moment when African populations reached the numbers and densities necessary for civilization" (p. 662).

In Baker and Darlington we are presented two hypotheses, each genetic, or at least biological. Baker infers cognitive differences from morphological ones in the typological format. Darlington, on the other hand, represents a kind of radical geneticism that includes many modern principles but in concept is as ancient as history, and probably more so. There is a lesson in the contrast in emphasis and conclusions between the two: We see for one thing how much there is room for maneuvering when addressing the same problems, which tells us for another how little is the solid evidence these scientists, or any scientists in the field, can call upon. We are again dealing with guesses and surmises.

CONCLUSIONS

Neither the genetic nor the environmental hypothesis with regard to racial differences in intelligence can be supported. The data are too sparse, too unreliable, too subject to alternative interpretation. We are thus left with the sense that there is more to be learned than is known on the matter, and that bias rather than even judgment has in many cases intervened.

If this is true we must reason on other grounds, and it would seem best, without asserting that the environmental explanation is correct, to act as if it is. This in itself is a bias and has no place in science. However, the issue of racial differences, although it has a long scientific history, is as it affects us chiefly not a scientific question. It is a social problem and has been since the beginning of humanity. It is therefore no good saying, as does Jensen (1973), that we should think of individuals rather than groups or, as does Baker (1974), that even though some ethnic taxa are superior to others, we should not behave as though each individual in a superior taxon is superior to each individual in an inferior one—for the evidence is that in everyday affairs we shall do so. We always have, even in the best of times, perhaps because we can scarcely help ourselves.

Therefore, although genes are and have been essential to the expression of differences in intelligence among *individuals,* we cannot conclude even that there are differences in intelligence among *races,* much less that such differences are genetically based. We have no choice but to seek to ameliorate past inequities in the best, most workable ways, without looking over our shoulders at science when the cost becomes great. Science is no help here. It may never be.

References

Adler, M. J., and Cain, S. *Ethics*. Chicago: Encyclopaedia Britannica, 1962.

Adler, M. J., and McGill, V. J. *Biology, Psychology, and Medicine*. Chicago: Encyclopaedia Britannica, 1963.

Adler, M. J., and McKeon, R. *Philosophy*. Chicago: Encyclopaedia Britannica, 1963.

Adler, M. J., and Wolff, P. *Foundations of Science and Mathematics*. Chicago: Encyclopaedia Britannica, 1960.

Alexander, F. The psychiatric aspects of war and peace. *American Journal of Sociology*, **46**, 504–520, 1941.

Allee, W. C. *Cooperation Among Animals*. New York: Abelard-Schuman, 1938.

Altman, J. *Organic Foundations of Animal Behavior*. New York: Holt, Rinehart and Winston, 1966.

Anastasi, A. Heredity, environment, and the question "how?" *Psychological Review*, **65**, 197–208, 1958.

Anastasi, A. *Individual Differences*. New York: Wiley, 1965.

Anton-Tay, F., Anton, S. M., and Wurtman, R. J. Mechanism of changes in brain norepinephrine metabolism after ovariectomy. *Neuroendocrinology*, **6**, 265–273, 1970.

Arén-Engelbrektsson, B., Larsson, K., Södersten, P., and Wilhelmsson, M.

The female lordosis pattern induced in male rats by estrogen. *Hormones and Behavior*, **1**, 181–188, 1970.

Arieti, S. *Interpretation of Schizophrenia*. New York: Basic Books, 1974.

Axelrod, J. Biogenic amines and their impact in psychiatry. *Seminars in Psychiatry*, **4**, 199–210, 1972.

Baker, J. R. *Race*. New York: Oxford University Press, 1974.

Bandura, A. *Aggression: A Social Learning Analysis*. Englewood Cliffs, N.J.: Prentice-Hall, 1973.

Bandura, A., Ross, D., and Ross, S. A. Imitation of film-mediated aggressive models. *Journal of Abnormal and Social Psychology*, **66**, 3–11, 1963.

Barfield, R. J. Activation of sexual and aggressive behavior by androgen implanted into the male ring dove brain. *Endocrinology*, **89**, 1470–1476, 1971.

Barnett, S. A. *A Study in Behavior*. London: Methuen, 1963.   (a)

Barnett, S. A. Instinct. *Proceedings of the American Academy of Arts and Sciences*, **92**, 564–580, 1963.   (b)

Beach, F. A. The snark was a boojum. *American Psychologist*, **5**, 115–124, 1950.

Beach, F. A. The descent of instinct. *Psychological Review*, **62**, 401–410, 1955.

Beadle, G. W., and Tatum, E. L. Experimental control of developmental reaction. *American Naturalist*, **75**, 107–116, 1941.

Benkert, O., Gluba, H., and Matussek, N. Dopamine, noradrenaline, and 5-hydroxytryptamine in relation to motor activity, fighting and mounting behaviour. *Neuropharmacology*, **12**, 177–186, 1963.

Bennett, J. H. (Ed.) English translation of Gregor Mendel, *Experiments in Plant Hybridisation*, with commentary and assessment by R. A. Fisher. London: Oliver & Boyd, 1965.

Benzer, S. The fine structure of the gene. *Scientific American*, **206**, 70–84, 1962.

Berkowitz, L. The contagion of violence: An S-R mediational analysis of some effects of observed aggression. In W. J. Arnold and M. M. Page, Eds., *Nebraska Symposium on Motivation*, **18**. Lincoln: University of Nebraska Press, 1970.

Berkowitz, L., and Le Page, A. Weapons as aggression eliciting stimuli. *Journal of Personality and Social Psychology*, **7**, 202–207, 1967.

Bigelow, R. The evolution of cooperation, aggression, and self-control. In J. K. Cole and D. D. Jensen, Eds., *Nebraska Symposium on Motivation*, **20**. Lincoln: University of Nebraska Press, 1972.

Bodmer, W. F., and Cavalli-Sforza, L. L. Intelligence and Race. *Scientific American*, **223**, 19–29, 1970.

Boring, E. G. *A History of Experimental Psychology*. New York: Appleton-Century-Crofts, 1950.

Breland, K., and Breland, M. The misbehavior of organisms. *American Psychologist*, **16**, 681–684, 1961.

Bridgman, P. W. The logic of modern physics. In H. Feigl and M. Brodbeck, Eds., *Readings in the Philosophy of Science*. New York: Appleton-Century-Crofts, 1953.

Bronson, F. H., and Desjardins, C. Aggression in adult mice: Modification by neonatal injections of gonodal hormones. *Science*, **161**, 705–706, 1968.

Bronson, F. H., and Desjardins, C. Neonatal androgen administration and adult aggressiveness in female mice. *General and Comparative Endocrinology*, **15**, 320–325, 1970.

Bronson, F. H., and Eleftheriou, B. E. Adrenal response to crowding in peromyscus and C57BL/10 mice. *Physiological Zoology*, **36**, 161–166, 1963.

Brown, J. L. The neural control of aggression. In C. H. Southwick, Ed., *Animal Aggression*. New York: Van Nostrand Reinhold, 1970.

Brown, R. M. Historical patterns of violence in America. In H. D. Graham and T. R. Gurr, Eds., *Violence in America: Historical and Comparative Perspectives*. New York: Bantam, 1969.

Cameron, N. *Personality Development and Psychopathology: A Dynamic Approach*. Boston: Houghton Mifflin, 1963.

Campbell, B. (Ed.) *Sexual Selection and the Descent of Man*. Chicago: Aldine, 1972.

Campbell, B. *Human Evolution*. Chicago: Aldine, 1974.

Campbell, N. *What Is Science?* New York: Dover, 1952.

Carlson, E. A. *The Gene: A Critical History*. Philadelphia: Saunders, 1966.

Carpenter, C. R. Aggressive behavior systems. In R. L. Holloway, Ed., *Primate Aggression, Territorality, and Xenophobia*. New York: Academic Press, 1974

Cavalli-Sforza, L. L., and Bodmer, W. F. *The Genetics of Human Populations*. San Francisco: Freeman, 1971.

Champlin, A. K., Blight, W. C., and McGill, T. E. The effects of varying levels of testosterone on the sexual behaviour of the male mouse. *Animal Behaviour*, **11**, 2–3, 1963.

Christian, J. J. The roles of endocrine and behavioral factors in the growth of mammalian populations. In A. Gorbman, Ed., *Comparative Endocrinology*. New York: Wiley, 1959.

Clayton, R. B., Kogura, J., and Kraemer, H. C. Sexual differentiation in the brain: Effects of testosterone on brain RNA metabolism in newborn female rats. *Nature*, **226**, 810–811, 1970.

Clemente, C. D., and Chase, M. M. Neurological substrates of aggressive behavior. *Annual Review of Physiology*, **35**, 329–356, 1973.

Collins, G. G. S., Sandler, M., Williams, E. D., and Youdim, M. B. H. Multiple forms of human brain mitochondrial monoamine oxidase. *Nature*, **225**, 817–820, 1970.

Cornford, F. M. *From Religion to Philosophy*. New York: Harper & Row, 1957.

Coulson, J. C. Differences in the quality of birds nesting in the centre and on the edges of a colony. *Nature*, **217**, 478–479, 1968.

Craig, W. Why do animals fight? *International Journal of Ethics*, **31**, 264–278, 1921.

Crick, F. H. C. The genetic code: III. *Scientific American*, **215**, 55–62, October 1966.

Cronbach, L. J. The two disciplines of scientific psychology. *American Psychologist*, **12**, 671–684, 1957.

Cronbach, L. J. Five decades of public controversy over mental testing. *American Psychologist*, **30**, 1–14, 1975.

Crook, J. H. Sexual selection, dimorphism, and social organization in the primates. In B. Campbell, Ed., *Sexual Selection and the Descent of Man*. Chicago: Aldine, 1972.

Darlington, C. D. *The Evolution of Man and Society*. New York: Simon & Schuster, 1969.

Darwin, C. *The Origin of Species by Means of Natural Selection or the Preservation of Favoured Races in the Struggle for Life* (1859). New York: New American Library (Mentor Books), 1958.

Darwin, C. *The Descent of Man and Selection in Relation to Sex*. London: Murray, 1871.

Darwin, C. *The Expression of the Emotions in Man and Animals* (1872). University of Chicago Press, 1965.

Darwin, C. *The Autobiography of Charles Darwin* (1882), F. Darwin, Ed. New York: Dover, 1958.

Davidson, E. H. *Gene Activity in Early Development*. New York: Academic Press, 1968.

DeFries, J. C. Quantitative aspects of genetics and environment in the determination of behavior. In L. Ehrman, G. S. Omenn, and E. Caspari, Eds., *Genetics, Environment, and Behavior*. New York: Academic Press, 1972.

Denenberg, V. H. The mother as motivator. In W. J. Arnold and M. M. Page, Eds., *Nebraska Symposium on Motivation*, **18**. Lincoln: University of Nebraska Press, 1970.

Dewey, J. The reflex arc concept in psychology. *Psychological Review*, **3**, 357–370, 1896.

Diamond, S. Four hundred years of the instinct controversy. *Behavior Genetics*, **4**, 237–252, 1974.

Dobzhansky, T. A review of some fundamental concepts and problems of population genetics. *Cold Spring Harbor Symposium on Quantitative Biology*, **20**, 1–15, 1955.

Dobzhansky, T. *Mankind Evolving*. New Haven: Yale University Press, 1962.

Dobzhansky, T. Changing man. *Science*, **155**, 409–414, 1967.

Dobzhansky, T. *Genetic Diversity and Human Equality*. New York: Basic Books, 1973.

Dolhinow, P. The living nonhuman primates. In P. Dolhinow and V. Sarich, Eds., *Background for Man*. Boston: Little, Brown, 1971.

Dollard, J. C., Doob, L., Miller, N., Mowrer, O., and Sears, R. *Frustration and Aggression*. New Haven: Yale University Press, 1939.

Donoso, A. O., de Gutierrez, F. N., Moyano, M. B., and Santolaya, R. C. Metabolism of noradrenaline in the hypothalamus of castrated rats. *Neuroendocrinology*, **4**, 12–19, 1969.

Dorfman, R. I. Mode of action of gonadotrophins. In G. Litwack and D. Kritchevsky, Eds., *Actions of Hormones on Molecular Processes*. New York: Wiley, 1964.

Dubos, R. Man's nature and social institutions. In A. Montagu, Ed., *Man and Aggression.* London: Oxford University Press, 1973.

Dugdale, R. L. *The Jukes.* New York: Putnam, 1877.

Dunlap, K. Are there any instincts? *Journal of Abnormal Psychology,* **14,** 35–50, 1920.

Dworkin, G. Two views on IQs. *American Psychologist,* **29,** 465–467, 1974.

Eaton, G. G., and Resko, L. A. Plasma testosterone and male dominance in a Japanese macaque (*Macaca fuscata*) troop compared with repeated measures of testosterone in laboratory males. *Hormones and Behavior,* **5,** 251–259, 1974.

Ebert, P. D. and Hyde, J. S. Selection for agonistic behavior in wild female *Mus musculus. Behavior Genetics,* **6,** 291–304, 1976.

Eiduson, S. Ontogentic development of monoamine oxidase. In E. Costa and M. Sandler, Eds., *Monoamine Oxidases—New vistas. Advances in Biochemical Psychopharmacology,* **5.** New York: Raven Press, 1972.

Eiseley, L. C. Charles Darwin. *Scientific American,* **195,** 62–72, February 1956.

Eiseley, L. C. Charles Lyell. *Scientific American,* **202,** 98–106, August 1959.

Eleftheriou, B. E., and Hancock, R. L. Hormonal regulation of sRNA methylase activity in regional brain areas in *Peromyscus maniculatus bairdii. Brain Research,* **28,** 311–316, 1971.

Elms, A. C. The crisis of confidence in social psychology. *American Psychologist,* **30,** 967–976, 1975.

Erlenmeyer-Kimling, L. Gene-environment interactions and the variability of behavior. In L. Ehrman, G. S. Omenn, and E. Caspari, Eds., *Genetics, Environment, and Behavior.* New York: Academic Press, 1972.

Erlenmeyer-Kimling, L. Schizophrenia: A bag of dilemmas. *Social Biology,* **23,** 123–134, 1976.

Estabrook, A. H. *The Jukes in 1915.* Washington, D.C.: Carnegie Institution, 1916.

Estes, W. K. Learning theory and the new "mental chemistry." *Psychological Review,* **67,** 207–223, 1960.

Everett, J. W., Holsinger, J. W., Zeilmaker, G. H., Redmond, W. C., and Quinn, D. L. Strain differences for preoptic stimulation of ovulation in cyclic, spontaneously persistent-estrous, and androgen-sterilized rats. *Neuroendocrinology,* **6,** 98–108, 1970.

Falconer, D. S. *Introduction to Quantitative Genetics.* New York: Ronald Press, 1960.

Falconer, D. S. The inheritance of liability to certain diseases, estimated from the incidence among relatives. *Annals of Human Genetics,* **29,** 51–76, 1965.

Feigl, H., and Brodbeck, M. (Eds.) *Readings in the Philosophy of Science.* New York: Appleton-Century-Crofts, 1953.

Feuer, L. S. *Einstein and the Generations of Science.* New York: Basic Books, 1974.

Fisher, R. A. The correlation between relatives on the supposition of Mendelian inheritance. *Transactions of the Royal Society of Edinburgh,* **52,** 399–433, 1918.

Fisher, R. A. *The Genetical Theory of Natural Selection.* New York: Oxford University Press (Clarendon Press), 1930.

Flynn, J. P. Patterning mechanisms, patterned reflexes, and attack behavior in cats. In J. K. Cole and D. D. Jensen, Eds., *Nebraska Symposium on Motivation,* **20.** Lincoln: University of Nebraska Press, 1972.

Freedman, L. Z., and Roe, A. Evolution and human behavior. In A. Roe and G. Simpson, Eds., *Behavior and Evolution.* New Haven: Yale University Press, 1958.

Freud, S. Why war? In T. Maple and D. W. Matheson, Eds., *Aggression, Hostility, and Violence.* New York: Holt, Rinehart and Winston, 1973.

Friedhoff, A. J., and Van Winkle, E. New developments in the investigation of the relationships of 3,4-dimethoxyphenylethylamine to schizophrenia. In H. E. Hernivich, S. S. Kety, and J. R. Smythes, Eds., *Amines and Schizophrenia.* New York: Pergamon Press, 1967.

Friedmann, T. Prenatal diagnosis of genetic disease. *Scientific American,* **225,** 34–42, November 1971.

Fromm, E. *The Anatomy of Human Destructiveness.* New York: Holt, Rinehart and Winston, 1973.

Fulker, D. W., Wilcock, J., and Broadhurst, P. L. Studies in genotype-environment interaction. I. Methodology and preliminary multivariate analysis of a diallel cross of eight strains of rat. *Behavior Genetics,* **2,** 261–287, 1972.

Fuller, J. L., and Thompson, W. R. *Behavior Genetics.* New York: Wiley, 1960.

Galton, F. *Hereditary Genius: An Inquiry into Its Laws and Consequences.* London: Macmillan, 1869.

Gandelman, R. Induction of pup killing in female mice by androgenization. *Physiology and Behavior,* **9,** 101–102, 1972.

Gardner, E. J. *Principles of Genetics.* New York: Wiley, 1964.

Gardner, R. A., and Gardner, B. T. Teaching sign language to a chimpanzee. *Science,* **165,** 664–672, 1969.

Garrod, A. *Inborn Errors of Metabolism.* London: Henry Frowe, 1909.

Geertz, C. *The Interpretation of Cultures.* New York: Basic Books, 1973.

Gerall, A. A., and Kenney, A. McM. Neonatally androgenized females' responsiveness to estrogen and progesterone. *Endocrinology,* **87,** 560–566, 1970.

Ginsburg, B. E. Genetic parameters in behavioral research. In J. Hirsch, Ed., *Behavior-Genetic Analysis.* New York: McGraw-Hill, 1967.

Ginsburg, B. E. Genotypic factors in the ontogeny of behavior. In C. H. Southwich, Ed., *Animal Aggression.* New York: Van Nostrand Reinhold, 1970.

Goddard, H. H. *The Kallikak Family.* New York: Macmillan, 1912.

Gorski, R. A. Steroid hormones and brain function: Progress, principles, and problems. In C. H. Sawyer and R. A. Gorski, Eds., *Steroid Hormones and Brain Function.* Berkeley: University of California Press, 1971.

Gorski, R. A. Perinatal effects of sex steroids on brain development and function. *Progress in Brain Research,* **39,** 149–163, 1973.

Goy, R. W., Bridson, W. E., and Young, W. C. Period of maximum susceptibility of prenatal female guinea pigs to masculinizing actions of testosterone propionate. *Journal of Comparative and Physiological Psychology*, **57**, 166–174, 1964.

Goy, R. W., and Phoenix, C. H. The effects of testosterone propionate administered before birth on the development of behavior in genetic
female rhesus monkeys. In C. H. Sawyer and R. A. Gorski, Eds., *Steroid
Hormones and Brain Function*. Berkeley: University of California Press,
1971.

Goy, R. W., and Resko, J. A. Gonadal hormones and behavior of normal and
pseudolohermaphroditic nonhuman female primates. *Recent Progress in
Hormone Research*, **28**, 707–733, 1972.

Grady, K. L., Phoenix, C. H., and Young, W. C. Role of the developing rat
testis in differentiation of the neural tissues mediating mating behavior.
Journal of Comparative and Physiological Psychology, **59**, 176–182,
1965.

Graham, H. D., and Gurr, T. R. (Eds.) Introduction. *The History of Violence in America*. New York: Bantam Books, 1969.

Grant, M. *The Passing of the Great Race, or the Racial Basis of European
History*. London: Bell, 1917.

Green, R. *Sexual Identity Conflict in Children and Adults*. New York: Basic
Books, 1974.

Griffiths, E. C., and Hooper, K. C. The effects of neonatal androgen on the
activity of certain enzymes in the rat hypothalamus. *Acta Endocri-
nologia*, **70**, 767–774, 1972.

Grunt, J. A., and Young, W. C. Differential reactivity of individuals and the
response of the male guinea pig to testosterone propionate. *Endocrinology*, **51**, 237–248, 1952.

Hailman, J. P. How an instinct is learned. *Scientific American*, **221**, 98–106,
1969.

Hall, C. S., and Lindzey, G. *Theories of Personality*. New York: Wiley,
1970.

Hall, K. L. R. Aggression in monkey and ape societies. In C. H. Southwick,
Ed., *Animal Aggression*. New York: Van Nostrand Reinhold, 1970.

Hamilton, T. H. Control by estrogen of genetic transcription and translation.
Science, **161**, 649–661, 1968.

Hardin, C. Sex differences and the effects of testosterone injections on biogenic amine levels of neonatal rat brain. *Brain Research*, **59**, 437–439,
1973.

Harrington, G. M. Genetic-environmental interaction in "intelligence." II.
Models of behavior, components of behavior, and research strategy.
Developmental Psychobiology, **1**, 245–253, 1969.

Harris, G. W. Sex hormones, brain development, and brain function. *Endocrinology*, **75**, 627–648, 1964.

Harrison, R. J., and Montagna, W. *Man*. New York: Appleton-Century-
Crofts, 1969.

Hart, B. L. Gonadal androgen and sociosexual behavior of male mammals:
A comparative analysis. *Psychological Bulletin*, **81**, 383–400, 1974.

Hebb, D. O. Heredity and environment in mammalian behaviour. *British Journal of Animal Behaviour*, **1**, 43–47, 1953.

Helmuth, H. Human behavior: aggression. In A. Montagu, Ed., *Man and Aggression*. New York: Oxford University Press, 1973.

Henderson, N. D. The confounding effects of genetic variables in early experience research: Can we ignore them? *Developmental Psychobiology*, **1**, 146–152, 1968.

Heston, L. L. Psychiatric disorders in foster home reared children of schizophrenic mothers. *British Journal of Psychiatry*, **112**, 819–825, 1966.

Hinde, R. A. *Animal Behavior*. New York: McGraw-Hill, 1969.

Hirsch, J. (Ed.) *Behavior-Genetic Analysis*. New York: McGraw-Hill, 1967.

Hitler, A. *Mein Kampf* (1925), Translated by R. Manheim. Boston: Houghton Mifflin, 1943.

Hoffmeister, F., and Wuttke, W. On the actions of psychotropic drugs on the attack- and aggressive-defensive behaviour of mice and cats. In S. Garattini and E. B. Sigg, Eds., *Aggressive Behaviour*. New York: Wiley, 1969.

Hofstadter, R. *Social Darwinism in American Thought*. Boston: Beacon Press, 1955.

Hsia, Y. E. Inherited disorders of amino acid, carbohydrate, and nucleic acid metabolism. In R. W. Albers, G. J. Siegel, R. Katzman, and B. W. Agranoff, Eds., *Basic Neurochemistry*. Boston: Little, Brown, 1972.

Hull, C. L. *Principles of Behavior*. New York: Appleton-Century-Crofts, 1943.

Hull, C. L. The place of innate individual and species differences in a natural-science theory of behavior. *Psychological Review*, **52**, 55–60, 1945.

Hull, C. L. *Essentials of Behavior*. New Haven: Yale University Press, 1951.

Hull, C. L. *A Behavior System*. New York: Wiley, 1952.

Hull, C. L., Hovland, C. I., Ross, R. T., Hall, M., Perkins, D. T., and Fitch, F. B. *Mathematico-Deductive Theory of Rote Learning*. New Haven: Yale University Press, 1940.

Huxley, J. Introduction to C. Darwin's *The Origin of Species*. New York: New American Library (Mentor Books), 1958.

Huxley, J. *Evolution: The Modern Synthesis*. New York: Wiley, 1964.

Hyde, J. S., and Ebert, P. D. Correlated responses in selection for aggressiveness in female mice. I. Male aggressiveness. *Behavior Genetics*, **6**, 421–427, 1976.

Ingram, V. M. How do genes act? *Scientific American*, **198**, 68–74, January 1958.

Inselman-Temkin, B. R., and Flynn, J. P. Sex-dependent effects of gonadotrophic hormones on centrally-elicited attack in cats. *Brain Research*, **60**, 393–410, 1973.

Jacob, F., and Monod, J. Genetic regulating mechanisms in the synthesis of proteins. *Journal of Molecular Biology*, **3**, 318–356, 1961.

James, W. The moral equivalent of war. In T. Maple and D. W. Matheson, Eds., *Aggression, Hostility, and Violence*. New York: Holt, Rinehart and Winston, 1973.

Jenkins, J. J., and Patterson, D. G. *Studies in Individual Differences*. New York: Appleton-Century-Crofts, 1961.

Jensen, A. How much can we boost IQ and scholastic achievement? *Harvard Educational Review*, **39**, 1–123, 1969.

Jensen, A. *Educability and Group Differences*. New York: Harper & Row, 1973.

Jensen, A. Kinship correlations reported by Sir Cyril Burt. *Behavior Genetics*, **4**, 1–28, 1974.

Jerison, H. J. Interpreting the evolution of the brain. *Human Biology*, **35**, 263–291, 1963.

Jinks, J. L., and Fulker, D. W. Comparison of the biometrical, genetical, MAVA, and classical approaches to the analysis of human behavior. *Psychological Bulletin*, **73**, 311–349, 1970.

Johnson, R. N. *Aggression in Man and Animals*. Philadelphia: Saunders, 1972.

Kalmus, H. Inherited sense defects. *Scientific American*, **186**, 64–70, May 1952.

Kaplan, A. *The Conduct of Inquiry*. San Francisco: Chandler, 1964.

Karli, P., Vergnes, M., Didiergeorges, F. Rat-mouse interspecific aggressive behaviour and its manipulation by brain ablation and by brain stimulation. In S. Garattini and E. B. Sigg, Eds., *Aggressive Behaviour*. New York: Wiley, 1969.

Karlsson, J. L. *The Biologic Basis of Schizophrenia*. Springfield, Ill.: Thomas, 1966.

Kety, S. S. Progress toward an understanding of the biological substrates of schizophrenia. In R. R. Fiene, D. Rosenthal, and H. Brile, Eds., *Genetic Research in Psychiatry*. Baltimore: The Johns Hopkins University Press, 1975.

Kety, S. S., Rosenthal, D., Wender, P. H., Schulsinger, F., and Jacobsen, B. Mental illness in the biological and adoptive families of adopted individuals who have become schizophrenic. *Behavior Genetics*, **6**, 219–225, 1976.

King, J. A. Intra- and inter-specific conflict of *Mus* and *Peromyscus*. *Ecology*, **38**, 355–357, 1957.

Koch, S. Psychology and emerging conceptions of knowledge as unitary. In T. W. Wann, Ed., *Behaviorism and Phenomenology*. Chicago: University of Chicago Press, 1964.

Kogan, B. R. *Darwin and His Critics: The Darwinian Revolution*. Belmont, Calif.: Wadsworth, 1960.

Kreuz, L. E., Rose, R. M., and Jennings, R. Suppression of plasma testosterone levels and psychological stress. *Archives of General Psychiatry*, **26**, 479–482, 1972.

Krutch, J. W. *The Great Chain of Life*. Boston: Houghton Mifflin, 1957.

Kuhn, T. *The Structure of Scientific Revolutions*. University of Chicago Press, 1962.

Kuo, Z. Y. A psychology without heredity. *Psychological Review*, **31**, 427–451, 1924.

Lagerspetz, K. M. Genetic and social causes of aggressive behaviour in mice. *Scandanavian Journal of Psychology*, **2**, 167–173, 1961.

Lagerspetz, K. M. Aggression and aggressiveness in laboratory mice. In S. Garattini and E. B. Sigg, Eds., *Aggressive Behaviour*. New York: Wiley, 1969.

Lagerspetz, K. Y., Tirri, R., and Lagerspetz, K. M. Neurochemical and endocrinological studies of mice selectively bred for aggressiveness. *Scandanavian Journal of Psychology*, **9**, 157–160, 1968.

Lane, R. Urbanization and criminal violence in the 19th century: Massachusetts as a test case. In H. D. Graham and T. R. Gurr, Eds., *The History of Violence in America*. New York: Bantam Books, 1969.

Larsson, K. Individual differences in reactivity to androgen in male rats. *Physiology and Behavior*, **1**, 255–258, 1966.

Laughlin, W. S., and Osborne, R. H. *Human Variation and Origins*. San Francisco: Freeman, 1968.

Lehrman, D. S. A critique of Konrad Lorenz's theory of instinctive behavior. *Quarterly Review of Biology*, **28**, 337–363, 1953.

Lerner, I. M. *Genetic Homeostasis*. London: Oliver & Boyd, 1954.

Lerner, I. M. *Heredity, Evolution, and Society*. San Francisco: Freeman, 1968.

Lindquist, E. F. *Design and Analysis of Experiments in Psychology and Education*. Boston: Houghton Mifflin, 1953.

Lindzey, G. General discussion. In S. G. Vandenberg, Ed., *Methods and Goals in Behavior Genetics*. New York: Academic Press, 1965.

Lodosky, W., and Gaziri, L. C. J. Brain serotonin and sexual differentiation of the nervous system. *Neuroendocrinology*, **6**, 168–174, 1970.

Loehlin, J. C., Lindzey, G., and Spuhler, J. N. *Race Differences in Intelligence*. San Francisco: Freeman, 1975.

Loomis, W. F. Skin pigment regulation of vitamin-D biosynthesis in man. *Science*, **157**, 501–506, 1967.

Lorenz, K. The function of colour in coral reef fishes. *Proceedings of the Royal Institute of Great Britain*, **39**, 282–296, 1962.

Lorenz, K. Ritualized fighting. In J. D. Carthy and F. J. Ebling, Eds., *The Natural History of Aggression*. New York: Academic Press, 1964.

Lorenz, K. *On Aggression*. New York: Bantam Books, 1967.

Lovejoy, A. O. *The Great Chain of Being: A Study of the History of an Idea*. Cambridge, Mass.: Harvard University Press, 1936.

Lunde, D. T., and Hamburg, D. A. Techniques for assessing the effects of sex hormones on affect, arousal, and aggression in humans. *Recent Progress in Hormone Research*, **28**, 627–663, 1972.

Luttge, W. G. The role of gonadal hormones in the sexual behavior of the rhesus monkey and human: A literature survey. *Archives of Sexual Behavior*, **1**, 61–88, 1971.

Luttge, W. G. Activation and inhibition of isolation induced inter-male fighting behavior in castrate male CD-1 mice treated with steroidal hormones. *Hormones and Behavior*, **3**, 71, 81, 1972.

Luttge, W. G., and Hall, N. R. Differential effectiveness of testosterone and its metabolites in the induction of male sexual behavior in two strains of mice. *Hormones and Behavior,* **4,** 31–43, 1973.

Maccoby, E. E., and Jacklin, C. N. *The Psychology of Sex Differences.* Stanford University Press, 1974.

MacCorquodale, K., and Meehl, P. E. On a distinction between hypothetical constructs and intervening variables. *Psychological Review,* **55,** 95–107, 1948.

Manning, A. *An Introduction to Animal Behaviour.* London: E. Arnold, 1967. (a)

Manning, A. Genes and the evolution of insect behavior. In J. Hirsch, Ed., *Behavior-Genetic Analysis.* New York: McGraw-Hill, 1967. (b)

Manning, A. and McGill, T. E. Neonatal androgen and sexual behavior in female house mice. *Hormones and Behavior,* **5,** 19–31, 1974.

Manosevitz, M. A. Note on genotype X environment interaction. *Texas Reports on Biology and Medicine,* **27,** 1969.

Mark, V. H., and Ervin, F. R. *Violence and the Brain.* New York: Harper & Row, 1970.

Marler, P. R. On animal aggression: The roles of strangeness and familiarity. *American Psychologist,* **31,** 239–246, 1976.

Marler, P. R., and Hamilton, W. J. *Mechanisms of Animal Behavior.* New York: Wiley, 1966.

Marx, M. (Ed.) *Theories in Contemporary Psychology.* New York: Macmillan, 1963.

Mayr, E. *Systematics and the Origin of Species.* New York: Dover, 1942, revised 1964.

McClearn, G. E., and DeFries, J. C. *Introduction to Behavioral Genetics.* San Francisco: Freeman, 1973.

McEwen, B. S., and Pfaff, D. W. Factors influencing sex hormone uptake by rat brain regions. I. Effects of neonatal treatment, hypophysectomy, and competing steroid on estradiol uptake. *Brain Research,* **21,** 1–16, 1970.

McEwen, B. S., Pfaff, D. W., and Zigmond, R. E. Factors influencing sex hormone uptake by rat brain regions. II. Effects of neonatal treatment and hypophysectomy on testosterone uptake. *Brain Research,* **21,** 17–28, 1970.

McGill, T. E. Sexual behavior in three inbred strains of mice. *Behavior,* **19,** 341–350, 1962.

McGill, T. E., and Blight, W. C. Effects of genotype on the recovery of sex drive in the male mouse. *Journal of Comparative and Physiological Psychology,* **56,** 887–888, 1963.

McGuire, L. S., Ryan, K. O., and Omenn, G. S. Congenital adrenal hyperplasia. II. Cognitive and behavioral studies. *Behavior Genetics,* **5,** 175–188, 1975.

McKinney, T. D., and Desjardins, C. Postnatal development of the testis, fighting behavior, and fertility in the house mouse. *Biology of Reproduction,* **9,** 279–294, 1973.

McKusick, V. A. *Human Genetics.* Englewood Cliffs, N.J.: Prentice-Hall, 1964.

Mead, M. Cultural determinants of behavior. In A. Roe and G. Simpson, Eds., *Behavior and Evolution*. New Haven: Yale University Press, 1958.

Meehl, P. E. Schizotaxia, schizotypy, and schizophrenia. *American Psychologist*, **17**, 827–838, 1962.

Michael, R. P., Zumpe, D., Keverne, E. B., and Bonsall, R. W. Neuroendocrine factors in the control of primate behavior. *Hormone Research*, **28**, 665–706, 1972.

Miller, N. E. The frustration-aggression hypothesis. *Psychological Review*, **48**, 337–342, 1941.

Money, J., and Ehrhardt, A. A. *Man and Woman, Boy and Girl*. Baltimore: The Johns Hopkins University Press, 1972.

Montagu, A. The new litany of "Innate Depravity, or Original Sin Revisited." In A. Montagu, Ed., *Man and Aggression*. New York: Oxford University Press, 1973.

Moyer, K. E. Kinds of aggression and their physiological basis. *Communications in Behavioral Biology*, **2**, 65–87, 1968.

Moyer, K. E. A preliminary model of aggressive behavior. In B. E. Eleftheriou and J. P. Scott, Eds., *The Physiology of Aggression and Defeat*. New York: Plenum, 1971. (a)

Moyer, K. E. The physiology of aggression and the implications for aggression control. In J. L. Singer, Ed., *The Control of Aggression and Violence*. New York: Academic Press, 1971. (b)

Moyer, K. E. *The Psychobiology of Aggression*. New York: Harper & Row, 1976.

Nakamura, C. Y., and Anderson, N. H. Avoidance behavior differences within and between strains of rats. *Journal of Comparative and Physiological Psychology*, **55**, 740–747, 1962.

Napier, J. The antiquity of human walking. *Scientific American*, **216**, 56–66, April 1967.

Nichols, P. L., and Anderson, V. E. Intellectual performance, race, and socioeconomic status. *Social Biology*, **20**, 367–374, 1973.

Omenn, G. S. and Motulsky, A. G. Biochemical genetics and the evolution of human behavior. In L. Ehrman, G. S. Omenn, and E. Caspari, Eds., *Genetics, Environment, and Behavior*. New York: Academic Press, 1972.

Orne, M. T. Hypnosis, motivation, and the ecological validity of the psychological experiment. In W. J. Arnold and M. M. Page, Eds., *Nebraska Symposium on Motivation*, **18**. Lincoln: University of Nebraska Press, 1970.

Orne, M. T., and Evans, F. J. Social control in the psychological experiment: Antisocial behavior and hypnosis. *Journal of Personality and Social Psychology*, **1**, 189–200, 1965.

Orne, M. T. and Schiebe, K. E. The contribution of nondeprivation factors in the production of sensory deprivation effects: The psychology of the "panic button." *Journal of Abnormal and Social Psychology*, **68**, 3–12, 1964.

Osmond, H. and Smythies, J. Schizophrenia: A new approach. *Journal of Mental Science*, **98**, 309–315, 1952.

Pfaff, D. W. Uptake of 3H-estradiol by the female rat brain. An autoradiographic study. *Endocrinology*, **82**, 1149–1155, 1968.

Philbeam, D. *The Ascent of Man*. New York: Macmillan, 1972.

Poirier, F. E. Colobine aggression: A review. In R. L. Holloway, Ed., *Primate Aggression, Territoriality, and Xenophobia*. New York: Academic Press, 1974.

Postman, L. (Ed.) *Psychology in the Making*. New York: Knopf, 1964.

Powers, J. B., and Valenstein, E. S. Individual differences in sexual responsiveness to estrogen and progesterone in ovariectomized rats. *Physiology and Behavior*, **8**, 673–676, 1972.

Premack, A. J., and Premack, D. Teaching language to an ape. *Scientific American*, **227**, 92–99, October 1972.

Reichenbach, H. *The Rise of Scientific Philosophy*. Berkeley: University of California Press, 1963.

Reis, D. J. Central monoamines as mediators of behavior. *Neurotransmitters*, **50**, 266–297, 1972.

Reppucci, N. D., and Saunders, J. T. Social psychology of behavior modification. *American Psychologist*, **29**, 649–660, 1974.

Resko, J. A., and Phoenix, C. H. Sexual behavior and testosterone concentrations in the plasma of the rhesus monkey before and after castration. *Endocrinology*, **91**, 499–503, 1972.

Riss, W., Valenstein, E. S., Sinks, J., and Young, W. C. Development of sexual behavior in male guinea pigs from genetically different stocks under controlled conditions of androgen treatment and caging. *Endocrinology*, **57**, 139–146, 1955.

Ritchie-Calder, P. R. *Leonardo and the Age of the Eye*. New York: Simon & Schuster, 1970.

Robinson, B. W. Aggression: Summary and overview. In B. E. Eleftheriou and J. P. Scott, Eds., *The Physiology of Aggression and Defeat*. New York: Plenum, 1971.

Rose, R. M., Holaday, J. W., and Bernstein, I. S. Plasma testosterone, dominance rank and aggressive behavior in male rhesus monkeys. *Nature*, **231**, 366–368, 1971.

Rosenthal, D. *Genetic Theory and Abnormal Behavior*. New York: McGraw-Hill, 1970.

Rosenthal, R. Interpersonal expectations: Effects of the experimenters hypothesis. In R. Rosenthal and R. Rosnow, Eds., *Artifact in Behavioral Research*. New York: Academic Press, 1969.

Rothenbuhler, W. C. Genetic and evolutionary considerations of social behavior of honeybees and some related insects. In J. Hirsch, Ed., *Behavior-Genetic Analysis*. New York: McGraw-Hill, 1967.

Rowe, D., and Plomin, R. The Burt controversy: A comparison of Burt's data on IQ with data from other studies. *Behavior Genetics*, **8**, 81–83, 1978.

Sar, M., and Stumpf, W. E. Autoradiographic localization of radioactivity in the rat brain after injection of 1, 2-3H-testosterone. *Endocrinology*, **92**, 251–256, 1973.

Sarason, S. B. Jewishness, blackishness, and the nature-nurture controversy. *American Psychologist*, **28**, 962–971, 1973.

Sarich, V. A molecular approach to the question of human origins. In P. Dolhinow and V. Sarich, Eds., *Background for Man*. Boston: Little, Brown, 1971.

Scarr, S., and Weinberg, R. A. IQ test performance of black children adopted by white families. *American Psychologist*, **31**, 726–739, 1976.

Scott, J. P. Theoretical issues concerning the origin and causes of fighting. In B. E. Eleftheriou and J. P. Scott, Eds., *The Physiology of Aggression and Defeat*. New York: Plenum, 1971.

Scott, J. P., and Fredericson, E. The causes of fighting in mice and rats. *Physiological Zoology*, **24**, 273–309, 1951.

Scriven, M. Views of human nature. In T. W. Wann, Ed., *Behaviorism and Phenomenology*. University of Chicago Press, 1964.

Sells, S. B. An interactionist looks at the environment. *American Psychologist*, **18**, 696–702, 1963.

Shaw, C. R. Electrophoretic variation in enzymes. *Science*, **149**, 936–943, 1965.

Shih, J.-H., C., and Eiduson, S. Multiple forms of monoamine oxidase in the developing brain. *Nature*, **224**, 1309–1310, 1969.

Shih, J.-H., C., and Eiduson, S. Multiple forms of monoamine oxidase in developing brain: Tissue and substrate specificities. *Journal of Neurochemistry*, **18**, 1221–1227, 1971.

Shimada, H., and Gorbman, A. Long lasting changes in RNA synthesis in the forebrains of female rats treated with testosterone soon after birth. *Biochemical and Biophysical Research Communications*, **38**, 423–430, 1970.

Shugar, D. (Ed.) *Enzymes and Isoenzymes*. New York: Academic Press, 1970.

Skinner, B. F. *Walden Two*. New York: Macmillan, 1948.

Skinner, B. F. Behaviorism at fifty. In T. W. Wann, Ed., *Behaviorism and Phenomenology*. University of Chicago Press, 1964.

Skinner, B. F. The phylogeny and ontogeny of behavior. *Science*, **153**, 1205–1213, 1966.

Skinner, B. F. *Beyond Freedom and Dignity*. New York: Knopf, 1971.

Skinner, B. F. *About Behaviorism*. New York: Knopf, 1974.

Smith, M. B. Criticism of a social science. *Science*, **180**, 610–612, 1973.

Södersten, P. Increased mounting behavior in the female rat following a single neonatal injection of testosterone propionate. *Hormones and Behavior*, **4**, 1–17, 1973.

Southwick, C. H. Genetic and environmental variables influencing animal aggression. In C. H. Southwick, Ed., *Animal Aggression*. New York: Van Nostrand Reinhold, 1970. (a)

Southwick, C. H. Conflict and violence in animal societies. In C. H. Southwick, Ed., *Animal Aggression*, New York: Van Nostrand Reinhold, 1970. (b)

Southwick, C. H., and Clark, L. H. Interstrain differences in aggressive behavior and exploratory activity of inbred mice. *Communications in Behavioral Biology*, **1**, 49–59, 1968.

Southwick, C. H., Farooqui, M. Y., Siddiqi, M. F., and Pal, B. C. Xenophobia among free-ranging rhesus groups in India. In R. L. Holloway, Ed., *Primate Aggression, Territoriality, and Xenophobia*. New York: Academic Press, 1974.

Sowell, T. The great IQ controversy. *Change*, 33–37, May 1973.

Splenger, O. The Decline of the West (Translated by C. F. Atkinson). London: G. Allen, 1959.

Spuhler, J. N., and Lindzey, G. Racial differences in behavior. In J. Hirsch, Ed., *Behavior-Genetic Analysis*. New York: McGraw-Hill, 1967.

Stanley, J. C. Interaction of organisms with experimental variables as a key to the integration of organismic and variable-manipulating research. In E. M. Huddleston, Ed., *17th Yearbook of the National Council on Measurement in Education*. East Lansing, Mich.: NCME, 1960.

Stein, L. and Wise, C. D. Possible etiology of schizophrenia: Progressive damage to the noradrenergic reward system by 6-hydroxydopamine. *Science*, **71**, 1032–1036, 1971.

Stern, C. *Principles of Human Genetics*, 3rd ed. San Francisco: Freeman, 1973.

Stoddard, L. *The Rising Tide of Colour Against White World Supremacy*. London: Chapman & Hall, 1920.

Storr, A. *Human Aggression*. New York: Bantam Books, 1970.

Stumpf, W. E. Hypophysiotrophic neurons in the periventricular brain: Typography of estradiol concentrating neurons. In C. H. Sawyer and R. A. Gorski, Eds., *Steroid Hormones and Brain Function*. Berkeley: University of California Press, 1971.

Sweet, W. H., Ervin, F. R., and Mark, V. H. The relationship of violent behaviour to focal cerebral disease. In S. Garattini and E. B. Sigg, Eds., *Aggressive Behaviour*. New York: Wiley, 1969.

Tata, J. R. Hormones and the synthesis and utilization of ribonucleic acids. *Progress in Nucleic Acid Research*, **5**, 191–250, 1966.

Teleki, G. The omnivorous chimpanzee. *Scientific American*, **228**, 32–42, January 1973.

Thiessen, D. D. A move toward species-specific analysis in behavior genetics. *Behavior Genetics*, **2**, 115–126, 1972.

Thiessen, D. D., and Rogers, D. A. Behavior genetics as the study of mechanism-specific behavior. In J. N. Spuhler, Ed., *Genetic Diversity and Human Behavior*. Chicago: Aldine, 1967.

Thompson, W. R. Some problems in the genetic study of personality and intelligence. In J. Hirsch, Ed., *Behavior-Genetic Analysis*. New York: McGraw-Hill, 1967.

Thorndike, E. L. Animal intelligence: an experimental study of the associative process in animals. *Psychological Monographs*, **2**, 1898.

Tiger, L. and Fox, R. *The Imperial Animal*. New York: Dell, 1971.

Tilly, C. Collective violence in European perspective. In H. D. Graham and T. R. Gurr, Eds., *The History of Violence in America*. New York: Bantam Books, 1969.

Tinbergen, N. On war and peace in animals and men. *Science*, **160**, 1411–1418, 1968.

Tinbergen, N. *The Animal in Its World*. Cambridge, Mass.: Harvard University Press, 1972.

Trivers, R. L. The evolution of reciprocal altruism. *Quarterly Review of Biology*, **46**, 35–57, 1971.

Tryon, R. C. Genetic differences in maze-learning ability in rats. *Yearbook of the National Society for the Study of Education*, **39**, 111–119, 1940.

Turner, M. B. *Philosophy and the Science of Behavior*. New York: Appleton-Century-Crofts, 1965.

Ulrich, R., and Azrin, N. H. Reflexive fighting in response to aversive stimulation. *Journal of the Experimental Analysis of Behavior*, **5**, 511–520, 1962.

Vale, J. R. and Ray, D. A diallel analysis of male mouse sex behavior. *Behavior Genetics*, **2**, 199–209, 1972.

Vale, J. R., Ray, D., and Vale, C. A. The interaction of genotype and exogenous neonatal androgen: Agonistic behavior in female mice. *Behavioral Biology*, **1**, 321–333, 1972.

Vale, J. R., Ray, D., and Vale, C. A. The interaction of genotype and exogenous neonatal androgen and estrogen: sex behavior in female mice. *Developmental Psychobiology*, **6**, 319–327, 1973.

Vale, J. R., Ray, D., and Vale, C. A. Neonatal androgen treatment and sexual behavior in males of three inbred strains of mice. *Developmental Psychobiology*, **7**, 483–488, 1974.

Vale, J. R. and Vale, C. A. Individual differences and general laws in psychology: A reconciliation. *American Psychologist*, **24**, 1093–1108, 1969.

Vale, J. R., Vale, C. A., and Harley, J. P. Interaction of genotype and population number with regard to aggressive behavior, social grooming, and adrenal and gonadal weight in male mice. *Communications in Behavioral Biology*, **6**, 209–221, 1971.

Valenstein, E. S. *Brain Control*. New York: Wiley, 1973.

Valenstein, E. S., Riss, W., and Young, W. C. Sex drive in genetically heterogeneous and highly inbred strains of male guinea pigs. *Journal of Comparative and Physiological Psychology*, **47**, 162–165, 1954.

Valenstein, E. S., Riss, W., and Young, W. C. Experential and genetic factors in the organization of sexual behavior in male guinea pigs. *Journal of Comparative and Experimental Psychology*, **48**, 397–403, 1955.

Valzelli, L., and Garattini, S. Biochemical and behavioral changes induced by isolation in rats. *Neuropharmacology*, **11**, 17–22, 1972.

Van Lawick-Goodall, J. The behaviour of free-living chimpanzees in the Gombe Stream Reserve. *Animal Behaviour Monographs*, **1**, 161–311, 1968.

Van Lawick-Goodall, J. *In the Shadow of Man*. New York: Dell, 1971.

Veblen, T. *The Theory of the Leisure Class*. New York: Macmillan, 1899.

Waddington, C. H. *The Strategy of the Genes*. London: G. Allen, 1957.

Wagner, R. P., and Mitchell, H. K. *Genetics and Metabolism*. New York: Wiley, 1964.

Wallis, D. I. Aggression in social insects. In C. H. Southwick, Ed., *Animal Aggression*. New York: Van Nostrand Reinhold, 1970.

Washburn, S. L. The study of race. *American Anthropologist*, **65**, 521–531, 1963.

Washburn, S. L., Jay, P. C., and Lancaster, J. B. Field studies of Old World monkeys and apes. In P. Dolhinow and V. Sarich, Eds. *Background for Man*. Boston: Little, Brown, 1971.

Watson, J. B. Psychology as the behaviorist views it. *Psychological Review*, **20**, 158–177, 1913.

Watson, J. B. *Behaviorism*. University of Chicago Press, 1930.

Watson, J. D., and Crick, F. H. C. A structure for deoxyribose nucleic acid. *Nature*, **171**, 737–738, 1953.

Watts, C. R., and Stokes, A. W. The social order of turkeys. *Scientific American*, **224**, 112–118, June, 1971.

Whalen, R. E. Comparative psychology. *American Psychologist*, **16**, 84, 1961. (a)

Whalen, R. E. Strain differences in sexual behavior of the male rat. *Behavior*, **18**, 199–204, 1961. (b)

Whalen, R. E., Beach, F. A., and Kuehn, R. E. Effects of exogenous androgen on sexually responsive and unresponsive male rats. *Endocrinology*, **69**, 373–380, 1961.

Whalen, R. E., Luttge, W. G., and Gorzalka, B. B. Neonatal androgenization and the development of estrogen responsivity in male and female rats. *Hormones and Behavior*, **2**, 83–90, 1971.

Wickler, W. *The Sexual Code*. Garden City, N.Y.: Doubleday (Anchor Books), 1973.

Wiesenfeld, S. L. Sickle-cell trait in human biological and cultural evolution. *Science*, **157**, 1134–1140, 1967.

Wilson, E. O. *Sociobiology*. Cambridge, Mass.: Harvard University Press, 1975. (a)

Wilson, E. O. Some central problems of sociobiology. *Social Science Information*, **14**, 5–18, 1975. (b)

Woolley, D. W., and Shaw, E. A biochemical and pharmacological suggestion about certain mental disorders. *Science*, **119**, 587–588, 1954.

Wynne-Edwards, V. C. Population control and social selection in animals. In D. C. Glass, Ed., *Biology and Behavior: Genetics*. New York: Rockefeller University Press and Russell Sage Foundation, 1968.

Young, W. C., Goy, R. R., and Phoenix, C. H. Hormones and sexual behavior. *Science*, **143**, 212–218, 1964.

Index

80   81   82   83   84   9   8   7   6   5   4   3   2   1